The PSI Handbook of Virtual Environments for Training and Education

The PSI Handbook of Virtual Environments for Training and Education

DEVELOPMENTS FOR THE MILITARY AND BEYOND

Volume 1
Learning, Requirements, and Metrics

Edited by Dylan Schmorrow, Joseph Cohn, and Denise Nicholson

Technology, Psychology, and Health

PRAEGER SECURITY INTERNATIONAL
Westport, Connecticut · London

Library of Congress Cataloging-in-Publication Data

The PSI handbook of virtual environments for training and education : developments for the military and beyond.

p. cm. – (Technology, psychology, and health, ISSN 1942–7573 ; v. 1-3)

Includes bibliographical references and index.

ISBN 978–0–313–35165–5 (set : alk. paper) – ISBN 978–0–313–35167–9 (v. 1 : alk. paper) – ISBN 978–0–313–35169–3 (v. 2 : alk. paper) – ISBN 978–0–313–35171–6 (v. 3 : alk. paper)

1. Military education–United States. 2. Human-computer interaction. 3. Computer-assisted instruction. 4. Virtual reality. I. Schmorrow, Dylan, 1967- II. Cohn, Joseph, 1969- III. Nicholson, Denise, 1967- IV. Praeger Security International. V. Title: Handbook of virtual environments for training and education. VI. Title: Praeger Security International handbook of virtual environments for training and education.

U408.3.P75 2009

355.0078'5–dc22 2008027367

British Library Cataloguing in Publication Data is available.

Library of Congress Catalog Card Number: 2008027367

ISBN-13: 978–0–313–35165–5 (set)
978–0–313–35167–9 (vol. 1)
978–0–313–35169–3 (vol. 2)
978–0–313–35171–6 (vol. 3)

ISSN: 1942–7573

First published in 2009

Praeger Security International, 88 Post Road West, Westport, CT 06881
An imprint of Greenwood Publishing Group, Inc.
www.praeger.com

Printed in the United States of America

The paper used in this book complies with the Permanent Paper Standard issued by the National Information Standards Organization (Z39.48–1984).

10 9 8 7 6 5 4 3 2 1

To our families, and to the men and women who have dedicated their lives to educate, train, and defend to keep them safe

CONTENTS

SERIES FOREWORD

LAUNCHING THE TECHNOLOGY, PSYCHOLOGY, AND HEALTH DEVELOPMENT SERIES

The escalating complexity and operational tempo of the twenty-first century requires that people in all walks of life acquire ever-increasing knowledge, skills, and abilities. Training and education strategies are dynamically changing toward delivery of more effective instruction and practice, wherever and whenever needed. In the last decade, the Department of Defense has made significant investments to advance the science and technology of virtual environments to meet this need. Throughout this time we have been privileged to collaborate with some of the brightest minds in science and technology. The intention of this three-volume handbook is to provide comprehensive coverage of the emerging theories, technologies, and integrated demonstrations of the state-of-the-art in virtual environments for training and education.

As Dr. G. Vincent Amico states in the Preface, an important lesson to draw from the history of modeling and simulation is the importance of *process*. The human systems engineering process requires highly multidisciplinary teams to integrate diverse disciplines from psychology, education, engineering, and computer science (see Nicholson and Lackey, Volume 3, Section 1, Chapter 1). This process drives the organization of the handbook. While other texts on virtual environments (VEs) focus heavily on technology, we have dedicated the first volume to a thorough investigation of learning theories, requirements definition, and performance measurement. The second volume provides the latest information on a range of virtual environment component technologies and a distinctive section on training support technologies. In the third volume, an extensive collection of integrated systems is discussed as virtual environment use-cases along with a section of training effectiveness evaluation methods and results. Volume 3, Section 3 highlights future applications of this evolving technology that span cognitive rehabilitation to the next generation of museum exhibitions. Finally, a glimpse into the potential future of VEs is provided as an original short story entitled "Into the Uncanny Valley" from Judith Singer and Hollywood director Alex Singer.

Through our research we have experienced rapid technological and scientific advancements, coinciding with a dramatic convergence of research achievements representing contributions from numerous fields, including neuroscience, cognitive psychology and engineering, biomedical engineering, computer science, and systems engineering. Historically, psychology and technology development were independent research areas practiced by scientists and engineers primarily trained in one of these disciplines. In recent years, however, individuals in these disciplines, such as the close to 200 authors of this handbook, have found themselves increasingly working within a unified framework that completely blurs the lines of these discrete research areas, creating an almost "metadisciplinary" (as opposed to multidisciplinary) form of science and technology. The strength of the confluence of these two disciplines lies in the complementary research and development approaches being employed and the interdependence that is required to achieve useful technological applications. Consequently, with this handbook we begin a new Praeger Security International Book Series entitled *Technology, Psychology, and Health* intended to capture the remarkable advances that will be achieved through the continued seamless integration of these disciplines, where unified and simultaneously executed approaches of psychology, engineering, and practice will result in more effective science and technology applications. Therefore, the esteemed contributors to the *Technology, Psychology, and Health Development Series* strive to capture such advancements and effectively convey both the practical and theoretical elements of the technological innovations they describe.

The *Technology, Psychology, and Health Development Series* will continue to address the general themes of requisite foundational knowledge, emergent scientific discoveries, and practical lessons learned, as well as cross-discipline standards, methodologies, metrics, techniques, practices, and visionary perspectives and developments. The series plans to showcase substantial advances in research and development methods and their resulting technologies and applications. Cross-disciplinary teams will provide detailed reports of their experiences applying technologies in diverse areas—from basic academic research to industrial and military fielded operational and training systems to everyday computing and entertainment devices.

A thorough and comprehensive consolidation and dissemination of psychology and technology development efforts is no longer a noble academic goal—it is a twenty-first century necessity dictated by the desire to ensure that our global economy and society realize their full scientific and technological potentials. Accordingly, this ongoing book series is intended to be an essential resource for a large international audience of professionals in industry, government, and academia.

We encourage future authors to contact us for more information or to submit a prospectus idea.

Dylan Schmorrow and Denise Nicholson
Technology, Psychology, and Health Development Series Editors
TPHSeries@ist.ucf.edu

PREFACE

G. Vincent Amico

It is indeed an honor and pleasure to write the preface to this valuable collection of articles on simulation for education and training. The fields of modeling and simulation are playing an increasingly important role in society.

You will note that the collection is titled virtual environments for *training and education*. I believe it is important to recognize the distinction between those two terms. Education is oriented to providing fundamental scientific and technical skills; these skills lay the groundwork for training. Simulations for training are designed to help operators of systems effectively learn how to operate those systems under a variety of conditions, both normal and emergency situations. Cognitive, psychomotor, and affective behaviors must all be addressed. Hence, psychologists play a dominant role within multidisciplinary teams of engineers and computer scientists for determining the effective use of simulation for training. Of course, the U.S. Department of Defense's Human Systems Research Agencies, that is, Office of the Secretary of Defense, Office of Naval Research, Air Force Research Lab, Army Research Laboratory, and Army Research Institute, also play a primary role—their budgets support many of the research activities in this important field.

Volume 1, Section 1 in this set addresses many of the foundational learning issues associated with the use of simulation for education and training. These chapters will certainly interest psychologists, but are also written so that technologists and other practitioners can glean some insight into the important science surrounding learning. Throughout the set, training technologies are explored in more detail. In particular, Volume 2, Sections 1 and 2 include several diverse chapters demonstrating how learning theory can be effectively applied to simulation for training.

The use of simulation for training goes back to the beginning of time. As early as 2500 B.C., ancient Egyptians used figurines to simulate warring factions. The precursors of modern robotic simulations can be traced back to ancient China, from which we have documented reports (circa 200 B.C.) of artisans constructing mechanical automata, elaborate mechanical simulations of people or animals.

These ancient "robots" included life-size mechanical humanoids, reportedly capable of movement and speech (Kurzweil, 1990; Needham, 1986). In those early days, these mechanical devices were used to train soldiers in various phases of combat, and military tacticians used war games to develop strategies. Simulation technology as we know it today became viable only in the early twentieth century.

Probably the most significant event was Ed Link's development of the Link Trainer (aka the "Blue Box") for pilot training. He applied for its patent in 1929. Yet, simulation did not play a major role in training until the start of World War II (in 1941), when Navy captain Luis de Florez established the Special Devices Desk at the Bureau of Aeronautics. His organization expanded significantly in the next few years as the value of simulation for training became recognized. Captain de Florez is also credited with the development of the first flight simulation that was driven by an analog computer. Developed in 1943, his simulator, called the operational flight trainer, modeled the PBM-3 aircraft. In the period after World War II, simulators and simulation science grew exponentially based upon the very successful programs initiated during the war.

There are two fundamental components of any modern simulation system. One is a sound mathematical understanding of the object to be simulated. The other is the real time implementation of those models in computational systems. In the late 1940s the primary computational systems were analog. Digital computers were very expensive, very slow, and could not solve equations in real time. It was not until the late 1950s and early 1960s that digital computation became viable. For instance, the first navy simulator to use a commercial digital computer was the Attack Center Trainer at the FBM Facility (New London, Connecticut) in 1959. Thus, it has been only for the past 50 years that simulation has made major advancements.

Even today, it is typical that user requirements for capability exceed the ability of available technology. There are many areas where this is particularly true, including rapid creation of visual simulation from actual terrain environment databases and human behavior representations spanning cognition to social networks. The dramatic increases in digital computer speed and capacity have significantly closed the gap. But there are still requirements that cannot be met; these gaps define the next generation of science and technology research questions.

In the past decade or so, a number of major simulation initiatives have developed, including distributed interactive simulation, advanced medical simulation, and augmented cognition supported simulation. Distributed simulation enables many different units to participate in a joint exercise, regardless of where the units are located. The requirements for individual simulations to engage in such exercises are mandated by Department of Defense standards, that is, high level architecture and distributed interactive simulation. An excellent example of the capabilities that have resulted are the unprecedented number of virtual environment simulations that have transitioned from the Office of Naval Research's Virtual Technologies and Environments (VIRTE) Program to actual military training applications discussed throughout this handbook. The second area of major growth is the field of medical simulation. The development of the human patient simulator clearly heralded this next phase of medical simulation based training, and the field of medical simulation will certainly expand during the next decade. Finally, the other exciting development in recent years is the exploration of augmented cognition, which may eventually enable system users to completely forgo standard computer interfaces and work seamlessly with their equipment through the utilization of neurophysiological sensing.

Now let us address some of the issues that occur during the development process of a simulator. The need for simulation usually begins when a customer experiences problems training operators in the use of certain equipment or procedures; this is particularly true in the military. The need must then be formalized into a requirements document, and naturally, the search for associated funding and development of a budget ensues. The requirements document must then be converted into a specification or a work statement. That then leads to an acquisition process, resulting in a contract. The contractor must then convert that specification into a hardware and software design. This process takes time and is subject to numerous changes in interpretation and direction. The proof of the pudding comes when the final product is evaluated to determine if the simulation meets the customer's needs.

One of the most critical aspects of any modeling and simulation project is to determine its effectiveness and whether it meets the original objectives. This may appear to be a rather straightforward task, but it is actually very complex. First, it is extremely important that checks are conducted at various stages of the development process. During the conceptual stages of a project, formal reviews are normally conducted to ensure that the requirements are properly stated; those same reviews are also conducted at the completion of the work statement or specification. During the actual development process, periodic reviews should be conducted at key stages. When the project is completed, tests should be conducted to determine if the simulation meets the design objectives and stated requirements. The final phase of testing is validation. The purpose of validation is to determine if the simulation meets the customer's needs. Why is this process of testing so important? The entire development process is lengthy, and during that process there is a very high probability that changes will be induced. The only way to manage the overall process is by performing careful inspections at each major phase of the project.

As the organization and content of this handbook make evident, this process has been the fundamental framework for conducting most of today's leading research and development initiatives. Following section to section, the reader is guided through the requirements, development, and evaluation cycle. The reader is then challenged to imagine the state of the possible in the final, Future Directions, section.

In summary, one can see that the future of simulation to support education and training is beyond our comprehension. That does not mean that care must not be taken in the development process. The key issues that must be addressed were cited earlier. There is one fact that one must keep in mind: no simulation is perfect. But through care, keeping the simulation objectives in line with the capabilities of modeling and implementation, success can be achieved. This is demonstrated by the number of simulations that are being used today in innovative settings to improve training for a wide range of applications.

REFERENCES

Kurzweil, R. (1990). *The age of intelligent machines.* Cambridge, MA: MIT Press.

Needham, J. (1986). *Science and civilization in China: Volume 2.* Cambridge, United Kingdom: Cambridge University Press.

ACKNOWLEDGMENTS

These volumes are the product of many contributors working together. Leading the coordination activities were a few key individuals whose efforts made this project a reality:

Associate Editor
Julie Drexler

Technical Writer
Kathleen Bartlett

Editing Assistants
Kimberly Sprouse and Sherry Ogreten

We would also like to thank our Editorial Board and Review Board members, as follows:

Editorial Board

John Anderson, Carnegie Mellon University; Kathleen Bartlett, Florida Institute of Technology; Clint Bowers, University of Central Florida, Institute for Simulation and Training; Gwendolyn Campbell, Naval Air Warfare Center, Training Systems Division; Janis Cannon-Bowers, University of Central Florida, Institute for Simulation and Training; Rudolph Darken, Naval Postgraduate School, The MOVES Institute; Julie Drexler, University of Central Florida, Institute for Simulation and Training; Neal Finkelstein, U.S. Army Research Development & Engineering Command; Bowen Loftin, Texas A&M University at Galveston; Eric Muth, Clemson University, Department of Psychology; Sherry Ogreten, University of Central Florida, Institute for Simulation and Training; Eduardo Salas, University of Central Florida, Institute for Simulation and Training and Department of Psychology, Kimberly Sprouse, University of Central Florida, Institute for Simulation and Training; Kay Stanney, Design Interactive,

Inc.; Mary Whitton, University of North Carolina at Chapel Hill, Department of Computer Science

Review Board (by affiliation)

Advanced Brain Monitoring, Inc.: Chris Berka; Alion Science and Tech.: Jeffery Moss; Arizona State University: Nancy Cooke; AuSIM, Inc.: William Chapin; Carlow International, Inc.: Tomas Malone; CHI Systems, Inc.: Wayne Zachary; Clemson University: Pat Raymark, Patrick Rosopa, Fred Switzer, Mary Anne Taylor; Creative Labs, Inc.: Edward Stein; Deakin University: Lemai Nguyen; Defense Acquisition University: Alicia Sanchez; Design Interactive, Inc.: David Jones; Embry-Riddle Aeronautical University: Elizabeth Blickensderfer, Jason Kring; Human Performance Architects: Richard Arnold; Iowa State University: Chris Harding; Lockheed Martin: Raegan Hoeft; Max Planck Institute: Betty Mohler; Michigan State University: J. Kevin Ford; NASA Langley Research Center: Danette Allen; Naval Air Warfare Center, Training Systems Division: Maureen Bergondy-Wilhelm, Curtis Conkey, Joan Johnston, Phillip Mangos, Carol Paris, James Pharmer, Ronald Wolff; Naval Postgraduate School: Barry Peterson, Perry McDowell, William Becker, Curtis Blais, Anthony Ciavarelli, Amela Sadagic, Mathias Kolsch; Occidental College: Brian Kim; Office of Naval Research: Harold Hawkins, Roy Stripling; Old Dominion University: James Bliss; Pearson Knowledge Tech.: Peter Foltz; PhaseSpace, Inc.: Tracy McSherry; Potomac Institute for Policy Studies: Paul Chatelier; Renee Stout, Inc.: Renee Stout; SA Technologies, Inc.: Haydee Cuevas, Jennifer Riley; Sensics, Inc.: Yuval Boger; Texas A&M University: Claudia McDonald; The Boeing Company: Elizabeth Biddle; The University of Iowa: Kenneth Brown; U.S. Air Force Academy: David Wells; U.S. Air Force Research Laboratory: Dee Andrews; U.S. Army Program Executive Office for Simulation, Training, & Instrumentation: Roger Smith; U.S. Army Research Development & Engineering Command: Neal Finkelstein, Timothy Roberts, Robert Sottilare; U.S. Army Research Institute: Steve Goldberg; U.S. Army Research Laboratory: Laurel Allender, Michael Barnes, Troy Kelley; U.S. Army TRADOC Analysis Center–Monterey: Michael Martin; U.S. MARCORSYSCOM Program Manager for Training Systems: Sherrie Jones, William W. Yates; University of Alabama in Huntsville: Mikel Petty; University of Central Florida: Glenda Gunter, Robert Kenny, Rudy McDaniel, Tim Kotnour, Barbara Fritzsche, Florian Jentsch, Kimberly Smith-Jentsch, Aldrin Sweeney, Karol Ross, Daniel Barber, Shawn Burke, Cali Fidopiastis, Brian Goldiez, Glenn Martin, Lee Sciarini, Peter Smith, Jennifer Vogel-Walcutt, Steve Fiore, Charles Hughes; University of Illinois: Tomas Coffin; University of North Carolina: Sharif Razzaque, Andrei State, Jason Coposky, Ray Idaszak; Virginia Tech.: Joseph Gabbard; Xavier University: Morrie Mullins

SECTION 1
LEARNING

SECTION PERSPECTIVE

Gwendolyn Campbell

Thirty years ago, as a discussant for the International Conference on Levels of Processing organized by Laird Cermak and Fergus Craik, Jenkins (1979) noted that perhaps one of the most surprising and promising developments in the field of memory research was that "no one around this table thinks that memory is simple anymore" (pp. 429–430). Apparently the speakers had backed off from making the kinds of sweeping assertions about the nature of memory that had been common in the past and were instead making reasonable claims that allowed for complex interactions between a host of variables on the processes associated with human learning and memory.

Jenkins went further in his discussion of the conference presentations and proposed an organizing framework to support memory researchers in both the design of their work and in the understanding of how their work fit into the larger body of memory research. This "Theorist's Tetrahedron" was formed with four vertices, each representing a category of variables that could be manipulated and studied within the context of memory research. The value of this type of organizational framework is that it reminds researchers, who are easily caught up in the excitement of the variables that they are manipulating, of the possibility that there are also impacts in their research from variables that they are ignoring.

Jenkins used one vertex of this tetrahedron, labeled "subjects," to represent variables associated with the population being studied—age, gender, ability, knowledge, motivation, and so forth. The vertex labeled "materials" represented variables associated with the stimuli that were presented to those subjects, including the nature of those stimuli (images, numbers, words, nonsense syllables, and so forth) and the way in which those stimuli were organized or sequenced. The "orienting tasks" vertex represented the nature of the mental and/or physical activities that the subjects were asked to conduct with the stimuli. A classic contrast from research around that time, for example, was to ask some subjects to process materials in a "meaningful" way (for example, generate a sentence using the word) and other subjects to process the material in a "nonmeaningful" way (for example, indicate whether or not the word is being presented in all capital letters).

Finally, the fourth vertex of the tetrahedron was labeled "criterial tasks" and contained variables associated with the nature of the post-test—was it immediate or delayed, recognition or free recall, and so forth.

Well, as is often said, the more things change, the more they stay the same. Looking over the chapters in this section and indeed, the sections in this volume and the volumes in this series, it is clear that, with some minor modifications, the Theorist's Tetrahedron is still a useful way to acknowledge and organize learning research. Most of the modifications would be in the nature of simply updating the terminology. For example, consider the vertex representing those mental and physical activities that are required of the student. Jenkins referred to these as "orienting tasks," but we might be more likely today to use the term "instructional activities." Similarly, it is likely that we would replace the "subject" label with the term "participant" or "student" and the term "criterial tasks" with "performance assessment."

One vertex, "materials," might require a bit more of an overhaul. At the time, Jenkins was dealing with research that had a common goal for the students—to remember something—and the variable that they often manipulated was the nature of that something (for example, images versus words). If we expect this model to handle learning research, then we need to acknowledge the fact that "remembering" is not the only goal of instruction. Oftentimes we want our students to be able to apply some process, make a decision, recognize a pattern, solve a problem, and so forth. Thus, it seems that a more appropriate category for this vertex might be "learning objectives."

Finally, given that the point of the model is to help researchers not lose sight of the variables that they are holding constant, I would propose the addition of a fifth vertex to represent the context in which the learning is taking place. Obviously, the focus of this series is on learning taking place within virtual environment (VE) based training systems, but there are other possible contexts, such as within a classroom or on the job. The addition of this fifth vertex, as illustrated in Figure SP1.1, creates a geometric solid that is officially known as a pentahedron, but is more commonly referred to as a pyramid.

With this modified model in hand, we can now examine the sections in this series of volumes and the chapters in this section to see how they are related to each other, and their coverage of the problem space. As befitting a section titled "Learning," a construct that takes place inside the student, the majority of the chapters in this section focus primarily on those aspects that are brought to the learning context by the student, or the "student characteristics" vertex. Our final two chapters begin to shift focus to the instructional activities and learning objectives vertices. Section 2 of this volume focuses on requirements analysis, which is, roughly speaking, the process for determining the learning objectives. Section 3 focuses on a third vertex, "performance assessment." The second volume of this series focuses on the components and technologies associated with our "learning context," virtual environments. Finally, the last volume of this series embeds this pyramid in a broader context that includes the work environment and makes projections for this pyramid into the future. Thus, at a high level this series

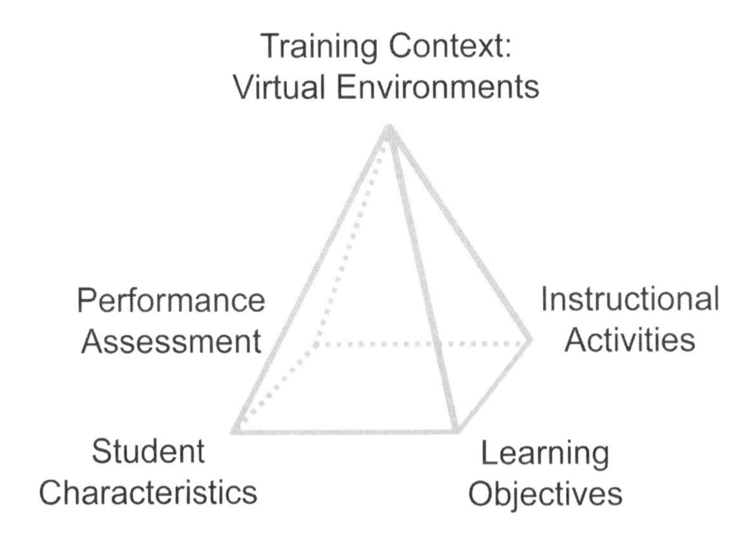

Figure SP1.1. Theorist's Pyramid

demonstrates both alignment with and coverage of the classes of variables that have long been recognized as important in the study of learning and memory.

Next consider this section. As mentioned earlier, the first four chapters in this section focus on those characteristics that the students bring with them to the learning context. Bowers, Vogel-Walcutt, and Cannon-Bowers (Chapter 2) provide one perspective on this topic by addressing those characteristics that vary from student to student, often referred to as individual differences. These authors cover many different research areas within the broad topic of individual differences and provide a nice organizing framework that distinguishes those characteristics that are relatively stable from those that are more malleable. In addition, they present open research questions regarding the relevance of these individual differences for VE training environments and provide some preliminary design guidance based on the existing literature.

Chapters 1 (Poulsen, Luu, and Tucker), 3 (Klein and Baxter), and 4 (Ross, Phillips, and Cohn), on the other hand, focus on an area that is generally thought of as common across students, by describing the nature of learning processes. At first glance, these chapters might seem quite disparate, as the topics range from the neurophysiology of animal learning to a conceptual model of expertise in complex, ill-structured domains such as firefighting or medical practice. In fact, what ties these chapters together is that each looks at learning as it occurs on a different level, or time scale, of human activity.

A useful way to see the relationship between these chapters is to place them within Newell's (1990) time scales of human activity. Newell proposed a series of four levels or bands, each containing units of time that cover approximately three orders of magnitude. Newell's biological band covers events that take place in the time frame of one-tenth of a millisecond to 10 milliseconds. As is obvious by both the time scale and the band label, these are primarily activities occurring at the neural level. Newell's cognitive band includes simple deliberate acts and unit tasks, activities that fall between 100 milliseconds and 10 seconds of time. More complex tasks, which often take between minutes and hours to complete, fall into Newell's rational band, and those human activities that stretch across days, weeks, months, and years fall into his social band.

Within the context of this framework, we can see that Poulsen, Luu, and Tucker (Chapter 1) are studying the learning process at the level of Newell's biological band by focusing on the neurophysiology of learning. Klein and Baxter (Chapter 3), on the other hand, present a characterization of the learning process that focuses on activities that are more likely to take place within Newell's rational band. More specifically, they present an alternative to the conceptualization that learning is a simple process of accumulating new information and argue that learning often requires replacing oversimplified and incorrect mental models. They focus on the challenges inherent in inducing change to a mental model and describe the roles that virtual environments might play in facilitating this process. Finally, Ross, Phillips, and Cohn (Chapter 4) describe the stages of becoming an expert in a field, an activity that takes years and falls squarely in Newell's social band. As with the other two chapters, these authors call out the instructional implications of this model of learning and make recommendations about the roles that virtual environments might play in facilitating the process.

One might wonder if it is really necessary to have chapters that address the learning process at these different levels. It is certainly an established heuristic that there is a "right" level of analysis for any given question. If a teacher wants to know how best to introduce the topic of fractions to young children, is it really helpful to discuss brain activation? And this levels-of-analysis issue is not restricted to psychology, but is a continuing debate among many communities. For example, sociologists have long been struggling with the question of whether or not you can understand societal phenomena by studying individual behavior (for example, Jepperson, 2007), and 10 years ago there was a growing trend for biology departments to reorganize according to the level of study, splitting molecular biologists from those who study ecosystems (Roush, 1997).

On the other hand, there are also many arguments in favor of taking a multilevel approach to understanding human behavior. At a minimum, it can easily be argued that principles identified at one level should not violate principles identified at other levels, and so a multilevel program of study may very well yield constraints and boundary conditions that flow between levels. It has also been argued at a relatively high level that the only way to truly understand a multilevel phenomenon such as human behavior is by analyzing it at multiple levels (for example, Hitt, Beamish, Jackson, & Mathieu, 2007). This argument has been made explicitly within the context of virtual environments, with regards to the construct of "presence" (IJsselsteijn, 2002). Anderson (2002) went beyond these general claims and assessed the empirical evidence available on the question of whether or not there is value in studying learning as it occurs over shorter time intervals when your interest is ultimately in understanding (and affecting) learning as it occurs across longer time intervals. He concluded that there is reasonably strong evidence that learning at the social band can be understood by decomposing it into cognition that occurs at lower bands (all the way down to the biological band), but more work needs to be done to demonstrate that attending to processes and events at the lowest level of Newell's hierarchy can yield improvements in outcomes at the social band.

Turning more specifically to the chapters in this section, it is interesting to note that, despite the huge disparity in time scales, they do, in fact, have a common thread. That thread has to do with the existence of two learning systems. Poulsen, Luu, and Tucker (Chapter 1) refer to those systems as the fast learning system and the slow learning system and provide evidence that these systems engage different areas of the brain. The fast learning system is typically in charge early on during the acquisition of knowledge in a new area; this system requires a lot of cognitive resources and explicit processing and results in relatively quick and discrete changes to memory. The slow learning system, on the other hand, typically takes over as a person's knowledge and skill in an area become rich, elaborated, and more and more compiled or automated. When this system is in charge, learning is gradual, may be unconscious, and repeated exposures are required to make even the smallest changes to a person's cognitive structures and processes.

An understanding of these two learning systems explicates a tension that is evident in the other two chapters, and more generally in the study of learning at any level of Newell's hierarchy. That is the tension between desiring to promote the fast, effortless, automatic performance that we associate with expertise and simultaneously desiring to promote the control and flexibility that allow people to deal with novel situations in new and inventive ways. This tension is, in fact, explicitly called out in the chapter by Ross, Phillips, and Cohn (Chapter 4), and this chapter presents some of the same instructional advice as that of Poulsen, Luu, and Tucker (Chapter 1) in regards to maintaining the proper balance between the two learning systems.

This thread is also evident in the discussion that Klein and Baxter (Chapter 3) present regarding the challenges in getting a student to unlearn or discard an established mental model. This challenge can be understood as a recasting of the challenge that Poulsen and colleagues describe of trying to get a person who has compiled knowledge, and thus is operating under the slow learning system, to shift back to using the fast learning system. Thus, while this is not by any means a definitive answer to the question of whether or not it is necessary to study learning at all levels of human activity, it does show one instance in which the principles at these levels are not isolated and independent, but rather inform, explicate, and illustrate each other.

The last two chapters can be seen as forming a bridge between this section and the following sections, as they begin to take on perspectives from other vertices. Biddle, McBride, and Malone (Chapter 5) begin by addressing the issue of how understanding biology (in this case, human maturation and neural mylenization) can contribute to our understanding of, and ability to facilitate, human learning. They continue by discussing the need to study interactions between two vertices, "student states" and "instructional activities," in order to optimize learning outcomes.

Finally, in the last chapter in this section (Chapter 6), Van Buskirk, Cornejo, Astwood, Russell, Dorsey, and Dalton explicitly address a leg rather than a vertex of the pyramid by considering the relationship between learning objectives and instructional activities. The basic premises of this chapter are that all instructional activities are not equally effective for tackling all learning objectives and that empirical research can provide guidance as to how to select and implement optimal instructional activities for a given learning objective. Like Jenkins and the other researchers at that international conference in the late 1970s, these authors do not think that the answers will be found in simple, sweeping assertions and instead present a framework of their own to help organize, integrate, and guide current and future instructional research. It will be interesting to see if, 30 years from now, their matrix (possibly with some minor modifications) provides a useful organizing framework for a new generation of researchers using a new generation of technology.

REFERENCES

Anderson, J. R. (2002). Spanning seven orders of magnitude: A challenge for cognitive modeling. *Cognitive Science, 26,* 85–112.

Hitt, M. A., Beamish, P. W., Jackson, S. E., & Mathieu, J. E. (2007). Building theoretical and empirical bridges across levels: Multilevel research in management. *Academy of Management Journal, 50*(6), 1385–1399.

IJsselsteijn, W. (2002, October). Elements of a multi-level theory of presence: Phenomenology, mental processing and neural correlates. *Proceedings of PRESENCE 2002* (pp. 245–249). Porto, Portugal.

Jenkins, J. (1979). Four points to remember: A tetrahedral model of memory experiments. In L. S. Cermak & F. I. M. Craik (Eds.), *Levels of processing in human memory* (pp. 429–446). Hillsdale, NJ: Lawrence Erlbaum.

Jepperson, R. L. (2007, August). Multilevel analysis versus doctrinal individualism: The use of the "Protestant Ethic Thesis" as intellectual idealogy. *Paper presented at the annual meeting of the American Sociological Association,* New York. Retrieved April 11, 2008, from http://www.allacademic.com/meta/ p177199.index.html

Newell, A. (1990). *Unified theories of cognition.* Cambridge, MA: Cambridge University Press.

Roush, W. (1997). News & Comment. *Science, 275* (5306), 1556.

Part I: Biological Band

Chapter 1

THE NEUROPHYSIOLOGY OF LEARNING AND MEMORY: IMPLICATIONS FOR TRAINING

Catherine Poulsen, Phan Luu, and Don Tucker

Learning is often considered to be a unitary phenomenon. However, neurophysiological evidence suggests that multiple systems regulate learning, thereby creating unique forms of memory. This chapter presents a model of learning and memory that integrates two complementary learning circuits into a coherent framework that can be used to understand the development of expertise. In this model, learning and memory arise out of primitive systems of motivation and action control. These systems are responsible for memory consolidation within corticolimbic networks, and they are thus critical to higher cognitive function. Yet they include unique motivational influences intrinsic to the control of learning within each system. Understanding these motivational biases may be important in designing effective training methods. For example, the optimistic mood in response to successful achievement may be particularly important to gaining a broad representation of situational awareness. In contrast, anxiety under stress may facilitate focused attention on obvious threats, but impair situational awareness for unexpected threats. Theoretical progress in understanding the neurophysiology of learning and memory may lead to new insights into why certain methods of education and training are effective, and it may suggest new strategies for effectively motivating the learning process in ways that translate to real world contexts.

INTRODUCTION

With a third of their careers spent in training, military personnel need training to be well motivated, efficient, and cost-effective. The standard approach to training is to conduct a task analysis that identifies required knowledge, motor skills, and cognitive skills and then to develop programs that target these skills. This is a logical approach, but it is not driven by an understanding of the brain's learning systems or by assessment of these systems during learning.

We propose that design of effective training should start with the theoretical principles of learning based on neuroanatomical and neurofunctional evidence.

Task analysis, although empirically derived, should be informed by these principles. Identification of skill sets, training, and learning protocols are interdependent and must be guided by empirical measurement of brain system dynamics within a coherent theoretical framework. The result of a neural systems analysis could be principled adjustments during training, with these adjustments optimized for each individual learner.

The model presented in this chapter emphasizes that animals as well as humans operate through goal-directed learning, an expectancy based process in which the discrepancy between the anticipated and the actual outcome of an action drives new learning and thereby consolidates context-adaptive performance. Understanding such goal-directed learning may be essential to developing training methods that make experts out of novices. The chapter presents a brief overview of the neural systems underlying expectancy based learning, provides examples of the neurophysiological signatures of expertise and learning in cognitive neuroscience experiments, and concludes by considering the implications of this approach for enhanced training and performance.

THE GOALS OF EDUCATION AND TRAINING: LEARNING VERSUS PERFORMANCE

A primary goal of education is to guide learning in its broadest sense, facilitating the acquisition of new knowledge, abilities, attitudes, and perspectives. Training typically has a more restricted objective, focusing on the directed acquisition and practice of a specific skill, from novice to expert level of performance. In either case, a major challenge is how best to facilitate learning. Specifically, how can the rate of learning be enhanced and the level of attainment be maximized? The development of effective interventions requires both sound theory and a sensitive, reliable measure of progress and attainment. This is particularly challenging for a process, such as learning, that is inherently hidden to the observer. In contrast to overt performance, learning is an internal process, sometimes referred to as a latent, unobservable state.

Educators and trainers traditionally use performance measures as indicators of this latent state. But, performance indicators alone may be absent, incomplete, or ambiguous. For example, an identical error may reflect a learner's incomplete knowledge of the task or simply a slip in performance (Campbell & Luu, 2007). Furthermore, considerable learning may take place before any overt change in performance occurs. Trainers need more reliable assessment of the learning *process.* Measurement of brain activity can provide a useful window on the process of learning, even in the absence of behavioral changes. Recent research indicates that novices differ from experts not just in task performance, but also in the very nature of the brain systems that mediate their learning and performance. Rather than relying solely on performance indicators, *neuroadaptive training* could directly monitor brain activity and thereby fine-tune information delivery and feedback to more directly support the neural systems engaged at different stages of learning.

NEUROPHYSIOLOGY OF LEARNING AND MEMORY

Advances in neuroscience have provided new insights into the specific mechanisms of learning and memory. These new insights apply not only to the process of memory consolidation, but also to the motivational influences that often determine success and failure.

All multicell organisms learn, yet scientists often neglect to recognize that the fundamental reason an animal learns is to anticipate effectively both internal homeostatic challenges and environmental constraints. The consequences of such neglect is that motivation is often viewed as separate from the learning process itself; motivational levels or affective reactions are merely considered to get the animal to "learn" a task; the motivation and affective mechanisms are often considered external to the core process of learning. Yet self-regulatory mechanisms may be inherent to the memory control processes that achieve learning. It is now well-known that an important aspect of animal learning involves forming and maintaining implicit or explicit cognitive expectancies for the hedonic regularities in the world and adapting these expectancies as they are adjusted or disconfirmed by the ongoing flow of events (for example, Rescorla & Wagner, 1972; Balleine & Ostlund, 2007). It is the internal regulation of information content (that is, the error between expectancies and outcomes) that significantly drives this form of learning. The memory systems supporting these expectancy and outcome representations are thus central to the learning process.

Memories as Learning Outcomes

As individuals learn through observation and interaction with their environments, traces of these experiences are retained, distributed across multiple neural networks. Conceptual knowledge is extracted from stimulus regularities (Potter, 1999), and stimulus-action-outcome contingencies are encoded and retrieved alongside contextual and hedonic features of the experience (Pribram, 1991). These processing traces, or memories, are thus formed in multiple neural systems as the outcomes of learning and experience.

Learning and Dual-Action Control Systems

The traditional view of learning as a process often divides it into two distinct stages: an early stage and a late stage. The early stage relies on executive control and short-term memory buffers. Task execution is carried out by slow and effortful control processes. These control processes are limited by the capacity of cognitive resources, require active attention, and can be directed consciously in new task situations (Schneider & Shiffrin, 1977; Shiffrin & Schneider, 1977). In contrast, the late stage is not dependent on executive control or temporary memory buffers, but rather on long-term memory stores. Routine task components, such as identifying task-relevant information in the environment, become *automatized* and reduce demands on limited cognitive resources. Learning reflects the progression through these stages.

Neurophysiological models of animal learning provide additional insights into the cognitive conceptualization of the learning process. We have argued that it is best to conceptualize learning as "action regulation" (Luu & Pederson, 2004; Luu & Tucker, 2003). Action regulation emphasizes the need to adjust behavior according to both internal states and external demands, which require different learning and memory systems; these systems reflect cybernetic constraints on action control. Learning and memory naturally arise from these action regulation processes. Our research on human memory and learning over the past decade has been informed by neuroanatomical findings on self-regulated learning in animals. These findings, and the resultant model, fit nicely with the separation of early and late stages of learning in the cognitive literature.

In the animal neurophysiology research, two complementary cortico-limbic-thalamic circuits have been distinguished, each providing a unique strategic control on the learning process (Gabriel, Burhans, Talk, & Scalf, 2002). The *ventral limbic* circuit is made up of the anterior cingulate cortex (ACC) and the dorsomedial nucleus of the thalamus, with input from the amygdala (see Figure 1.1). This ACC based circuit is triggered by exogenous feedback and leads to rapid changes in learning in response to new information that is discrepant with expectations. This circuit is involved in the early stages of learning, whenever new tasks must be learned, or when routine actions and a priori knowledge are no longer appropriate for current demands (Gabriel, 1990; Gabriel, Sparenborg, & Stolar, 1986;

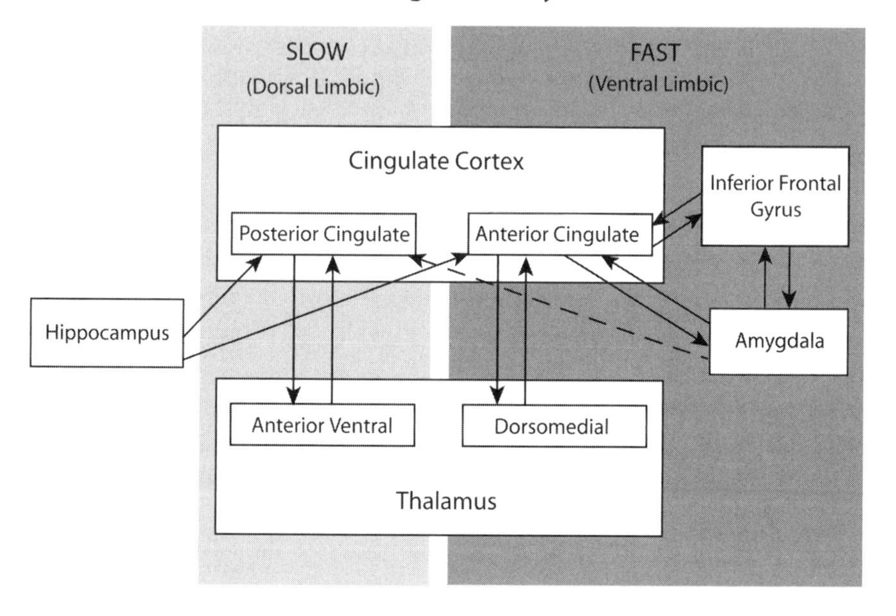

Figure 1.1. Slow (Dorsal) and Fast (Ventral) Cortico-Limbic-Thalamic Learning Circuits

Poremba & Gabriel, 2001). At the psychological level, information in this circuit is deliberately held in short-term memory, existing for only a few seconds. The unique properties of this fast learning system, for example, its contribution to overcoming habitual responses, led Gabriel and colleagues (Gabriel et al., 2002) to suggest this circuit is integral to what has been called the executive control of cognition (Posner & DiGirolamo, 1998).

Although numerous modern neuroimaging studies have repeatedly shown the involvement of the ACC early in learning (for example, Chein & Schneider, 2005; Toni, Ramnani, Josephs, Ashburner, & Passingham, 2001), it has been noted that lesions to the ACC, traditionally, do not change scores on intelligence tests or tests of executive control (Devinsky & Luciano, 1993), implying that this region may not be important to learning in humans. The most consistent observation after cingulate lesions in humans is alteration of affect. Patients are often described as being either apathetic or laissez-faire in their attitudes (Cohen, Kaplan, Moser, Jenkins, & Wilkinson, 1999); that is, they are less concerned about daily life events. For example, Rylander (1947) noted that patients report not being concerned when they make mistakes. In more recent studies, ACC lesions have been shown to affect both performance on traditional tests of executive control (such as the Stroop task; Janer & Pardo, 1991; Cohen et al., 1999) and error monitoring (Swick & Turken, 2002).

The second circuit is centered on the posterior cingulate cortex (PCC) and the anterior ventral nucleus of the thalamus, with input from the hippocampus (Tucker & Luu, 2006). This *dorsal limbic* circuit is involved in the later stages of learning (Keng & Gabriel, 1998), when consolidation of information into long-term memory becomes important (Gabriel, 1990). It functions in an automated manner, shaping the context model with small adjustments, requiring little or no effort. In the later stages of learning, a contextual model is fully formed, and discrepancies with expectations result in minor changes that are largely consistent with the internal model and can be made with minimal attentional demands. The PCC based system then applies a feed-forward bias to action regulation, in which action is controlled endogenously and learning is slowly and incrementally updated (Tucker & Luu, 2006). The late stage learning process has been described as context updating (Donchin & Coles, 1988).

This neurophysiological model indicates that goal-directed learning is an active process achieved by circuits with qualitative strategic biases. One bias, emerging from feedback control from the viscerosensory regulation of the ventral limbic pathway, leads to rapid, focused changes of associations under conditions of context violation or threat. A second bias, emerging from feed-forward control inherent to the visceromotor function in the dorsal limbic pathway, leads to endogenous, hedonic expectancies for action and a gradual updating of a valued context model (Tucker & Luu, 2006). Noninvasive neurophysiological measures of brain activity, including dense-array electroencephalography (EEG), near-infrared spectroscopy, and functional magnetic resonance imagery (fMRI) now allow unprecedented access to investigation of these learning mechanisms during learning and performance in humans. We focus here on recent EEG research examining the operation of these two action regulation circuits.

Neural Signatures of Learning and Skill Development

Characteristic changes in brain activity during early learning, in contrast to changes that occur late in learning, have been identified with both fMRI and EEG. Chein and Schneider's (2005) fMRI based analysis of the neural components of performance suggested that the effortful control required early in learning engages brain regions characterized as regulating attention (PCC and parietal), comparison (ACC), and task control (dorsolateral prefrontal cortex). As skilled performers became more automatic, activity in frontal regions declined, presumably due to a reduced need for executive control.

Similar effects were obtained in a dense-array EEG study of task switching (Poulsen, Luu, Davey, & Tucker, 2005; Poulsen et al., 2003).[1] Subjects performed one of two alternative tasks, either a letter (vowel/consonant) or digit (even/odd) judgment, in response to bivalent (for example, G5) or univalent (for example, &3, G#) stimuli. A cue indicated which task to perform, and trials were sequenced to require task repetition or task switch. Consistent with effortful, controlled processing, performance was slower on the challenging switch trials as compared to repeat trials and evidenced greater engagement of ACC and lateral anterior prefrontal regions (Figure 1.2). A reduction in these control processes with task experience was indicated by behavioral and brain measures. Reaction time decreased linearly early in learning (half 1) to asymptotic levels that were maintained later in learning (half 2). Reaction-time variability, a behavioral performance index of automaticity (Segalowitz, Poulsen, & Segalowitz, 1999; Segalowitz & Segalowitz, 1993) also decreased linearly during half 1, but fluctuated in half 2. This suggests that control processing was reengaged intermittently in half 2 in order to maintain high levels of performance, particularly for the most difficult, bivalent switch trials. This interpretation was further supported by EEG evidence of greater prefrontal cortex and ACC involvement in half 2 (Figure 1.2). Amplitude of the P300, an EEG component associated with context updating that source localized to the PCC and related parietal cortex, was larger in half 2 than in half 1. This suggests greater involvement of the PCC based circuit late in learning, with emphasis on memory consolidation and incremental context updating. The results of this study thus illustrate not only how these two learning circuits characteristically come into play early and late in learning, but also how the relative balance of these two circuits can be dynamically adjusted to meet the challenges of variable task demands.

Learning within this system is regulated by a simple cognitive phenomenon: violations of expectancy. EEG recorded during expectancy violations reveals brain activity consistent with the ACC learning circuit. Specifically, when expectancies are violated, such as when an error is committed, a negative deflection in the ongoing EEG is observed over medial frontal scalp regions (Falkenstein, Hohnsbein, Hoormann, & Blanke, 1991; Gehring, Goss, Coles, Meyer, & Donchin, 1993; Luu, Flaisch, & Tucker, 2000). This negative deflection is

[1]All results reported were statistically significant. Figures represent grand averages of experimental groups and/or conditions.

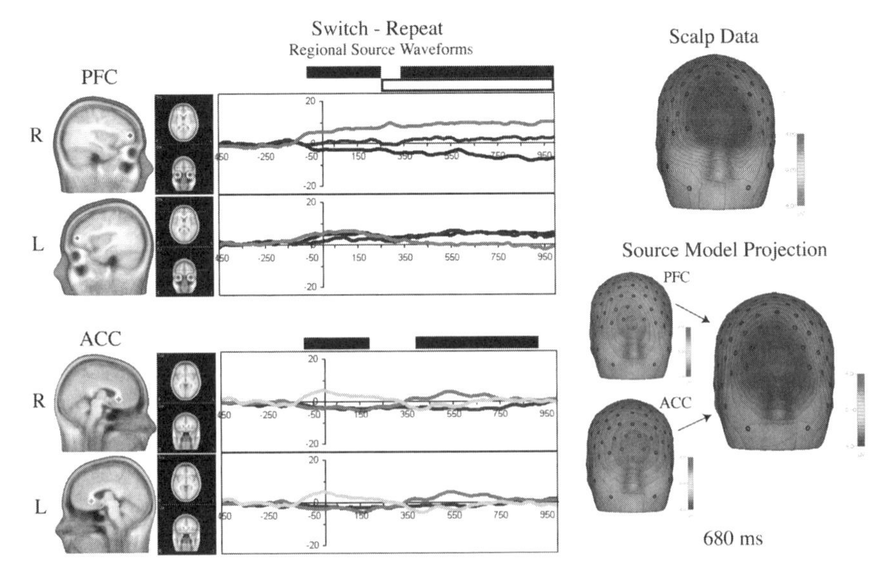

Figure 1.2. (Left) Regional dipole locations for the right and left prefrontal cortex (PFC) and anterior cingulate (ACC) sources. (Middle) Switch-repeat difference waveforms (nAm) of regional source activity (x, y, z vectors) for the right and left PFC and ACC sources. Differences reveal greater activity in preparation for, and during execution of, a switch trial than a repeat trial. Periods of significant effects are indicated with superposed bars (solid bar: trial-type main effect; open bar: trial-type x laterality interaction). (Right) Anterior medial frontal effect in half 2 at 680 ms: topographic map of the difference amplitude for switch-repeat scalp data on bivalent trials (top), followed by the forward topographic projection of the ACC and PFC sources illustrating their contribution to modeled source activity.

referred to as the error-related negativity and has been source localized to the ACC (see Figure 1.3) (Luu, Tucker, Derryberry, Reed, & Poulsen, 2003; Miltner et al., 2003). Brain responses similar to the error-related negativity are observed in other situations, when errors are not committed but when expectancies are violated in other ways. For example, as subjects learn a repeated sequence of stimuli, when a position within that sequence is changed, medial frontal negativity (MFN) is also observed in response to this violation (for a review, see Luu & Pederson, 2004).

TRACKING THE DEVELOPMENT OF EXPERTISE

The above studies suggest that brain activity typically shifts from the anterior, ACC based, early learning system with reliance on control processes, to the posterior, PCC based, late learning system with increased automaticity and memory consolidation. We sought to examine, more specifically, the differences between novices and experts and the transition from novice to expert performance.

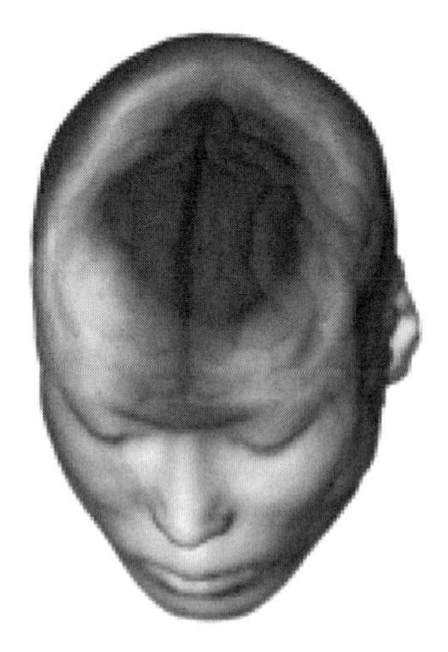

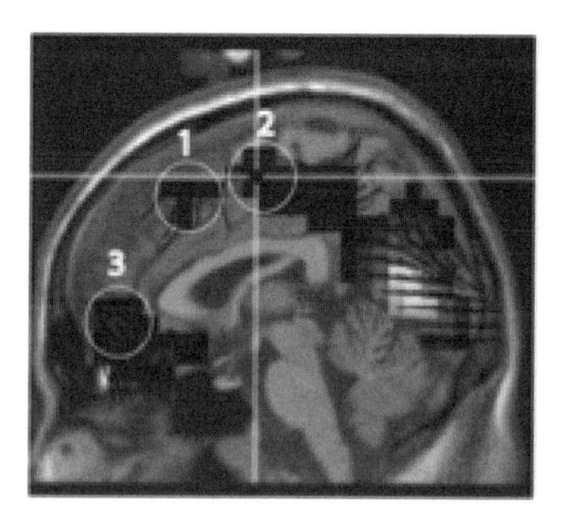

Figure 1.3. (Left) Medial frontal scalp distribution of the error-related negativity. (Right) Cortical source generators of the error-related negativity in anterior cingulate and mid-cingulate regions (plus residual occipital activity for this visual decision task).

Tanaka and Curran (2001) studied bird and dog experts and found that signatures over the visual cortex (a component referred to as the N170, which occurs ~170 ms [milliseconds] after stimulus onset) were largest for images that contained domain-specific expertise content. That is, bird experts exhibited larger activity for images that contained birds as opposed to dogs, and vice versa for the dog experts. Moreover, when trained to identify certain types of targets, amplitude of the N170 increased (Scott, Tanaka, Sheinberg, & Curran, 2006). The time course, location, and pattern of this effect suggest it indexes what has been described in cognitive research as conceptual short-term memory (Potter, 1999). Conceptual short-term memory is a construct that describes the rapid process (100–300 ms post stimulus) by which conceptual information is *extracted* from visual and auditory stimuli and *selected* for further processing (200–500 ms post stimulus). Conceptual short-term memory appropriately belongs to the late learning system. It is through learning and experience that concepts are abstracted and consolidated by the hippocampus into long-term memory traces stored in high order sensory (such as visual) cortices (Squire, 1998). These traces in turn permit rapid and automated extraction of goal-relevant information by these higher order sensory cortices from sensory input. The implication is that, for experts, experience reaches out to perception (via reentrant processes) such that what an expert sees is inherently different, at least neurophysiologically if not also phenomenologically, from what is perceived by the novice.

We attempted to understand these findings with respect to satellite image analysts and their performance in detecting targets for which they have extensive experience (Crane, Luu, Tucker, & Poolman, 2007). Experts and novices viewed satellite images, with or without targets of military interest, delivered in rapid

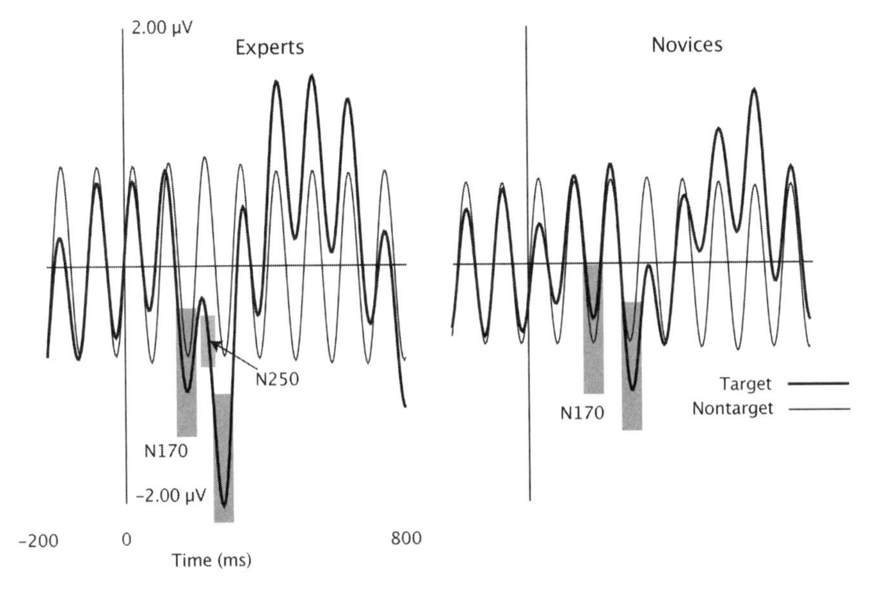

Figure 1.4. Grand-average waveform plots for a channel located over the midline of the occipital region for expert (left) and novice (right) subjects. First gray rectangle identifies the N170, the second gray rectangle identifies the N250 (missing in novices), and the third gray rectangle identifies the continuation of the N250 to a peak at 300 ms. The oscillatory appearance of the waveform reflects overlapping visual P1-N1 responses to each image presented in rapid serial visual presentation at a rate of 10 Hz.

serial visual presentation format at a rate of 10 Hz (hertz). We observed that for experts, but not novices, the N170 (see Figure 1.4) was enhanced for target images compared to nontarget images. Furthermore, compared to novices, the experts' N170 responses to targets were substantially enhanced. These findings are consistent with the notion that learning influences conceptual extraction by conceptual short-term memory. That is, the enhanced N170 observed for the experts likely reflects their domain-specific training, which allows for the separation of targets from nontarget brain responses.

We also observed another difference between the target and nontarget waveforms of experts at approximately 250 ms after stimulus onset, an effect that was absent for novices. In Figure 1.4, the second rectangle in the experts' waveform marks the time that a clear N250 can be seen for target images (see Figure 1.5). This cortical signature is absent in both the waveform (Figure 1.4) and topographic map (Figure 1.5) of novice subjects. When we estimated the cortical source of this scalp-recorded potential, sources were identified along the extent of the posterior temporal lobe, including the fusiform gyrus (see Figure 1.5, right), a region that has been implicated in expert visual processing (Gauthier, Skudlarski, Gore, & Anderson, 2000; but see also McKone, Kanwisher, & Duchaine, 2007). Previous research has shown that when subjects are trained to

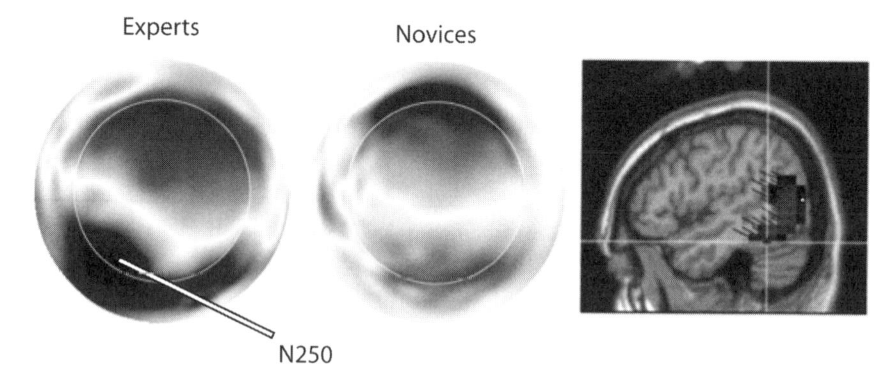

Figure 1.5. Topographic maps at 250 ms after target stimulus onset (left and middle figures), and the cortical source of the left-lateralized posterior negativity (N250) in experts (right).

discriminate objects at subordinate levels (for example, robins, starlings, and so forth), recognition of these subordinate-level images evoked larger N250 amplitudes than recognition of basic-level images (for example, birds; Scott et al., 2006), suggesting that the N250 reflects cortical activity related to conceptual extraction that is specific for a particular domain of expertise.

At approximately 300 ms after stimulus onset, both novices and experts show a negative peak that differentiates target images from nontarget images (see Figure 1.4, gray rectangle around the 300 ms peak). This component may reflect the additional processing of the target images, but is potentiated by the N170 and the N250.

The emerging view from these data is that conceptual extraction of information is experience dependent and occurs very early in visual processing. In experts, it is enhanced and automated (seen in the N170 and the N250), contributing to more accurate and rapid selection of target images.

Beyond 300 ms after stimulus onset, we observed additional differences in regional brain activity between experts and novices. Most remarkable was a centromedial negativity (N350) observed in experts only (see Figure 1.6). This effect resembles mediofrontal negativities (MFNs) observed in other studies (for example, Tucker et al., 2003), but for the expert intelligence analysts in this experiment there was engagement of the precuneus (posterior midline) and orbital frontal regions, as well as the ACC. Although MFNs are believed to reflect action monitoring functions (such as error monitoring and/or conflict monitoring), it is plausible that the activity here reflects the selection stage of conceptual short-term memory. Further research with expert performers may clarify the shifts in neural processing that can explain the efficiencies of cognition that are gained with domain expertise.

These studies illustrate how noninvasive measurement of neural signatures can distinguish expert from novice performers. As the following studies demonstrate, they can also be used to track changes in the learning state.

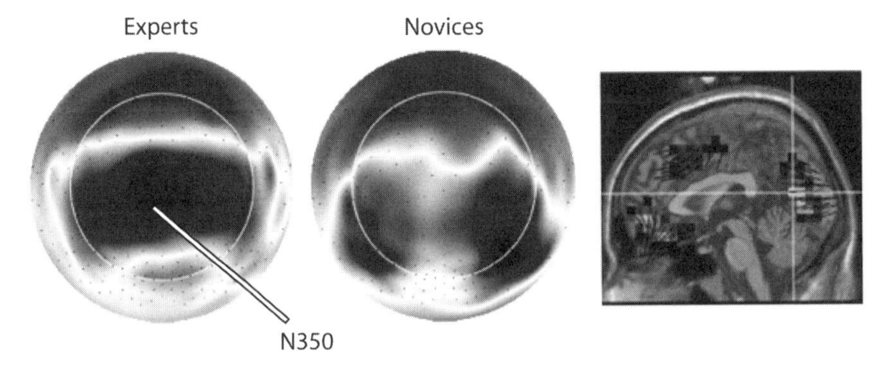

Figure 1.6. Topographic maps at 350 ms after target stimulus onset (left and middle figures), and the cortical source of the mediofrontal negativity (N350) in experts (right).

Brain Changes Associated with Learning and Practice

In the series of experiments described below, dense-array EEG was used to track neural activity during verbal and spatial associative learning. Subjects were required to discover by trial and error which arbitrary key press was associated with a two-digit code (verbal task) or with a dot location (spatial task; Luu, Tucker, & Stripling, 2007). Based on models of early and late learning systems, greater activity in the ventrolateral aspects of the inferior frontal lobes and the ACC was predicted early in learning, and greater activity in the hippocampus, the PCC, and the parietal lobes was predicted later in learning.

Figure 1.7 shows the topography, at about 400 ms after stimulus onset, of the scalp potentials (inferior frontal focus) for the digit and spatial learning task, as well as the associated cortical generators of those scalp potentials. We refer to this waveform feature as the lateral inferior anterior negativity (LIAN). As predicted, this activity was lateralized according to the nature of the task (left for digits and right for spatial) and decreased, for the spatial task, as subjects learned the task (for the digit task, the activity remained steady throughout learning).

We also found an ACC source that increased with learning (contrary to predictions). The ACC source was seen at the scalp as a mediofrontal negativity (MFN; see Figure 1.8). We interpret involvement of the inferior frontal source as reflecting memory encoding processes, and the increasing involvement of the ACC source as reflecting action monitoring relative to task demands (that is, with increased task knowledge there is a corresponding increase in response conflict). Note that although the time course of the MFN is similar to the negativity observed for expert image analysts (see Figure 1.6), a somewhat different pattern of cortical sources contributes to the MFN in the learning study. Although the pattern of results across several studies emphasizes the importance of midline corticolimbic networks to increasing expertise, it remains to be seen whether skill gained in a few training sessions will be associated with neural mechanisms that explain expertise gained over many years of training.

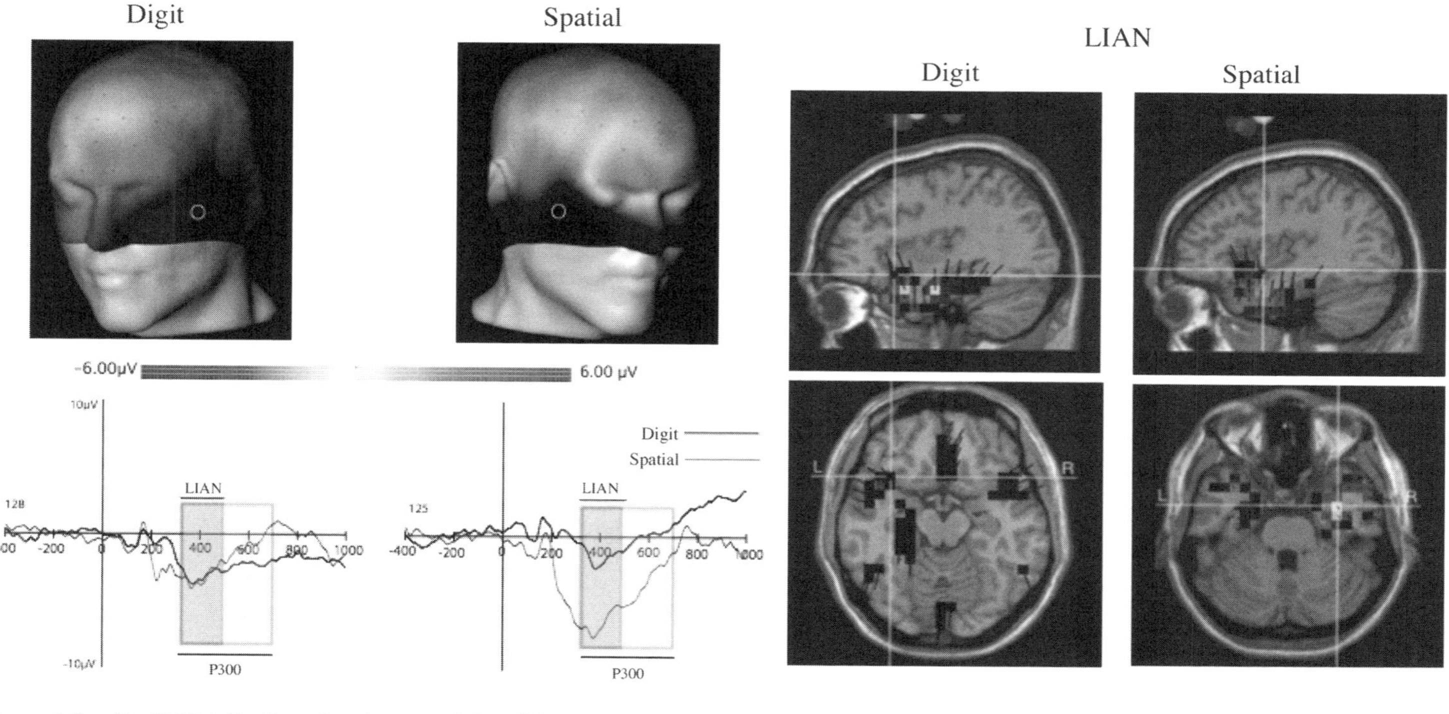

Figure 1.7. **(Left) Distribution of scalp potential at 400 ms post stimulus associated with the digit and spatial memory task. In the code-learning task, early effortful processing was associated with hemisphere-specific inferior frontotemporal negativities (the lateral inferior anterior negativity; LIAN), with activity greater on the left side in learning the digit code, and on the right side in learning the spatial code. Dots on the topographic maps indicate the channel locations of the waveform plots below. The time window to quantify the LIAN (overlapping with the centroparietal P3 window) is indicated by the narrow gray box. (Right) Cortical sources of the LIAN.**

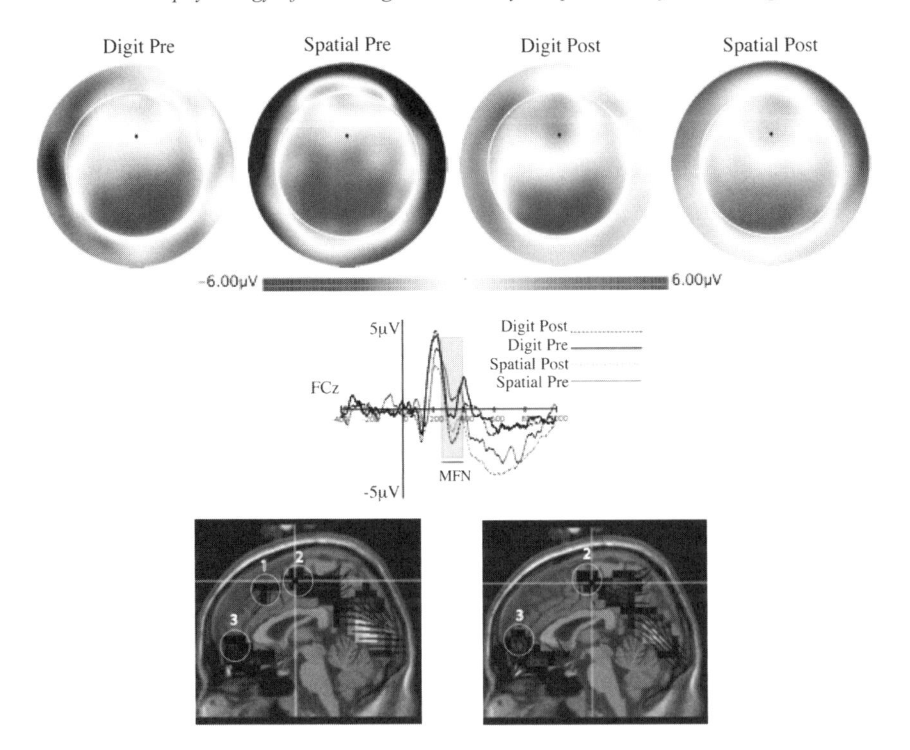

Figure 1.8. (Top and Middle) Scalp topography of contextual learning, as indexed by an increase in medial frontal negativity (MFN) post learning. Dots on the topographic maps indicate the channel location of the waveform plot below. (Bottom) Cortical sources of the MFN component for digit (left) and spatial (right) targets post learning.

From the learning model we also predicted that there would be hippocampal, PCC, and parietal lobe activity late in learning. As in an earlier study (Poulsen et al., 2005), the increase in the P300 closely tracked the increase in demonstrated task learning (see Figure 1.9). Source analysis indicated activity in the hippocampus, the PCC, and the parietal cortex, consistent with memory-updating processes and memory consolidation.

The Luu et al. (2007) study showed that source-localized EEG effects can track brain changes related to trial-and-error learning that are predicted from animal learning models, as well as from human fMRI studies. The results from this study provide unique clues about the time course of regional brain activation associated with the early learning process, as well as changes that occur with task acquisition. However, this study did not address the brain changes associated with extensive practice. Therefore, a second study was conducted using the digit version of the learning task. In this study participants completed four sessions, over four separate days. In the first session, they learned the digit-response mappings through trial and error. The second and third sessions provided additional practice

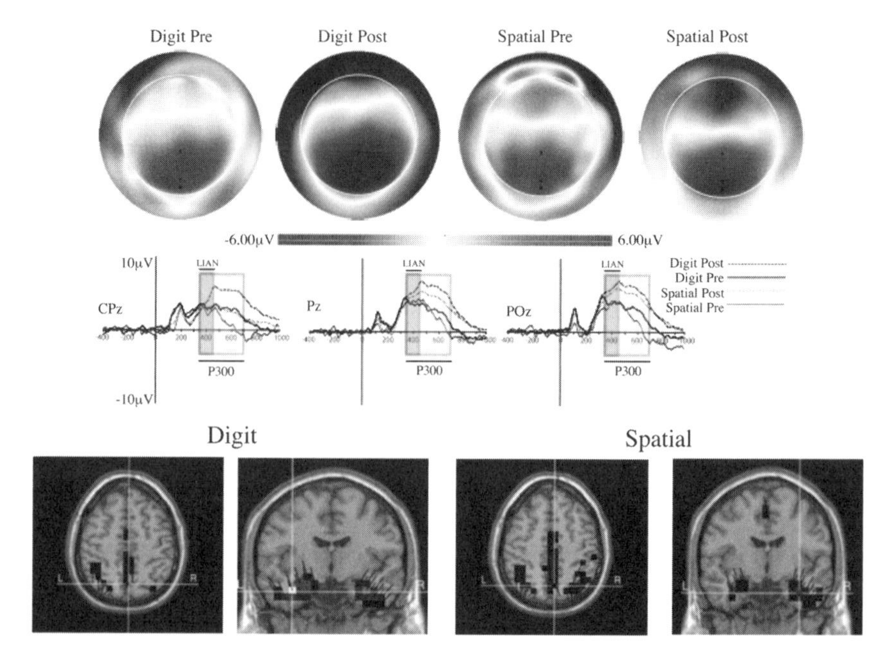

Figure 1.9. (Top and Middle) Scalp topography of the P300. Dots on the topographic maps indicate the channel location of the waveform plots below. The time window to quantify the P300 (overlapping with the anterior LIAN window) is indicated by the full box width. (Bottom) Cortical sources (including the hippocampus, the PCC, and the parietal cortex) of P300.

to reinforce these digit-response mappings. In the fourth session, subjects were required to learn new digit-response mappings.

Based on results obtained with the image analyst data, we anticipated learning-related changes over the occipital cortex at approximately 170 ms after stimulus onset. Data presented in Figure 1.10 confirm this prediction. As subjects learned the task, amplitude of the N170 increased. The most prominent increase was during the first session as subjects progressed from the prelearned stage to the learned stage. The N170 from subsequent sessions (second and third) displayed similar amplitudes to the learned stage of the first session. When subjects had to learn a new digit-response mapping, N170 amplitude decreased to a similar level as the first prelearned state. These results confirm a stimulus-specific learning effect.

The N170 progressed bilaterally along the temporal lobes, including the fusiform gyrus, and engaged the anterior temporal lobes by 250 ms. In the target detection task, only the experts had shown this temporal lobe engagement. In the extended practice experiment, however, this engagement was observed for all conditions, including the prelearning period. This was perhaps due to the nature of the stimuli (digits) in this experiment; all subjects have extensive experience with numbers, and activity along the fusiform gyrus may reflect this experience with familiar perceptual objects.

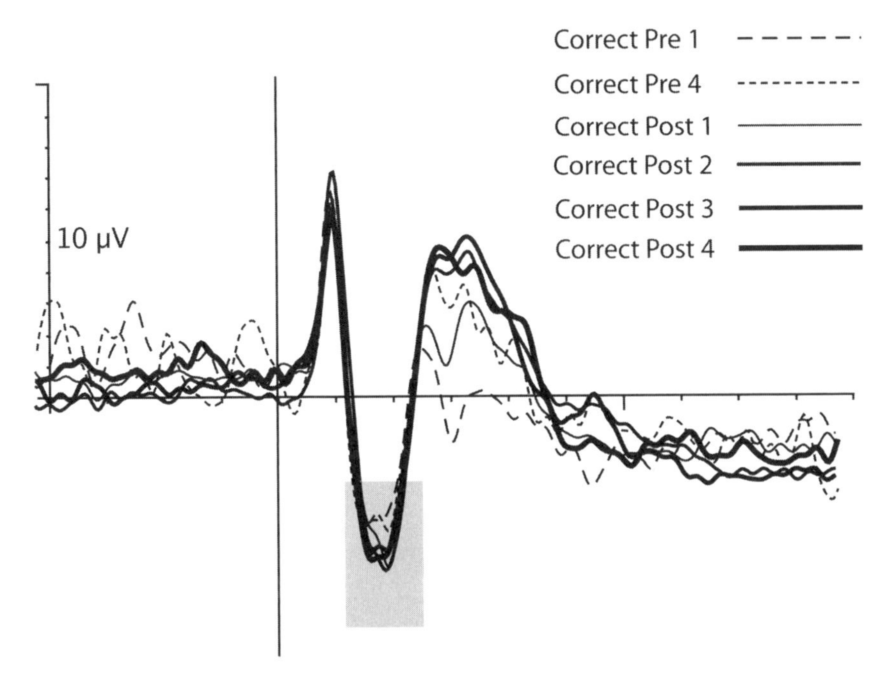

Figure 1.10. Waveform plots for a channel located over the left occipital region for correct responses in the four sessions. The prelearned conditions are illustrated for sessions 1 and 4. The gray rectangle identifies the N170.

Following this sequence of electrical events, we observed the development of a midline negativity that peaked at approximately 350 ms post stimulus. This negativity was absent in the prelearned state during the first session (see Figure 1.11), but increased across practice sessions 2 and 3, replicating and extending previous findings on the MFN (see Figure 1.8). Surprisingly, in the fourth session, when subjects were required to learn a new stimulus-response mapping, the enhanced negativity did not disappear, but rather remained, being of similar amplitude to the postlearned state of the first session. This suggests that the N350 reflects experience with the task and not the specific stimulus-response requirements of the task (compare with the N170 effect). Note the similarity between Figures 1.11 and 1.6.

Cortical sources for the midline negativity associated with extensive practice are shown in Figure 1.12. The figure on the left shows the cortical activity at 300 ms, when the midline negativity begins to develop. This is similar to the MFN we observed in the first learning study (see Figure 1.8), particularly for the ACC, medial occipital, and orbitofrontal sources. At 350 ms, the sources are much more posterior (PCC, Figure 1.12, middle; left parietal, Figure 1.12, right). The PCC source may be associated with the contextual representation of task parameters, whereas the parietal lobe activity may reflect the actual action-stimulus representation (Goodale & Milner, 1992).

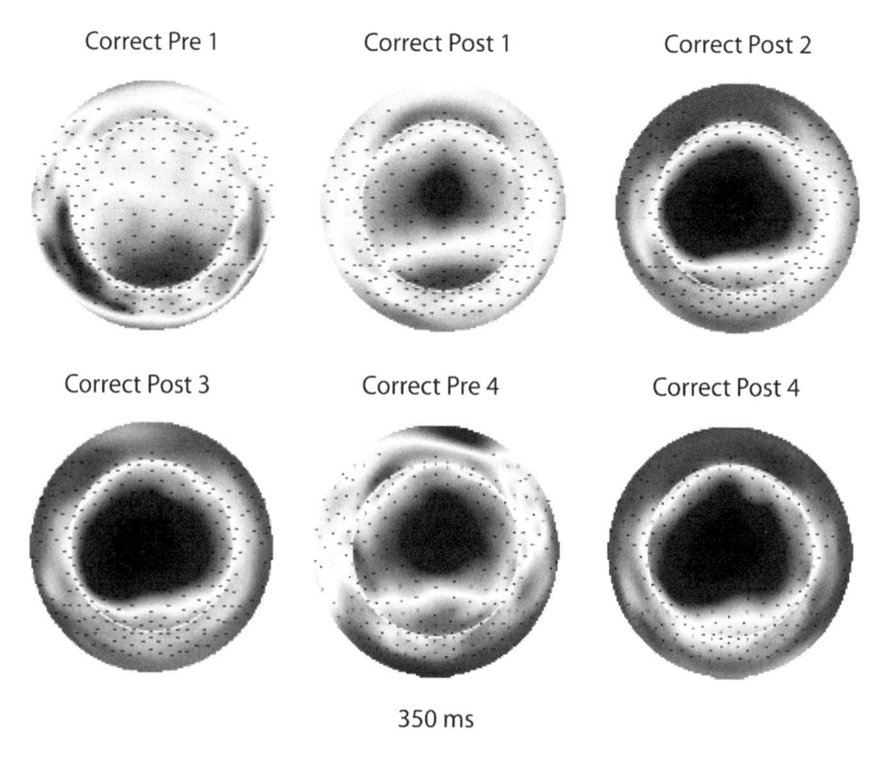

Figure 1.11. Topographic maps illustrating the MFN response at 350 ms post stimulus for correct responses before learning (sessions 1 and 4, pre), and after learning (sessions 1 through 4, post).

These studies revealed important learning-related brain activation that can be related to differences observed between expert image analysts and novices. The important findings are summarized here. First, from the learning studies, brain activity associated with both early and late learning systems was observed. The components of the early learning system, the inferior frontal lobe and the ACC, were engaged early in learning, whereas components of the late learning system, the PCC, the hippocampus, and sensory cortices, were engaged in the late stages of learning.

Second, several learning-related changes consistent with the observed differences between image analysts and novice subjects were observed within the late learning system, including changes in the N170 and the appearance of the centromedial negativity at 350 ms. The N170 indexes stimulus-specific experience, as new learning is associated with a decrease of its amplitude. In contrast, the centromedial negativity reflects the acquisition of general task parameters because it persists even when subjects are required to learn new stimulus-response mappings in the same task. The similarities between these two learning-related markers and expert-novice differences suggest that certain markers of expertise reflect experience-dependent changes that are a consequence of learning generally.

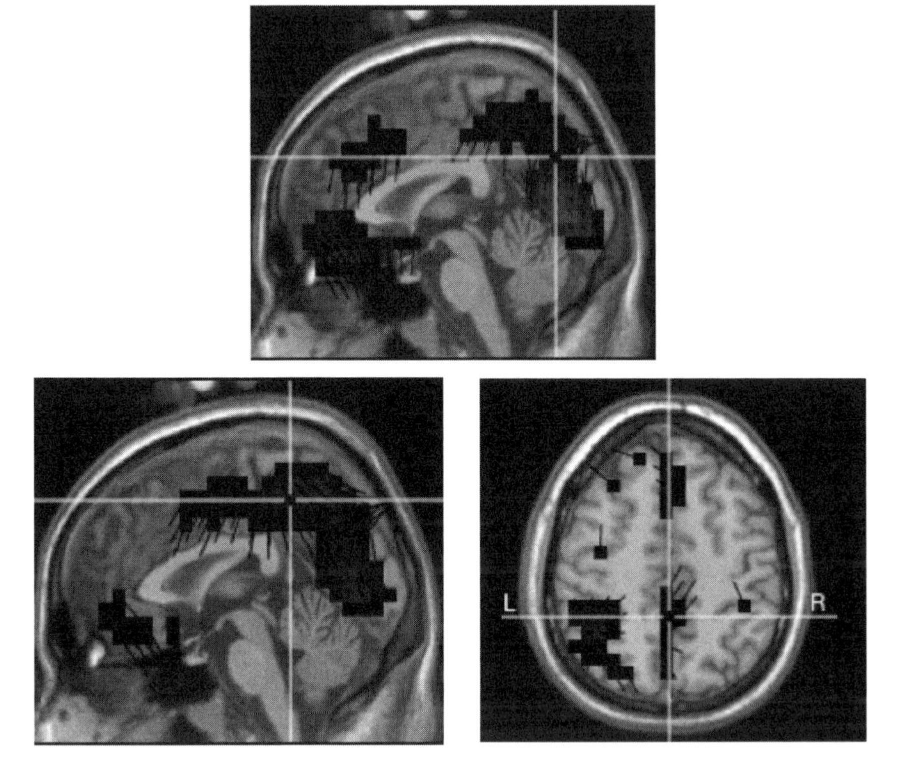

Figure 1.12. Sources of the MFN. (Top) 300 ms post stimulus. (Bottom) 350 ms post stimulus.

Third, there were expert-related neural signatures that were specific to image analysts. An N250 was present in image analysts' brain responses, but was absent in novices. In these learning studies, the N250 was observed in both the pre- and postlearned states. These results suggest that for the task of image analysis, which involves acquiring expertise in analysis of uncommon images (at least from the perspective of the experience of the general population), there may be specific changes to the higher order visual cortex. These changes may contribute to rapid conceptual extraction of information from image data, marking the attainment of expertise for image analysts.

EDUCATION AND TRAINING PROTOCOLS

We have theorized that the intrinsic motive biases of the dorsal (contextual feed-forward) and ventral (item based feedback) human memory systems cause human learning to be controlled for specific purposes of adaptive cognition (Luu et al., 2007; Tucker & Luu, 2006). Such learning is thus strategic, and it is tailored to the processing demands of performance. We reviewed EEG studies of skilled performance that have identified electrophysiological signatures of the engagement of these two circuits in response to processing demands early and late in learning. In translating both the theoretical rationale and the empirical measures into training implications, we apply three principles. First, effective training should provide instruction appropriate to the learner's current level. This applies not just to content knowledge and performance, but especially to underlying neural processing. Second, effective training should engage processes during learning and practice that match those later required in real world performance environments. This transfer-appropriate training entails introducing not simply the external task features during learning, but significantly the internal, neural dynamics underlying targeted performance. Third, effective training should monitor and adapt to the operation of, and interaction between, both these action control systems.

Emphasis Change: Emphasis on Strategic Control

The involvement of the ventral limbic circuit at the early stage of learning indicates the importance of feedback-guided instruction. Very early in learning, while stimulus-response representations are still being acquired, feedback carries a large informational load. As the task is acquired, knowledge of the correct response leads to the generation of expectancies, allowing the learner to employ greater endogenous (feed-forward) control and self-monitoring of performance. Recording of electrical brain responses to errors and feedback thus provides a sensitive index of the development of expectancies through learning. In our research, this progression from prelearned to learned behavior seems to be reflected in a reduction in the feedback-related negativity (exogenous guidance), concurrent with an increase in the error-related negativity (endogenous guidance). During this period, the learner is also developing conceptual short-term representations of the stimuli and their task relevancy (indicated by an increase in the amplitude of the N170) alongside the context of occurrence (indicated by an increase in P300) and the tracking of global task parameters (indexed by the N350). For task context and parameters to be accurately encoded in brain activity for later transfer to performance, the learner should, therefore, experience the task in its full context during training.

An existing approach to training, the emphasis-change protocol (Gopher, 1996, 2007) is compatible with these objectives. Emphasis-change training comprises a collection of methods, including training with variable priorities, emphasis change, secondary tasks, and task switching (Gopher, 2007). Common to all these methods is the introduction of systematic variability into training, guided task exploration with performance feedback, and experience with the task in its entirety throughout training. Based on cognitive task analysis, emphasis-change protocols attempt to match processing requirements during training to those required in a real world context. This approach is particularly well suited to the training of complex skills and has been applied effectively in several high performance training programs, including the training of air force pilots (Gopher, Weil,

& Bareket, 1994) and helicopter navigation with helmet-mounted displays (Seagull & Gopher, 1997). Whereas emphasis-change methods lead to slower initial learning rates, research indicates they result in high levels of attainment with superior transfer to the real world. Research into the neural mechanisms of learning as described in this chapter may provide additional insight into the effectiveness of emphasis-change programs and how they may be further adapted to optimize learning and transfer to operational performance.

Rote Training: Emphasis on Automatization

When learners become highly skilled in a given task domain, routine task components become *automatized*. As noted earlier, automaticity improves the efficiency of performance by reducing demands on cognitive resources and increasing the speed, fluency, and reliability of execution. The N170 and N250 components appear to index the development of automatized stimulus recognition during the later stages of learning and extended practice.

Cognitive experimental research indicates that automaticity is most effectively developed through repeated, consistent mapping between stimulus and response (Schneider & Fisk, 1982; Strayer & Kramer, 1994). A common element of many training protocols, therefore, is to provide rote practice to enhance automaticity. Automaticity, however, may also come at a cost. Automatic processes are triggered by external stimuli and are ballistic; that is, they are difficult to stop once initiated. In the absence of concurrent strategic control, automaticity can lead to rigid performance and potentially critical errors of commission. Moreover, automaticity achieved through rote practice is often abstracted from the real task context and associated goal-directed processing; consequently, rote training will typically transfer poorly to real performance situations. One challenge of training programs, therefore, is to provide extended practice environments that develop automaticity within a meaningful context that maintains the goal-directed nature of performance and strategic control dynamics (for example, Gatbonton & Segalowitz, 1988; Gopher, 1996).

Training for Complex, Demanding Environments

Complex environments create highly variable processing demands that require dynamic self-regulation. Although we have stressed the relative contribution of the ACC and PCC circuits in early versus late learning, respectively, variable environments require dynamic reengagement of the ACC circuit with its feedback control bias—even after learning has been achieved. A similar concept has been put forth by Goschke (2002), who characterized the demands of adaptive behavior as a dynamic balance between competing control dilemmas: persistence versus flexibility, stability versus plasticity, and selection versus monitoring. In cognitive terms, this behavior entails balancing automatization with cognitive flexibility and is the hallmark of an individual who can rapidly perceive, comprehend, and take action in a complex environment. We refer to this capability as flexible expertise. We highlight the importance of this dynamic balance in relation to two challenges of flexible expertise in complex environments—performance under stress and the maintenance of situational awareness.

High levels of stress in performance situations can arise from multiple sources, including cognitive overload, fatigue, threats to self-preservation, and even performance anxiety. The improved efficiency afforded by automaticity increases the resilience of performance to stress, fatigue, and distraction. It also frees up cognitive resources for allocation to situational awareness and strategic processing, such as monitoring the environment for critical events, assessing the outcomes of action, and coordinating the performance with others. The development of automaticity alone, however, does not ensure that essential, strategic attention processes will be engaged. Prior research on complex, semi-automated environments has found that most errors stem from lapses in situational awareness, particularly the failure to perceive and attend to critical information, to integrate this information, and to revise one's mental model of the situation (Endsley, 1995; Jones & Endsley, 1996). Most of these incidents have involved highly trained personnel. Although the automaticity of expertise affords many benefits, performance risks getting "stuck in set" making it difficult to recognize novelty and change one's dominant mode of response—despite the greater availability of resources. From the perspective of action-control theory, such suboptimal performance by experts in complex environments reflects an imbalance between the two control circuits, with overreliance on the feed-forward bias of the PCC system.

Training for optimal performance in complex environments must, therefore, exercise strategic switching between these action-control systems by, for example, introducing unpredictability and variability alongside continued practice for automaticity. In addition to guiding new learning, expectancy violation may be essential for maintaining attentional awareness and cognitive flexibility along with the benefits of automaticity. The introduction of variability during practice will trigger the engagement of the expectancy violation mechanism and will train learners to assess the relevancy of this violation and to adaptively adjust their response strategies. Such a protocol will promote an optimal, dynamic balance between the feed-forward PCC circuit and the feedback ACC circuit.

INTEGRATION OF NEURAL MEASURES INTO TRAINING AND EDUCATION

Coupled with sophisticated brain monitoring technologies, the neurophysiological model advanced in this chapter offers a new window into adaptive learning and performance processes. It provides sensitive neurometrics for developing online, tailored training protocols and for measuring learner progress and training program success. As described in the preceding sections, we have made initial progress in mapping these learning systems in humans. We think it is now feasible to apply the dense-array EEG technology to monitor learning in real time and thereby guide the training process.

Our near-term objective is to develop and test neural based assessment and training tools that will increase the rate and efficiency of domain-specific learning through real time adaptive feedback and information delivery, as well as augment domain-general performance. Our protocol targets three interrelated components of cognitive enhancement: automaticity, situational awareness, and cognitive flexibility. Neuroadaptive training will adjust task parameters (such as stimulus and feedback presentations and required responses) to the learning state and cognitive capacity of the learner. The training protocols will specifically target the engagement of the ACC and PCC based learning and memory systems and monitor their responses through online electrophysiological recording. Along with behavioral performance indicators, these neurophysiological markers of flexible expertise and feedback processing can be used as metrics to (1) assess a learner's current level of development in each of these cognitive capacities, (2) select candidates for targeted adaptive training, and (3) predict success in actual performance environments.

REFERENCES

Balleine, B. W., & Ostlund, S. B. (2007). Still at the choice-point: Action selection and initiation in instrumental conditioning. *Annals of the New York Academy of Sciences, 1104,* 147–171.

Campbell, G. E., & Luu, P. (2007, October). *A preliminary comparison of statistical and neurophysiological techniques to assess the reliability of performance data.* Paper presented at the 4th Augmented Cognition International Conference, Baltimore, MD.

Chein, J. M., & Schneider, W. (2005). Neuroimaging studies of practice-related change: fMRI and meta-analytic evidence of a domain-general control network for learning. *Cognitive Brain Research, 25,* 607–623.

Cohen, R. A., Kaplan, R. F., Moser, D. J., Jenkins, M. A., & Wilkinson, H. (1999). Impairments of attention after cingulotomy. *Neurology, 53,* 819–824.

Crane, S. M., Luu, P., Tucker, D. M., & Poolman, P. (2007). *Expertise in the human visual system: Neural target detection without conscious detection.* Manuscript in preparation.

Devinsky, O., & Luciano, D. (1993). The contributions of cingulate cortex to human behavior. In B. A. Vogt & M. Gabriel (Eds.), *Neurobiology of the cingulate cortex and limbic thalamus* (pp. 427–556). Boston: Birkhauser.

Donchin, E., & Coles, M. G. (1988). Is the P300 component a manifestation of context updating? *Behavioral and Brain Sciences, 11,* 357–374.

Endsley, M. R. (1995). Toward a theory of situation awareness in dynamic systems. *Human Factors, 37,* 32–64.

Falkenstein, M., Hohnsbein, J., Hoormann, J., & Blanke, L. (1991). Effects of crossmodal divided attention on late ERP components. II. Error processing in choice reaction tasks. *Electroencephalography and Clinical Neurophysiology, 78,* 447–455.

Gabriel, M. (1990). Functions of anterior and posterior cingulate cortex during avoidance learning in rabbits. *Progress in Brain Research, 85,* 467–482.

Gabriel, M., Burhans, L., Talk, A., & Scalf, P. (2002). Cingulate cortex. In V. Ramachandran (Ed.), *Encyclopedia of the human brain* (Vol. 1, pp. 775–791). San Diego, CA: Academic.

Gabriel, M., Sparenborg, S. P., & Stolar, N. (1986). An executive function of the hippocampus: Pathway selection for thalamic neuronal significance code. In R. L. Isaacson & K. H. Pribram (Eds.), *The hippocampus* (Vol. 4, pp. 1–39). New York: Plenum.

Gatbonton, E., & Segalowitz, N. (1988). Creative automatization: Principles for promoting fluency within a communicative framework. *TESOL Quarterly, 22,* 473–492.

Gauthier, I., Skudlarski, P., Gore, J. C., & Anderson, A. W. (2000). Expertise for cars and birds recruits brain areas involved in face recognition. *Nature Neuroscience, 3,* 191–197.

Gehring, W. J., Goss, B., Coles, M. G. H., Meyer, D. E., & Donchin, E. (1993). A neural system for error detection and compensation. *Psychological Science, 4,* 385–390.

Goodale, M. A., & Milner, D. A. (1992). Separate visual pathways for perception and action. *Trends in Neurosciences, 15,* 20–25.

Gopher, D. (1996). Attention control: Explorations of the work of an executive controller. *Cognitive Brain Research, 5*(1–2), 23–38.

Gopher, D. (2007). Emphasis change as a training protocol for high-demand tasks. In A. F. Kramer, D. A. Wiegmann, & A. Kirlik (Eds.), *Attention: From theory to practice* (pp. 209–224). New York: Oxford University Press.

Gopher, D., Weil, M., & Bareket, T. (1994). Transfer of skill from a computer game trainer to flight. *Human Factors, 36,* 387–405.

Goschke, T. (2002). Voluntary action and cognitive control from a cognitive neuroscience perspective. In S. Maasen, W. Prinz, & G. Roth (Eds.), *Voluntary action: An issue at the interface of nature and culture* (pp. 49–85). Oxford, United Kingdom: Oxford University Press.

Janer, K. W., & Pardo, J. V. (1991). Deficits in selective attention following bilateral anterior cingulotomy. *Journal of Cognitive Neuroscience, 3,* 231–241.

Jones, D. G., & Endsley, M. R. (1996). Sources of situation awareness errors in aviation. *Aviation, Space and Environmental Medicine, 67,* 507–512.

Keng, E., & Gabriel, M. (1998). Hippocampal modulation of cingulo-thalamic neuronal activity and discriminative avoidance learning in rabbits. *Hippocampus, 8,* 491–510.

Luu, P., Flaisch, T., & Tucker, D. M. (2000). Medial frontal cortex in action monitoring. *Journal of Neuroscience, 20*(1), 464–469.

Luu, P., & Pederson, S. M. (2004). The anterior cingulate cortex: Regulating actions in context. In M. I. Posner (Ed.), *Cognitive neuroscience of attention* (pp. 232–244). New York: Guilford Press.

Luu, P., & Tucker, D. M. (2003). Self-regulation and the executive functions: Electrophysiological clues. In A. Zani & A. M. Proverbio (Eds.), *The cognitive electrophysiology of mind and brain* (pp. 199–223). San Diego, CA: Academic Press.

Luu, P., Tucker, D. M., Derryberry, D., Reed, M., & Poulsen, C. (2003). Electrophysiological responses to errors and feedback in the process of action regulation. *Psychological Science, 14,* 47–53.

Luu, P., Tucker, D. M., & Stripling, R. (2007). Neural mechanisms for learning actions in context. *Brain Research, 1179,* 89–105.

McKone, E., Kanwisher, N., & Duchaine, B. C. (2007). Can generic expertise explain special processing for faces? *Trends in Cognitive Sciences, 11*(1), 8–15.

Miltner, W. H., Lemke, U., Weiss, T., Holroyd, C., Scheffers, M. K., & Coles, M. G. (2003). Implementation of error-processing in the human anterior cingulate cortex: A source analysis of the magnetic equivalent of the error-related negativity. *Biological Psychology, 64*(1–2), 157–166.

Poremba, A., & Gabriel, M. (2001). Amygdalar efferents initiate auditory thalamic discriminative training-induced neuronal activity. *Journal of Neuroscience, 21,* 270–278.

Posner, M. I., & DiGirolamo, G. J. (1998). Executive attention: Conflict, target detection, and cognitive control. In R. Parasuraman (Ed.), *The attentive brain* (pp. 401–423). Cambridge, MA: MIT Press.

Potter, M. C. (1999). Understanding sentences and scenes: The role of conceptual short-term memory. In V. Coltheart (Ed.), *Fleeting memories: Cognition of brief visual stimuli* (pp. 13–46). Cambridge, MA: MIT Press.

Poulsen, C., Luu, P., Davey, C., & Tucker, D. M. (2005). Dynamics of task sets: Evidence from dense-array event-related potentials. *Cognitive Brain Research, 24,* 133–154.

Poulsen, C., Luu, P., Tucker, D. M., Scherg, M., Davey, C., & Frishkoff, G. (2003, March). *Electrical brain activity during task switching: Neural source localization and underlying dynamics of scalp-recorded potentials.* Poster presented at the Tenth Annual Meeting of the Cognitive Neuroscience Society, New York.

Pribram, K. H. (1991). *Brain and perception: Holonomy and structure in figural processing.* Hillsdale, NJ: Erlbaum.

Rescorla, R. A., & Wagner, A. R. (1972). A theory of Pavlovian conditioning: Variations in the effectiveness of reinforcement and nonreinforcement. In A. H. Black & W. F. Prokasy (Eds.), *Classical conditioning II: Current research and theory* (pp. 64–99). New York: Appleton-Century-Crofts.

Rylander, G. (1947). Personality analysis before and after frontal lobotomy. In J. F. Fulton, C. D. Aring, & B. S. Wortis (Eds.), *Research publications—Association for research in nervous and mental disease: The frontal lobes* (pp. 691–705). Baltimore, MD: Williams & Wilkins.

Schneider, W., & Fisk, A. D. (1982). Degree of consistent training. Improvements in search performance and automatic process development. *Perception and Psychophysics, 31,* 160–168.

Schneider, W., & Shiffrin, R. M. (1977). Controlled and automatic human information processing: 1. Detection, search, and attention. *Psychological Review, 84,* 1–66.

Scott, L. S., Tanaka, J. W., Sheinberg, D. L., & Curran, T. (2006). A reevaluation of the electrophysiological correlates of expert object processing. *Journal of Cognitive Neuroscience, 18,* 1453–1465.

Seagull, F. J., & Gopher, D. (1997). Training head movement in visual scanning: An embedded approach to the development of piloting skills with helmet mounted displays. *Journal of Experimental Psychology: Applied, 3,* 463–480.

Segalowitz, N., Poulsen, C., & Segalowitz, S. (1999). RT coefficient of variation is differentially sensitive to executive control involvement in an attention switching task. *Brain and Cognition, 40,* 255–258.

Segalowitz, N., & Segalowitz, S. (1993). Skilled performance, practice, and the differentiation of speed-up from automatization effects: Evidence from second language word recognition. *Applied Psycholinguistics, 14,* 369–385.

Shiffrin, R. M., & Schneider, W. (1977). Controlled and automatic human information processing: II. Perceptual learning, automatic attending, and a general theory. *Psychological Review, 84,* 127–190.

Squire, L. R. (1998). Memory systems. *Comptes Rendus de l'Academie des Sciences. Serie III, Sciences de la Vie, 321,* 153–156.

Strayer, D. L., & Kramer, A. F. (1994). Strategies and automaticity: I. Basic findings and conceptual framework. *Journal of Experimental Psychology: Learning, Memory, and Cognition, 20,* 318–341.

Swick, D., & Turken, A. U. (2002). Dissociation between conflict detection and error monitoring in the human anterior cingulate cortex. *Proceedings of the National Academy of Science, 99,* 16354–16359.

Tanaka, J. W., & Curran, T. (2001). A neural basis for expert object recognition. *Psychological Science, 12,* 43–47.

Toni, I., Ramnani, N., Josephs, O., Ashburner, J., & Passingham, R. E. (2001). Learning arbitrary visuomotor associations: Temporal dynamic of brain activity. *Neuroimage, 14,* 1048–1057.

Tucker, D. M., & Luu, P. (Eds.). (2006). *Adaptive binding.* New York: Oxford University Press.

Tucker, D. M., Luu, P., Desmond, R. E., Jr., Hartry-Speiser, A., Davey, C., & Flaisch, T. (2003). Corticolimbic mechanisms in emotional decisions. *Emotion, 3,* 127–149.

Part II: Cognitive/Rational Band

Chapter 2

THE ROLE OF INDIVIDUAL DIFFERENCES IN VIRTUAL ENVIRONMENT BASED TRAINING

Clint Bowers, Jennifer Vogel-Walcutt, and Jan Cannon-Bowers

Most modern theories of learning and instructional design converge on the conclusion that the attributes that learners bring to the instructional environment are important ingredients in the learning process. In fact, the notion that learners bring a unique set of knowledge, skills, aptitudes, abilities, preferences, and experiences to a learning environment is captured by a popular approach known as learner-centered instruction (for example, see Cognition and Technology Group at Vanderbilt [CTGV], 2000; Bransford, Brown, & Cocking, 1999; Kanfer & McCombs, 2000; Clark & Wittrock, 2000). Essentially, proponents of this approach argue that characteristics of the learner must be taken into account in the design and delivery of instruction and that an explicit attempt must be made to build on the strengths of the student. Variables that have been implicated in this regard include prior knowledge, prior skill, prior experience, misconceptions, and interests, among others. In addition, a number of other personal attributes have been shown to affect learning. These include motivation (CTGV, 2000; Clark, 2002; Clark & Wittrock, 2000; Bransford, Brown, & Cocking, 1999), personal agency/self-efficacy (Kanfer & McCombs, 2000; Bandura, 1977, 1986; Gist, 1992), goal orientation (Dweck, 1986; Dweck & Legget, 1988; Bransford, Brown, & Cocking, 1999; Kanfer & McCombs, 2000), goal commitment (Kanfer & McCombs, 2000), emotional state (Clark, 2002), self-regulation (Kanfer & McCombs, 2000), misconceptions (Gentner, 1983), interest (Kanfer & McCombs, 2000), instrumentality (Tannenbaum, Mathieu, Salas, & Cannon-Bowers, 1991), ability (Mayer, 2001), and spatial ability (Mayer, 2001).

Clearly, the findings cited above (as well as other work) justify the study of learner characteristics as an important variable in technology-enabled learning. Interestingly, the popular notion that people have unique learning styles—that is, some people learn differently than others—has not been substantiated by the empirical literature (for example, see Cronbach & Snow, 1977). Nonetheless, a host of learner characteristics clearly do affect learning; in fact, it is quite likely these learner attributes will interact with instructional approaches, such that some interventions may be more effective for some learners than others.

At the same time, researchers have also discussed the degree to which individual differences might influence an individual's experience in virtual environments. For example, Kaber and his colleagues discuss this problem from the viewpoint of virtual environment (VE) design in their recent chapter (Kaber, Draper, & Usher, 2002). They discuss a variety of factors that might determine the degree to which individuals perceive or experience the elements of the virtual environment as "real." They include a review of variables such as user experience, spatial aptitude, age, and so forth. They describe the effect that these variables may have in the degree of immersion and other experiential variables associated with virtual environments and how designers might consider them.

In this chapter, we seek to combine these two bodies of literature described above to discuss those *individual differences that are most likely to influence learning outcomes in virtual environments.* As such, we will focus on the subset of variables mentioned above. In so doing, we will discuss the state of the existing literature and also the research that is required to more fully understand these important relationships.

TYPES OF INDIVIDUAL DIFFERENCES THAT AFFECT LEARNING IN VIRTUAL ENVIRONMENTS

In reviewing the many variables that might be included in this review, it became apparent there were two distinct classes of individual characteristics that might affect learning in a virtual environment. One of these classes includes a set of immutable characteristics, such as gender, age, and cognitive ability. The second class involves more alterable characteristics, such as attitudes, expectations, and experiences.

According to Figure 2.1, both classes of variables are predicted to affect learning in a virtual environment. Moreover, at least some of these variables (for example, cognitive ability) may also have a direct impact on learning outcomes. The following sections review what is known about the influence of individual differences in learning, with specific emphasis on virtual environment based learning systems.

IMMUTABLE CHARACTERISTICS

As noted, there are a number of individual variables that are fairly consistent in learners over time, and probably not amenable to much alteration. It is important to understand how these features operate since they can affect the degree of learning expected. Furthermore, it may be worthwhile designing variations in the learning system's design to accommodate various users. Obviously, concerns such as cost and configuration management will have implications on how much the VE can be altered for different users; however, if learning is affected significantly, the cost may be justified. The following sections summarize what is known about the impact of stable individual differences in VE based learning.

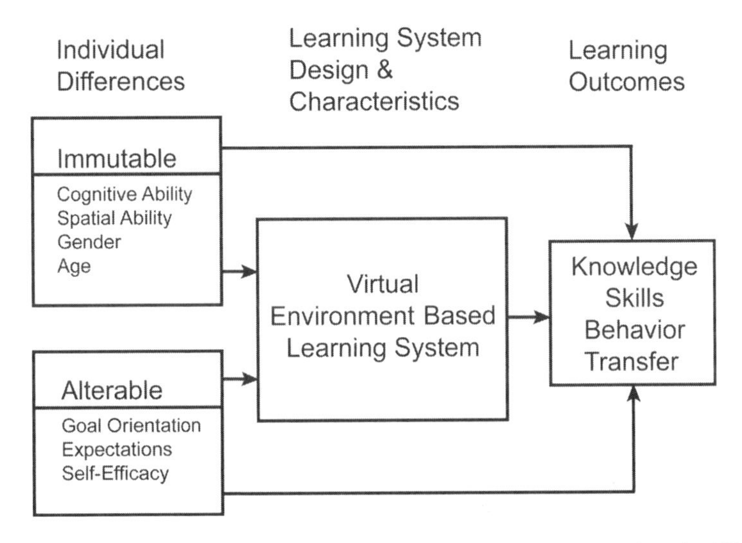

Figure 2.1. Model of Individual Characteristics on Learning in Virtual Environments

Age

Perceptual and cognitive skills decline as people age; consequently, these changes must be considered when creating training for VE. Specifically, Paas, Van Gerven, and Tabbers (2005) note that there is a reduction in working memory, cognitive speed, inhibition, and integration likely negatively influencing working and learning in dynamic, complex computer systems. Several researchers have investigated these issues using principles derived from cognitive load theory, multimedia theory, and the human-computer interaction literature and will be summarized below.

Age is a general predictor of learning to use computers (Egan, 1988). Because the aging process impacts perceptual and cognitive skills, it may also impact interactions with VE systems (Birren & Livingson, 1985; Fisk & Rogers, 1991; Hertzog, 1989; Salthouse, 1992, as cited in Kaber, Draper, & Usher, 2002). Given that elderly learners are dependent on a more limited pool of cognitive resources, they appear to learn at a slower pace than younger learners, and this becomes more pronounced in complex tasks requiring substantial mental processing (Fisk & Warr, 1996; Salthouse, 1994, 1996, as cited in Paas, Van Gerven, & Tabbers, 2005). Those areas affected by the decline of resources include working memory, cognitive speed, inhibition, and integration. Reduced working memory capacity means that there is essentially less space available to hold thoughts for assimilation before being discarded or filed in long-term memory. To address this problem, Paas, Van Gerven, and Tabbers suggest that, in an effort to maximize the efficiency of the available cognitive resources in older adults, efforts should be made to reduce extraneous cognitive load and increase or focus the learner's attention on the germane cognitive load. Because aging adults experience more limited capacity for cognitive load and slower mental processing speeds, the cognitive space available may be diminished enough to significantly affect learning in VEs (Clark, Nguyen, & Sweller, 2006). To accomplish this, Paas, Van Gerven, and Tabbers suggest using bimodal presentations (utilizing both the auditory and visual channels), eliminating redundant information (extraneous cognitive load), and, to a lesser degree, using enhanced timing, attention scaffolding, and enhanced layout. However, it is not clear whether the results of this approach would result in transfer of learning in virtual environments. Additional research in this regard is needed.

Communication Skills

While there have been many personality factors implicated in learning and (particularly) performance, we focus attention here on the collaborative aspects of VEs. Collaborative virtual environments (CVEs) allow people to work together in a shared virtual space and communicate both verbally and nonverbally in real time (Bailenson, Beall, Loomis, Blascovich, & Turk, 2004). When communicating online, people display less social anxiety, fewer inhibitions, and reduced self-awareness. These differences may help to reduce stress, promote a more relaxed atmosphere for learning compared to a classroom setting (Roed, 2003), and support social awareness (Prasolova-Forland & Divitini, 2003a, 2003b).

Self-concept and self-representation may also play a role in VE performance. For example, using an avatar (self-representation) may impact virtual group dynamics and often reflects the true personalities of the users. Thus, leaders would generally take on leadership roles in collaborative virtual environments, while followers would take on a more passive role. Other important characteristics that impact performance in collaborative virtual environments include user dedication in virtual tasks, sense of group cohesiveness, and willingness to collaborate (Kaber, Draper, & Usher, 2002). Further, performance in collaborative virtual environments is not only dependent on the user, but also on the feeling of immersion and group acceptance in the VE (Slater & Wilber, 1997). Those who are more open to the experience generally perform better in these environments. On the other hand, those who do not effectively collaborate with others in the VE may actually degrade group performance by acting as a distracter from individual task performance (Romano, Drna, & Self, 1998).

Recommendations regarding these issues focus on display transparency, realistic avatars, and individual differences in group work. Attention should be paid to these areas of concern in an effort to improve performance in collaborative virtual environments and decrease the extraneous cognitive load for the participants. Such reductions will help improve the focus on the pertinent or germane information for training and improve group dynamics for learning.

Spatial Ability and Memory

As noted, spatial ability appears to be an important potential variable in the effectiveness of virtual environments as agents of learning. Both Garg, Norman, and Sperotable (2001) and Waller (2000) found mitigating effects of spatial ability in learning. Accordingly, Mayer (2001) suggests that the design of multimedia knowledge presentations should be adjusted according to the learner's prior knowledge and spatial ability. Clearly, the demands imposed on learners in an interactive, three-dimensional VE are different from those associated with more passive forms of instruction (for example, listening to a lecture or reading a textbook). Precisely what these differences translate to in terms of learning effects and design implications remains to be seen.

Several studies have investigated the relationship between spatial abilities and the ability to navigate in VEs and/or acquire spatial knowledge from VEs. For example, it has been found that novice users experience more difficulties using VEs, especially when they have to navigate through small spaces such as virtual doorways (Kaber, Draper, & Usher, 2002). It is likely that these difficulties derive from the difficulty of learning how to use the controls. As such, designers should attempt to make control interfaces as natural as possible to promote time to proficiency for novices (Kaber, Draper, & Usher, 2002).

Previously, variable results have been found with regard to the connection between spatial abilities in the real world and the ability of the user to acquire spatial knowledge from a VE (Waller, 2000). However, in Waller's research, there is support for the connection. Specifically, it was found that spatial visualization and spatial orientation skills correlated with the ability to acquire spatial information from a VE. In agreement, scores on field independence tests and figure rotation ability correlate with task performance in a VE (Oman et al., 2000). Thus, it would seem one could test users with paper and pencil tests of spatial abilities and better predict their abilities to function in VEs. Not surprisingly, Chen, Czerwinski, and Macredie (2000) found that those users with lower spatial abilities were slower and produced more errors than the high spatial abilities group. It is hypothesized that the difference is likely due to reduced navigation skills. Interestingly though, Tracey and Lathan (2001) report that those users with lower spatial skills demonstrated a significant positive transfer from simulation based training to real world tasks compared to users with high spatial abilities. Although it appears counterintuitive that those users with lower spatial abilities would perform worse in the VE but then better transfer the skills to the real world, it is possible that by moving slowly through the system, these users would be better able to master the skills learned and thus allow for better transfer of those skills.

However, in a meta-analysis done by Chen and Rada (1996), it was determined that those users with higher spatial abilities may be able to create structural mental models more quickly than users with lower spatial abilities as evidenced by the reduced need to review the table of contents. In other words, those with better spatial skills are able to more quickly learn the layout of the VE system and thus perform better and quicker. Additionally, this increased speed of acquisition of knowledge may act to reduce the extraneous cognitive load of navigation skills allowing the user to focus attention on the training materials. More research in this area is needed to better clarify these connections.

Cockburn and McKenzie (2002) found a negative correlation between freedom to locate objects in VEs and performance with users reporting that environments with more dimensions were perceived as more "cluttered" and less efficient. These findings may be explained through the theory of cognitive load in that the additional information acts as a distracter from the germane information needed to navigate and perform in the environment. As such, for training development purposes attention should be paid to balancing interest in having a high level of fidelity or functionality in the VE and the possibility of cognitive overload in the user. Information and functionality choices should be made meaningfully and with the intention of supporting the training goals to achieve maximum performance.

Spatial Abilities and Gender

The differential impact of gender on training in virtual environments is an area not widely studied in the current literature. However, three main themes have been addressed intermittently and are discussed widely in similar literatures such as human-computer interaction, interactive and serious games, and online or distance learning. From these areas of research, we draw out these themes, namely, differences in spatial abilities, language proficiency, and differential levels of computer usage. Generally speaking, females exhibit not only a lower level of spatial ability (Waller, 2000), but research has shown that they actually use different parts of the brain to attend to spatial tasks (Vogel, Bowers, & Vogel, 2003). Specifically, while males showed a right hemisphere preference for spatial tasks; females failed to show a hemispheric preference, meaning that they use both hemispheres equally. Implications of the use of the left hemisphere by females during spatial tasks may be a result of language usage during spatial tasks. The increased level of language proficiency by females may impact how they interact with the system and their interpretation of the verbal components of the training. Finally, the amount of time spent on the computer has a general positive correlation with the level of comfort using the system thereby influencing the learner's perceived cognitive task load during activities (Egan, 1988). Because the average female uses computers less frequently than males (Busch, 1995), this difference may influence cognitive task load and has the potential to indirectly impact training in virtual environments.

Differences by gender in spatial abilities may influence the ability to effectively function in a VE and the strategies used to acquire spatial information in a VE (Sandstrom, Kaufman, & Huettel, 1998). Research has shown that individuals with higher spatial skills are generally better able to perform in graphic or spatially oriented interfaces (Borgman, 1989); as such, we would expect that females would be at a disadvantage in virtual environments. Because the role of spatial skills is even more important in virtual environments than it is in the real world (Waller, 2000), they may directly impact females' ability to navigate the environment, with males significantly outperforming females (Tan & Czerwinsk, 2006). To mitigate these differences, females are often given extended time to work within the system. Clearly, however, there is a need to see if this extra practice results in an effective learning outcome. Alternatively, research has shown that increasing the size of the display not only increases spatial abilities across genders, allowing them to better perform in the system, but additionally acts to attenuate the differences between the genders in performance (Tan & Czerwinsk, 2006). This might be another strategy for equating the genders in terms of learning outcomes.

Finally, gender may interact with other important individual characteristics, such as computer experience. Not only do males use computers more often (Busch, 1995), research also shows that males spend more time playing video games than females (Colwell, Grady, & Rhaiti, 2006). This increased experience level has the eventual effect of greater familiarity of VEs and consequently lessens the extraneous cognitive load placed on the user. Therefore, females may become overwhelmed more quickly in virtual environments and might benefit from easier scenarios in early training trials.

Clearly, this is a topic worthy of further study, since it is crucial to understand whether females can benefit equally from VE based training. It is also worth noting that gender differences may eventually fade as girls are exposed to computers, virtual worlds, and computer games at an earlier age. In fact, it may be that gap between the genders, in terms of use of technology, that may be closing.

Cognitive Ability

There is a vast literature implicating trainee cognitive ability as an important consideration in instructional design (see Hunter, 1986; National Research Council, 1999; Ree, Carretta, & Teachout, 1995; Ree & Earles, 1991; Colquitt, LePine, & Noe, 2000; Randel, Morris, Wetzel, & Whitehill, 1992; Quick, Joplin, Nelson, Mangelsdorff, & Fiedler, 1996; Kaemar, Wright, & McMahan, 1997; Warr & Bunce, 1995), so we will not review that here. Suffice it to say that learners with higher cognitive ability tend to fare better in training, but many other variables are also involved. We do not have reason to believe this will be different for virtual environments. However, an interesting question posed with respect to cognitive ability and virtual environments is that they may have the unique characteristic of being able to mitigate the impact of low cognitive ability in learning. For example, as part of their work in anchored instruction, the Cognition and Technology Group at Vanderbilt (CTGV) discovered that material presented with dynamic visual support (as opposed to only linguistic support) was beneficial to young learners with lower linguistic competencies (Sharp et al., 1995). It is quite possible that the flexibility of presentation of content afforded by virtual environments can enable them to be tailored to the needs of the learner or compensate for any incoming weaknesses.

However, it is clear that specific cognitive abilities might enable learners to gain the maximum from virtual environment based training (and vice versa). For example, Yeh (2007) reported that teachers with specific abilities, such as interpersonal intelligence, obtained more benefit from a simulation based training experience to train classroom management skills. Kizony, Katz, and Weiss (2004) report that several cognitive abilities correlated with performance by stroke patients on virtual psychomotor tasks. The cognitive abilities included attention, visual search, and visual memory. Each of these was related to at least one aspect of performance.

Obviously, various cognitive abilities are predictive of performance in virtual environments. What is not known, however, is *which* abilities are important in these environments universally, and which are related to the elements of the learning task. Thus, there is a need for theory development, as well as the consequent empirical research, to explore this important area more fully.

Alterable States

As mentioned, there is also a class of individual differences that is considered to be states, rather than traits. Hence, these individual differences can be modified prior to training. In fact, understanding how to best prepare a learner for training can have important implications for learning outcomes. For example, if a learner has poor self-efficacy prior to training, it may be relatively easy (and cheap) to raise it before commencing training. It is likely to be more cost-effective to modify the attitude than potentially wasting time in a sophisticated virtual environment. Hence, efforts to understand and modify these so-called "alterable" differences should be made.

Prior Knowledge and Experience

Several studies have found that the learner's prior knowledge and experience in a domain affect his or her ability to learn new material. In particular, it has been found that learners with less knowledge and experience need a more structured learning environment (Hannafin, Oliver, & Sharma, 2003) and more direct scaffolding (CTGV, 2000). One goal of virtual environment based education, therefore, should be to activate prior knowledge in the creation of new knowledge—an application of the scaffolding approach to training (Kirkley et al., 2004).

Another approach would be to provide "pretraining" to ensure that all trainees have an adequate pool of knowledge to enable the virtual experience to be effective (Bingener et al., 2008). Yet another approach involves assessing the prior knowledge of trainees and selecting, or creating, scenarios that are designed to take the trainee to the next level of knowledge. This approach has been shown to be effective in a variety of settings using simulation based training (for example, Birch et al., 2007). However, this approach requires not only a large library of scenarios, but a selection approach to choose scenarios to fit a given level of prior knowledge.

It is also important to consider the experience of trainees with the technology and the interface that links that technology to the user. Researchers have shown that previous experience with a virtual environment is a significant predictor of subsequent performance in that environment (Draper, Fujita, & Herndon, 1987). Further, it might be that the process of developing a mental model of the virtual environment interferes with the development of a mental model of the actual learning material (Kaber, Draper, & Usher, 2002). It may be that there is a requirement to "learn" the interface before a trainee is comfortable enough to have a positive learning experience. However, in one case, differences between those who were familiar with the interface and those who were not were made equivalent with a few practice trials (Frey, Hartig, Ketzel, Zinkernagel, & Moosbrugger, 2007). It is not clear whether this type of intervention is routinely effective; further research is needed to better understand this phenomenon.

Goal Orientation

It has been suggested that the experience that a learner might have in a virtual environment can be affected by a variety of attitudes that the learner brings to the situation. One attitude that seems particularly important in this regard is goal orientation. Goal orientation refers to the learner's motivation when approaching the learning task. Researchers in this area argue that learners can be either performance or mastery oriented (Brett & VandeValle, 1999; Ford, Weissbein, Smith, Gully, & Salas, 1998; Phillips & Gully, 1997; Dweck, 1986; Dweck & Leggett, 1988; Elliot & Harackiewicz, 1994; Grieve, Whelan, Kottke, & Meyers, 1994; Kozlowski, Gully, McHugh, Salas, & Cannon-Bowers, 1996; Covington, 2000). Performance-oriented individuals are concerned with their performance outcomes (for example, maximizing their scores in a game), while mastery-oriented individuals are concerned more about their own learning process by which task competence is achieved (Kozlowski, Gully, Salas, & Cannon-Bowers, 1996). Further, mastery goals are self-referenced and focus on improving individual skills over past performance on a task (Harackiewicz & Elliot, 1993). Evidence also suggests that mastery goals may lead to faster task acquisition and task self-efficacy; improve perceptions of competence, mood, and motivation; and provide higher levels of enjoyment (see Cannon-Bowers & Salas, 1998). Performance goals, on the other hand, are more immediate and outcome oriented. They focus attention on achieving specific performance targets, but not on the strategies used to obtain them.

There has been some evidence to suggest that performance-oriented individuals are more likely to perform better in training, but worse on transfer tasks, while the opposite appears to be true for mastery-oriented learners (Dweck, 1986). This is typically explained as being due to the fact that a performance orientation leads learners to figure out one or two strategies that maximize immediate performance, but that they fail to establish more flexible, generalizable strategies. In virtual environments, this variable may be problematic if it is found that typical gaming features (such as score keeping and competition) exacerbate a trainee's tendency toward performance goals. To ameliorate this, it may be necessary for designers to explicitly develop early content that emphasizes mastery, or strategies embedded in game play that require trainees to explore more generalizable strategies.

Interestingly, it seems that goal orientation might also influence the manner in which trainees approach virtual environment based training (Hwang & Li, 2002;

Schmidt & Ford, 2003). For example, it has been demonstrated that learners with a mastery orientation achieved better learning outcomes in a computer-mediated learning environment than did those with a performance orientation (Klein, Noe, & Wang, 2006). However, the mechanism that underlies this difference is not well understood. It has been suggested that trainees with a performance orientation are more likely to possess an "entity theory" of the learning situation. That is, they are more likely to see ability as a fixed, inflexible quantity. It has also been argued that these individuals are also more likely to attribute their failures or frustrations to elements of the virtual environments (Hwang & Li, 2002). Furthermore, their data suggest that trainees with a mastery orientation are more likely to enjoy virtual training environments more than those with a performance orientation (Yi & Hwang, 2003). One might suggest that a similar relationship would exist between the constructs of goal orientation and presence, mediating the impact on learning. This is a relationship that requires additional study.

It should also be noted that the effectiveness of any virtual learning environment is dependent upon the scenario that is deployed in that system. An understanding of the impact of learner control on the effectiveness of training might help in the creation of these scenarios. For example, it might be advisable to provide trainees with a performance orientation with scenarios that provide less challenge and risk for error and to increase this challenge very gradually. More efficient training might be accomplished with mastery-oriented individuals using more challenging scenarios (Heimbeck, Frese, Sonnentag, & Keith, 2003).

Expectations for Training

Researchers have suggested that preconceived attitudes toward training are effective predictors of subsequent training outcomes (Cannon-Bowers, Salas, Tannenbaum, & Mathieu, 1995; Smith-Jentsch, Jentsch, Payne, & Salas, 1996). This is potentially a troublesome finding for scientists interested in using virtual environments for training. Specifically, it might be that the virtual environment is so different from the trainees' experience that it colors their expectations about how effective it can be, adversely influencing downstream learning. Indeed, some trainees may not readily accept the use of a virtual environment to train "serious" knowledge and skills. While there is virtually no research with which to evaluate this hypothesis, it is clear that some students are at least dubious about being educated in computer based environments (Chiou, 1995; Hunt & Bohlin, 1991). For example, it has been demonstrated that trainees who participated in a "simulation" designed to train decision-making skills had higher expectations, and learned more, than trainees who experience a "game" with the same learning content (Baxter, Ross, Phillips, & Shafer, 2004). Further, it has been suggested that the fidelity of training needs to be increased along with the experience of the trainee in order to keep trainees' expectations appropriately high (Merriam & Leahy, 2005).

Similarly, even if the trainee has positive expectations about the outcomes of this type of learning, the effects might also be mitigated by negative attitudes of instructors or coaches who are adjunct to the learning experience (MacArthur & Malouf, 1991). Consequently, there may be benefit in developing training interventions to improve the attitudes of instructors toward virtual environments.

Self-Efficacy

Much work has focused on the contribution of the student's self-efficacy in training. Self-efficacy is defined generally as the learner's belief that he or she has the necessary competence to accomplish the task (Bandura, 1982, 1989; Gist, Schwoerer, & Rosen, 1989; Gist, Stevens, & Bavetta, 1991; Cole & Latham, 1997; Eden & Aviram, 1993; Ford et al., 1998; Mathieu, Martineau, & Tannenbaum, 1993; Martocchio, 1994; Martocchio & Webster, 1992; Mathieu, Tannenbaum, & Salas, 1992; Quiñones, 1995; Mitchell, Hopper, Daniels, & George-Falvy, 1994; Phillips & Gully, 1997; Stevens & Gist, 1997; Stajkovic & Luthans, 1998). In general, studies converge on the conclusion that high self-efficacy learners perform better in training than low self-efficacy learners. In addition, self-efficacy appears to affect a host of other motivational variables such as goal setting, self-regulation, self-attribution, and others (see Kanfer & McCombs, 2000; Pintrich & Schunk, 1996).

Given these results, the question for the use of virtual environment design is not necessarily whether preexisting self-efficacy will affect training outcomes (evidence suggests that it will), but how the experience might be structured to assess and raise self-efficacy early in training. Although there is little research in this area, there are some findings to suggest that some elements of scenario design can be used to increase a trainee's self-efficacy. For example, Holladay and Quiñones (2003) demonstrated that varying aspects of practice were effective in increasing trainee self-efficacy. This might indicate that a variety of scenarios with varying challenges would be most effective for virtual environment based training.

Further, it is important to realize that self-efficacy with the technology of virtual environment itself might influence the training outcome. There are data to suggest that psychological states such as computer self-efficacy might mediate the relationship between a computer based education environment and the eventual training outcome from the virtual experience. This mediating relationship has been demonstrated by Spence and Usher (2007). However, it is interesting to note that increasing computer self-efficacy is not as easy as merely providing experience with the system (Wilfong, 2006). Indeed, even a simple training program designed to increase this type of self-efficacy did not have the expected effect on improving this attitude (Ross & Starling, 2005). Consequently, there is a need for further study of how to improve this apparently important mediating variable.

SUMMARY: IMPLICATIONS FOR RESEARCH AND DESIGN

Clearly, much is known about how individual differences operate in learning environments. However, specific investigations of individual differences in VE based learning systems are not common. Hence, in Table 2.1 we summarize by

Table 2.1. Individual Differences in VE Based Learning Systems

Learner Attribute	Definition/Description	Sample Research Question(s)	Preliminary Design Guidance
Immutable			
Gender	Degree to which gender directly or indirectly affects learning outcomes in VE based learning systems	Do males benefit more from training in VEs than females? Will early exposure to computers close any gender gaps?	Provide additional pretraining practice for females; use common interface metaphors
Age	Degree to which age-related changes affect learning outcomes in VE based learning systems	How do age-related cognitive changes affect learning in VEs?	Reduce extraneous stimuli in the virtual environment; create optimized interfaces
Personality—Collaboration	Degree to which users are inclined to work cooperatively with others in a VE	Does the tendency to work cooperatively affect learning in a shared virtual space?	Select for collaborative learning; provide pre-exposure team training
Personality—Immersive Tendencies	Possible individual difference in a user's tendency to feel immersed in a virtual environment	Is immersive tendency a reliable individual difference? If so, does it have an impact on learning in a VE?	Consider non-VE training for low tendency individuals
Cognitive Ability	General class of aptitudes associated with general intelligence (for example, reasoning, verbal comprehension, and working memory capacity)	How does cognitive ability impact the design of VEs for learning? What VE features are needed to accommodate learners of varying cognitive ability? Can VEs compensate for weaknesses in ability?	Provide greater visual support
Spatial Ability	Ability to generate, maintain, and manipulate mental visual images	How does spatial ability impact the design of VEs? What VE features are needed to accommodate learners of varying spatial ability?	Use 2-D graphics for low spatial ability individuals; provide spatial memory aids

Prior Knowledge and Experience	Domain-specific knowledge that a learner brings to the learning task	How can prior knowledge be automatically assessed in VEs? How can VE design be optimized to accommodate learners with different levels of prior knowledge?	Provide pretraining for low experience individuals; design training scenarios to scaffold learning
Alterable			
Goal Orientation	Nature of goals (mastery versus performance) set by trainees that influences their learning strategy and what they emphasize in training	How does goal orientation interact with VE design? How can VEs be designed to trigger mastery orientation in learners?	Reinforce mastery orientation behaviors in the virtual environment; provide mastery-oriented feedback
Expectations for Training	Beliefs about what the training system/ game will be like; can be past experience with instruction and/or past experience with virtual environments	How do ingoing expectations regarding gaming affect the success of VEs? Do trainees' incoming expectations for training affect their reactions to the VE?	Share "success stories" or other orientations before training; demonstrate applicability of training to the job
Self-Efficacy	Belief that one has the necessary capability to accomplish a certain level of performance in a given task	How must VE design be modified to accommodate learners with varying levels of self-efficacy?	Scaffold learning scenarios to ensure early successes; provide feedback designed to increase efficacy

offering a set of design recommendations that were gleaned from existing literature and also by providing a set of research questions to guide future work.

REFERENCES

Bailenson, J. N., Beall, A. C., Loomis, J., Blascovich, M., & Turk, M. A. (2004). Transformed social interaction: Decoupling representation from behavior and form in collaborative virtual environments. *Presence, 13*(4), 428–441.

Bandura, A. (1977). Self-efficacy: Toward a unifying theory of behavioral change. *Psychological Review, 84*(2), 191–215.

Bandura, A. (1982). The assessment and predictive generality of self-percepts of efficacy. *Journal of Behavior Therapy and Experimental Psychiatry, 13,* 195–199.

Bandura, A. (1986). From thought to action: Mechanisms of personal agency. *New Zealand Journal of Psychology, 15,* 1–17.

Bandura, A. (1989). Regulation of cognitive processes through perceived self-efficacy. *Developmental Psychology, 25,* 725–739.

Baxter, H. C., Ross, J. K., Phillips, J., & Shafer, J. E. (2004, December). *Framework for assessment of tactical decision making simulations.* Paper presented at the Interservice/Industry Training, Simulation & Education Conference (I/ITSEC), Orlando, FL.

Bingener, J., Boyd, T., Van Sickle, K., Jung, I., Saha, A., Winston, J., Lopez, P., Ojeda, H., Schwesinger, W., & Anastakis, D. (2008). Randomized double-blinded trial investigating the impact of a curriculum focused on error recognition on laparoscopic suturing training. *The American Journal of Surgery, 195*(2), 179–182.

Birch, L., Jones, N., Doyle, P., Green, P., McLaughlin, A., Champney, C., Williams, D., Gibbon, K., & Taylor, K. (2007). Obstetric skills drills: Evaluation of teaching methods. *Nurse Education Today, 27*(8), 915–922.

Birren, J. E., & Livingston, J. (1985). *Cognition, stress, and aging.* Englewood Cliffs, NJ: Prentice-Hall.

Borgman, C. L. (1989). All users of information retrieval systems are not created equal: An exploration into individual differences. *Information Processing & Management, 25,* 237–251.

Bransford, J. D., Brown, A. L., & Cocking, R. R. (1999). *How people learn: Brain, mind, experience, and school.* Washington, DC: National Academies Press.

Brett, J. F., & VandeValle, D. (1999). Goal orientation and goal content as predictors of performance in a training program. *Journal of Applied Psychology, 84,* 863–873.

Busch, T. (1995). Gender differences in self-efficacy and attitudes toward computers. *Journal of Educational Computing Research, 12,* 147–158.

Cannon-Bowers, J. A., & Salas, E. (Eds.). (1998). *Making decisions under stress: Implications for individual and team training.* Washington, DC: American Psychological Association.

Cannon-Bowers, J. A., Salas, E., Tannenbaum, S. I., & Mathieu, J. E. (1995). Toward theoretically-based principles of trainee effectiveness: A model and initial empirical investigation. *Military Psychology, 7,* 141–164.

Chen, C., Czerwinski, M., & Macredie, R. (2000). Individual differences in virtual environments—Introduction and overview. *Journal of the American Society for Information Science, 51*(6), 499–507.

Chen, C., & Rada, R. (1996). Interacting with hypertext: A meta-analysis of experimental studies. *Human-Computer Interaction, 11*(2), 125–156.

Chiou, G. F. (1995). Reader interface of computer-based reading environment. *International Journal of Instructional Media, 22*(2), 121–133.

Clark, R. (2002). Learning outcomes: The bottom line. *Communication Education, 51*(4), 396–404.

Clark, R., Nguyen, F., & Sweller, J. (2006). *Efficiency in learning: Evidence-based guidelines to manage cognitive load.* San Francisco, CA: Pfeiffer.

Clark, R., & Wittrock, M. C. (2000). Psychological principles in training. In S. Tobias & J. D. Fletcher (Eds.), *Training and retraining: A handbook for business, industry, government, and the military* (pp. 51–84). New York: Macmillan.

Cockburn, A., & McKenzie, B. (2002). Evaluating the effectiveness of spatial memory in 2D and 3D physical and virtual environments. *Proceedings of the Conference on Human Factors in Computing Systems* (pp. 203–210). New York: ACM.

Cognition and Technology Group at Vanderbilt (CTGV). (2000). Connecting learning theory and instructional practice: Leveraging some powerful affordances of technology. In H. F. O'Neil, Jr., & R. S. Perez (Eds.), *Technology applications in education: A learning view* (pp. 173–209). Mahwah, NJ: Lawrence Erlbaum.

Cole, N. D., & Latham, G. P. (1997). Effects of training in procedural justice on perceptions of disciplinary fairness by unionized employees and disciplinary subject matter experts. *Journal of Applied Psychology, 82*(5), 699–705.

Colquitt, J. A., LePine, J. A., & Noe, R. A. (2000). Toward an integrative theory of training motivation: A meta-analytic path analysis of 20 years of research. *Journal of Applied Psychology, 85*(5), 678–707.

Colwell, J., Grady, C., & Rhaiti, S. (2006). Computer games, self-esteem and gratification of needs in adolescents. *Journal of Community and Applied Social Psychology, 5*(3), 195–206.

Covington, M. V. (2000). Goal theory, motivation, and school achievement: An integrative review. *Annual Review of Psychology, 51,* 171–200.

Cronbach, L. J., & Snow, R. E. (1977). *Aptitudes and instructional methods: A handbook for research on aptitude-treatment interactions.* New York: Irvington.

Draper, J. V., Fujita, Y., & Herndon, J. N. (1987). *Evaluation of high-definition television for remote task performance* (Rep. No. ORNL/TM-10303). Blacksburg, VA: Virginia Polytechnic Institute and State University.

Dweck, C. (1986). Motivational processes affecting learning. *American Psychologist, 41* (10), 1040–1048.

Dweck, C., & Legget, E. (1988). A social-cognitive approach to motivation and personality. *Psychological Review, 95,* 256–273.

Eden, D., & Aviram, A. (1993). Self-efficacy training to speed reemployment: Helping people to help themselves. *Journal of Applied Psychology, 78*(3), 325–360.

Egan, D. (1988). Individual differences in human-computer interaction. In M. Helander (Ed.), *Handbook of human-computer interaction* (pp. 541–568). North-Holland: Elsevier.

Elliot, A. J., & Harackiewicz, J. M. (1994). Goal setting, achievement orientation, and intrinsic motivation: A mediational analysis. *Journal of Personality and Social Psychology, 66,* 968–980.

Fisk, A. D., & Rogers, W. A. (1991). Toward an understanding of age-related memory and visual search effects. *Journal of Experimental Psychology, 120,* 131–149.

Fisk, J. E., & Warr, P. (1996). Age and working memory: The role of perceptual speed, the central executive, and the phonological loop. *Psychology and Aging, 11,* 316–323.

Ford, J. K., Weissbein D. A., Smith, E. M., Gully, S. M., & Salas, E. (1998). Relationships of goal orientation, metacognitive activity, and practice strategies with learning outcomes and transfer. *Journal of Applied Psychology, 83*(2), 218–233.

Frey, A., Hartig, J., Ketzel, A., Zinkernagel, A., & Moosbrugger, H. (2007). The use of virtual environments based on a modification of the computer game Quake III Arena® in psychological experimenting. *Computers in Human Behavior, 23*(4), 2026–2039.

Garg, A., Norman, G., & Sperotable, L. (2001). How medical students learn spatial anatomy. *The Lancet, 357*(9253), 363–364.

Gentner, D. (1983). Structure-mapping: A theoretical framework for analogy. *Cognitive Science, 77,* 155–170.

Gist, M. E. (1992). Self-efficacy: A theoretical analysis of its determinants and malleability. *Academy of Management Review, 17,* 183–211.

Gist, M. E., Schwoerer, C., & Rosen, B. (1989). Effects of alternative training methods on self-efficacy and performance in computer software training. *Journal of Applied Psychology, 74*(5), 884–891.

Gist, M. E., Stevens, C. K., & Bavetta, A. G. (1991). Effects of self-efficacy and post-training intervention on the acquisition and maintenance of complex interpersonal skills. *Personnel Psychology, 44,* 837–861.

Grieve, F. G., Whelan, J. P., Kottke, R., & Meyers, A. W. (1994). Manipulating adults' achievement goals in a sport task: Effects on cognitive, affective and behavioral variables. *Journal of Sport Behavior, 17*(4), 227–246.

Hannafin, M., Oliver, K., & Sharma, P. (2003). Cognitive and learning factors in web-based distance learning environments. In M. G. Moore & W. G. Anderson (Eds.), *Handbook of distance education* (pp. 245–260). Mahwah, NJ: Erlbaum.

Harackiewicz, J. M., & Elliot, A. J. (1993). Achievement goals and intrinsic motivation. *Journal of Personality and Social Psychology, 65,* 904–915.

Heimbeck, D., Frese, M., Sonnentag, S., & Keith, N. (2003). Integrating errors into the training process: The function of error management instructions and the role of goal orientation. *Personnel Psychology, 56*(2), 333–361.

Hertzog, C. (1989). Influences of cognitive slowing in age differences in intelligence. *Developmental Psychology, 25,* 636–651.

Holladay, C. L., & Quiñones, M. A. (2003). Practice variability and transfer of training: The role of self-efficacy generality. *Journal of Applied Psychology, 88*(6), 1094–1103.

Hunt, N. P., & Bohlin, R. M. (1991). Entry attitudes of students towards using computers. Paper presented at the 70th Annual Meeting of California Education Research Association, Fresno, CA.

Hunter, J. E. (1986). Cognitive ability, cognitive aptitudes, job knowledge, and job performance. *Journal of Vocational Behavior, 29*(3), 340–346.

Hwang, W. Y., & Li, C. C. (2002). What the user log shows based on learning time distribution. *Journal of Computer Assisted Learning, 18,* 232–236.

Kaber, D. B., Draper, J. V., & Usher, J. M. (2002). Influence of individual differences on application design for individual and collaborative immersive virtual environments. In K. M. Stanney (Ed.), *Handbook of virtual environments: Design, implementation, and applications* (pp. 379–402). Mahwah, NJ: Lawrence Erlbaum.

Kaemar, K. M., Wright, P. M., & McMahan G. C. (1997). The effects of individual differences on technological training. *Journal of Management Issues, 9,* 104–120.

Kanfer, R., & McCombs, B. L. (2000). Motivation: Applying current theory to critical issues in training. In S. Tobias & J. D. Fletcher (Eds.), *Training and retraining: A handbook for business, industry, government, and the military* (pp. 85–108). New York: MacMillan.

Kirkley, J. R., Kirkley, S. E., Swan, B., Myers, T. E., Sherwood, D., & Singer, M. J. (2004, December). *Developing an embedded scaffolding framework to support problem-based embedded training (PBET) using mixed and virtual reality simulations.* Paper presented at the Interservice/Industry Training, Simulation & Education Conference (I/ITSEC), Orlando, FL.

Kizony, R., Katz, N., & Weiss, P. L. (2004). Virtual reality based intervention in rehabilitation: Relationship between motor and cognitive abilities and performance within virtual environments for patients with stroke. *Proceedings of the 5th International Conference on Disability, Virtual Reality & Associated Technologies* (pp. 19–26). Oxford, United Kingdom: University of Reading.

Klein, H. J., Noe, R. A., & Wang, C. (2006). Motivation to learn and course outcomes: The impact of delivery mode, learning goal orientation, and perceived barriers and enablers. *Personnel Psychology, 59*(3), 665–702.

Kozlowski, S. W. J., Gully, S. M., McHugh, P. P., Salas, E., & Cannon-Bowers, J. A. (1996). A dynamic theory of leadership and team effectiveness: Developmental and task contingent leader roles. In G. R. Ferris (Ed.), *Research in personnel and human resource management* (Vol. 14, pp. 235–305). Greenwich, CT: JAI Press.

Kozlowski, S. W. J., Gully, S. M., Salas, E., & Cannon-Bowers, J. A. (1996). Team leadership and development: Theory, principles, and guidelines for training leaders and teams. In M. Beyerlein, D. Johnson, & S. Beyerlein (Eds.), *Advances in interdisciplinary studies of work teams: Team leadership* (Vol. 3, pp. 251–289). Greenwich, CT: JAI Press.

MacArthur, C. A., & Malouf, D. B. (1991). Teacher beliefs, plans and decisions about computer-based instruction. *Journal of Special Education, 25,* 44–72.

Martocchio, J. J. (1994). Effects of conception of ability on anxiety, self-efficacy, and learning in training. *Journal of Applied Psychology, 79*(6), 819–825.

Martocchio, J. J., & Webster, J. (1992). Effects of feedback and cognitive playfulness on performance in microcomputer software training. *Personnel Psychology, 45*(3), 553–578.

Mathieu, J. E., Martineau, J. W., & Tannenbaum, S. I. (1993). Individual and situational influences on the development of self-efficacy: Implication for training effectiveness. *Personnel Psychology, 46,* 125–147.

Mathieu, J. E., Tannenbaum, S. I., & Salas, E. (1992). Influences of individual and situational characteristics on measures of training effectiveness. *Academy of Management, 35,* 828–847.

Mayer, R. E. (2001). *Multimedia learning.* New York: Cambridge University Press.

Merriam, S. B., & Leahy, B. (2005). Learning transfer: A review of the research in adult education and training. *PAACE Journal of Lifelong Learning, 14,* 1–24.

Mitchell, T. R., Hopper, H., Daniels, D., & George-Falvy, J. (1994). Predicting self-efficacy and performance during skill acquisition. *Journal of Applied Psychology, 79* (4), 506–517.

National Research Council. (1999). *How people learn: Bridging research and practice.* Washington, DC: National Academy Press.

Oman, C. M., Shebilske, W. L., Richards, J. T., Tubre, T. C., Beall, A. C., & Natapoff, A. (2000). Three dimensional spatial memory and learning in real and virtual environments. *Spatial Cognition and Computation, 2*(4), 355–372.

Paas, F., Van Gerven, P. W. M., & Tabbers, H. K. (2005). The cognitive aging principle in multimedia learning. In R. E. Mayer (Ed.), *The Cambridge handbook of multimedia learning* (pp. 339–354). New York: Cambridge University Press.

Phillips, J. M., & Gully, S. M. (1997). Role of goal orientation, ability, need for achievement, and locus of control in the self-efficacy and goal-setting process. *Journal of Applied Psychology, 82*(5), 792–802.

Pintrich, P. R., & Schunk, D. H. (1996). *Motivation in education: Theory, research, and applications.* Englewood Cliffs, NJ: Prentice Hall Merrill.

Prasolova-Forland, E., & Divitini, M. (2003a). Collaborative virtual environments for supporting learning communities: An experience of use. *Proceedings of the 2003 International ACM SIGGROUP Conference on Supporting Group Work* (pp. 58–67). New York: ACM.

Prasolova-Forland, E., & Divitini, M. (2003b). Supporting social awareness: Requirements for educational CVE. *Proceedings of the 3rd IEEE International Conference on Advanced Learning Technologies* (pp. 366–367). Los Alamitos, CA: IEEE Computer Society.

Quick, J. C., Joplin, J. R., Nelson, D. L., Mangelsdorff, A. D., & Fiedler, E. (1996). Self-reliance and military service training outcomes. *Military Psychology, 8,* 279–293.

Quiñones, M. A. (1995). Pretraining context effects: Training assignment as feedback. *Journal of Applied Psychology, 80,* 226–238.

Randel, J. M., Morris, B. A., Wetzel, C. D., & Whitehill, B. V. (1992). The effectiveness of games for educational purposes: A review of recent research. *Simulation and Gaming, 23,* 261–276.

Ree, M. J., Carretta, R. T., & Teachout, S. M. (1995). Role of ability and prior job knowledge in complex training performance. *Journal of Applied Psychology, 80*(6), 721–730.

Ree, M. J., & Earles, J. A. (1991). Predicting training success: Not much more than g. *Personnel Psychology, 44,* 321–332.

Roed, J. (2003). Language learner behaviour in a virtual environment. *Computer Assisted Language Learning, 16*(2), 155–172.

Romano, D. M., Drna, P., & Self, J. A. (1998). *Influence of collaboration and presence on task performance in shared virtual environments.* Paper presented at the 1998 United Kingdom Virtual Reality Special Interest Group (UKVRSIG) Conference, Exeter, England.

Ross, J. A., & Starling, M. (2005, April). *Achievement and self-efficacy effects of self-evaluation training in a computer-supported learning environment.* Paper presented at the annual meeting of the American Educational Research Association, Montreal, Canada.

Sandstrom, N. J., Kaufman, J., & Huettel, S. A. (1998). Males and females use different distal cues in a virtual environment navigation task. *Cognitive Brain Research, 6,* 351–360.

Schmidt, A. M., & Ford, J. K (2003). Learning within a learner control training environment: The interactive effects of goal orientation and metacognitive instruction on learning outcomes. *Personnel Psychology, 56*(2), 405–429.

Sharp, D. L. M., Bransford, J. D., Goldman, S. R., Risko, V. J., Kinzer, C. K., & Vye, N. J. (1995). Dynamic visual support for story comprehension and mental model building by young, at-risk children. *Educational Technology Research and Development, 43*(4), 1042–1629.

Slater, M., & Wilber, S. (1997). A framework for immersive virtual environments (FIVE): Speculations on the role of presence in virtual environments. *Presence, 6*(6), 603–617.

Smith-Jentsch, K. A., Jentsch, F. G., Payne, S. C., & Salas, E. (1996). Can pretraining experiences explain individual differences in learning? *Journal of Applied Psychology, 81,* 909–936.

Spence, D. J., & Usher, E. L. (2007). Engagement with mathematics courseware in traditional and online remedial learning environments: Relationship to self-efficacy and achievement. *Journal of Educational Computing Research, 37*(3), 267–288.

Stajkovic, A. D., & Luthans, F. (1998). Self-efficacy and work-related performance: A meta-analysis. *Psychological Bulletin, 124,* 240–261.

Stevens, C. K., & Gist, M. E. (1997). Effects of self-efficacy and goal-orientation training on negotiation skill maintenance: What are the mechanisms? *Personnel Psychology, 50,* 955–978.

Tan, D. S., & Czerwinsk, M. P. (2006). Large displays enhance optical flow cues and narrow the gender gap in 3-D virtual navigation. *Human Factors, 48*(2), 318–333.

Tannenbaum, S. I., Mathieu, J. E., Salas, E., & Cannon-Bowers, J. A. (1991). Meeting trainees' expectations: The influence of training fulfillment on the development of commitment, self-efficacy, and motivation. *Journal of Applied Psychology, 76,* 759–769.

Tracey, M. R., & Lathan, C. E. (2001). The interaction of spatial ability and motor learning in the transfer of training from virtual to a real task. In J. D. Westwood, H. M. Hoffman, G. T. Mogel, D. Stredney, & R. A. Robb (Eds.), *Medicine meets virtual reality* (pp. 521–527). Amsterdam: IOS Press.

Vogel, J. J., Bowers, C. A., & Vogel, D. S. (2003). Cerebral lateralization of spatial abilities: A meta-analysis. *Brain and Cognition, 52*(2), 197–204.

Waller, D. (2000). Individual differences in spatial learning from computer-simulated environments. *Journal of Experimental Psychology, 6*(4), 307–321.

Warr, P., & Bunce, D. (1995). Trainee characteristics and the outcomes of open learning. *Personnel Psychology, 48,* 347–376.

Wilfong, J. D. (2006). Computer anxiety and anger: the impact of computer use, computer experience, and self-efficacy beliefs. *Computers in Human Behavior, 22*(6), 1001–1011.

Yeh, Y. C. (2007). Aptitude-treatment interactions in preservice teachers' behavior change during computer-simulated teaching. *Computers & Education, 48*(3), 495–507.

Yi, M. Y., & Hwang, Y. (2003). Predicting the use of web-based information systems: Self-efficacy, enjoyment, learning goal orientation, and the technology acceptance model. *International Journal of Human-Computer Studies, 59*(4), 431–449.

Chapter 3

COGNITIVE TRANSFORMATION THEORY: CONTRASTING COGNITIVE AND BEHAVIORAL LEARNING

Gary Klein and Holly C. Baxter

The traditional approach to learning is to define the objectives (the gap between the knowledge a person has and the knowledge the person needs to perform the task), establish the regimen for practice, and provide feedback. Learning procedures and factual data are seen as adding more information and skills to the person's storehouse of knowledge. However, this storehouse metaphor is poorly suited for cognitive skills and does not address the differing learning needs of novices and experts. Teaching cognitive skills requires the diagnosis of the problem in terms of flaws in existing mental models, not gaps in knowledge. It requires learning objectives that are linked to the person's current mental models, practice regimens that may have to result in "unlearning" that enables the person to abandon the current, flawed mental models, and it requires feedback that promotes sensemaking. We propose a Cognitive Transformation Theory to guide the development of cognitive skills. We also present several strategies that might be useful in overcoming barriers to understanding and to revising mental models. Finally, we show the implications of Cognitive Transformation Theory for using virtual environments (VEs; where a "live" student interacts with a "simulated" environment) in training.

INTRODUCTION

How can cognitive skills be improved? The conventional mechanisms of practice, feedback, and accumulation of knowledge rarely apply to cognitive skills in the same way they apply to behavioral skills. In this chapter we argue that cognitive learning requires a different concept of the learning process.

Traditional approaches to learning seem clear-cut: (1) identify what you want the student to learn; (2) provide the knowledge and present an opportunity to practice the skill or concept; (3) give feedback so the student can gauge whether the learning has succeeded. Educating students in behavioral skills appears to simply be a matter of practice and feedback.

This approach to learning relies on a storehouse metaphor. It assumes that the learner is missing some critical form of knowledge—factual information or procedures. The learner or the instructor defines what knowledge is missing. Together, they add this knowledge via a course, a practice regimen, or through simple study. Instructors provide feedback to the learner. Then, they test whether the new knowledge was successfully added to the storehouse.

We believe that this storehouse metaphor is insufficient to describe learning of cognitive skills. The storehouse metaphor may be useful for learning factual information or for learning simple procedures. But cognitive learning should help people discover new ways to understand events. We can distinguish different forms of knowledge that people need in order to gain expertise: declarative knowledge, routines and procedures, recognition of familiar patterns, perceptual discrimination skills, and mental models.

The storehouse metaphor seems best suited for acquiring declarative knowledge and for learning new routines/procedures. It may be less apt for building pattern-recognition skills. It is least appropriate for teaching people to make perceptual discriminations and for improving the quality of their mental models.

When people build a larger repertoire of patterns and prototypes, they are not simply adding new items to their lists. They are learning how to categorize the new items and are changing categories and redefining the patterns and prototypes as they gain new experience. The storehouse metaphor implies a simple additive process, which would lead to confusion rather than to growth. We encounter this kind of confusion when we set up a new filing system for an unfamiliar type of project and quickly realize that adding more files is creating only more confusion—the initial categories have to be changed.

When people develop perceptual discrimination skills through training in VEs or other methods, they are learning to make distinctions that they previously did not notice. They are learning to "see the invisible" (Klein & Hoffman, 1993) in the sense that they can now make discriminations they previously did not notice. Perceptual learning depends on refashioning the way we attend and the way we see, rather than just adding additional facts to our knowledge base.

Cognitive skills depend heavily on mental models. We define a mental model as a cluster of causal beliefs about how things happen. We have mental models for how our car starts when we turn our key in the ignition, for how water is forced out of a garden hose when the spigot is turned on, and for why one sports team has beaten another. In steering a simple sailboat, we have a mental model of why the nose of the boat will turn to the left when we press the tiller to the right. We believe that the water will press against the rudder in a way that swings the back of the boat to the right, creating a counterclockwise rotation in the boat's heading. Therefore, the slower the boat moves, the less the water pressure on the rudder and the less pronounced this effect should be. According to Glaser and Chi (1988), mental models are used to organize knowledge. Mental models are also described as knowledge structures and schemata.

Cognitive learning is not simply a matter of adding additional beliefs into the existing mental models. Rather, we have to revise our belief systems and our mental models as experience shows the inadequacy of our current ways of thinking. We discover ways to extend or even reject our existing beliefs in favor of more sophisticated beliefs.

The scientist metaphor is much more suited to cognitive learning. This metaphor views a learner as a scientist engaged in making discoveries, wrestling with anomalies, and finding ways to restructure beliefs and mental models (Carey, 1986). The scientist metaphor is consistent with the field of science education, where students are taught to replace their flawed mental models with better concepts about how physical, chemical, and biological processes actually work. The scientist metaphor emphasizes conceptual change, not accumulation of declarative information. Within psychology, the scientist metaphor is epitomized by Piaget (1929) who described conceptual change as a process of *accommodation.* Posner, Strike, Hewson, and Gertzog (1982) point out that within the philosophy of science, the empiricist tradition that evaluated a theory's success in generating confirmed predictions has been superseded by views that emphasize a theory's resources for solving problems. This replacement fits better within the Piagetian process of accommodation than does the empiricist approach. Posner et al. have described some of the conditions necessary for accommodation to take place: dissatisfaction with existing conceptions, including the difficulties created by anomalies; the intelligibility of new concepts, perhaps by linkage with analogies and metaphors; and the initial plausibility of new conceptions.

Although our own approach is firmly within the scientist metaphor, we should note some disconnects. The field of science education assumes a knowledgeable teacher attempting to convince students to accept scientifically acceptable theories. In contrast, many cognitive learning situations do not come equipped with knowledgeable teachers, and the learners have to discover for themselves where their mental models are wrong and how to replace them with more effective ones.

The next section describes the kinds of sensemaking needed for cognitive learning. Following that, we present the concept of cognitive transformation as an alternative to the storehouse metaphor, and as an elaboration of the scientist metaphor. Finally, we offer some implications for achieving cognitive learning in virtual environments.

SENSEMAKING REQUIREMENTS FOR LEARNING COGNITIVE SKILLS

What is hard about learning cognitive skills is that none of the traditional components of learning—diagnosis, practice, feedback, or training objectives—are straightforward. Each of them depends heavily on sensemaking (for example, Weick, 1995). Bloom's (1956) taxonomy includes a component of synthesis—building a structure or pattern from diverse elements; and putting parts together to form a whole, with an emphasis on creating a new meaning or structure. This corresponds to the process of sensemaking.

We treat cognitive learning as a *sensemaking* activity that includes four components: diagnosis, learning objectives, practice, and feedback. These components of sensemaking must be the up-front focus of any VE development in order for effective training transfer to occur.

Diagnosis

Diagnosing the reasons for weak performance depends on sensemaking. The instructor, whether in person or virtual, has to ferret out the reasons why the student is confused and making errors. Sometimes trainees do not even notice errors or weaknesses and may resist suggestions to overcome problems they do not realize they have. Even if trainees do realize something is wrong, the cause/effect mechanisms are subtle and complex. Outcome feedback, the type of feedback that is most often available in the technologies associated with virtual environments, usually does not provide any clues about what to do differently. That is why instructors and technologies need to be able to provide process feedback as the trainee progresses through the learning process, but they first must diagnose what is wrong with the trainee's thinking. Diagnosing the reason for poor performance is a challenge to trainees. It is also a challenge to the instructors who may not be able to figure out the nature of the problem and who have no technologies capable of providing a diagnosis at this level.

Diagnosis is difficult for instructional developers. The classical systems approach to instructional design is to subtract the existing knowledge, skills, and abilities (KSAs) from the needed KSAs. But for cognitive skills, instructional developers need to understand why the students are struggling. The goal of diagnosis goes beyond establishing learning objectives—it depends on discovering what flaw in a mental model needs to be corrected.

For cognitive skills, it is very difficult to determine and define the existing problem. Cognitive Task Analysis (for example, Crandall, Klein, & Hoffman, 2006) methods may be needed to diagnose subtle aspects of cognitive skills.

Within the framework of science education, Chi, Glaser, and Rees (1982) have discussed the use of misconceptions to understand why students are confused. Similarly, Shuell (1986) described how a student's "buggy algorithms" could lead to misconceptions and how analysis of mistakes can provide educators with insights into how to repair the flaws.

Learning Objectives

With the storehouse metaphor, learning objectives are clear and succinct—the additional declarative or procedural knowledge to be imparted and the changes in performance that reflect whether the student has acquired the new material.

But for cognitive learning, the objectives may be to help the students revise their mental models and perhaps to reorganize the way they categorize events. Some learning theorists emphasize the importance of integrating new learning with the concepts that are already known. For example, both Kolb (1984) and Dewey (1938) focus on learning through experience. What is important in Kolb's reflective observation stage is how the learner transforms an experience into learning through reflection. During reflection, the student compares the new learning to what is already known and tries to make it fit with existing knowledge and sees how to leverage this new knowledge for additional learning.

For Dewey, the key is what the learner does with experience. Not all experiences are equal and not all experiences are educational. According to Dewey, individuals reflect on their experiences to learn what thoughts and actions can change real world conditions that need improving. Dewey thought that people were constantly trying to resolve perplexing intellectual situations and difficult moral situations.

Theorists such as Kolb and Dewey do not view accumulating or storing knowledge as an end state. Instead, knowledge accumulation kicks off a series of cognitive activities by the individual to figure out ways to test the "goodness" of the new learning through active experimentation or to use the new learning to change an unsatisfactory situation.

The field of science education describes this process as "restructuring" (Chi et al., 1982; Shuell, 1986). Carey (1986) draws on the philosophy of science and, in particular, the work of Kuhn (1962), Feyerabend (1962), and Toulmin (1953) to describe conceptual change. When theories change, successive conceptual systems will differ in the phenomena they address, the kinds of explanations they offer, and the concepts they employ. Carey uses the example of theories of mechanics, which historically used different meanings for the terms force, velocity, time, and mass. Thus, Aristotle did not distinguish between average velocity and instantaneous velocity, whereas Galileo highlighted this difference.

Carey distinguishes weak restructuring, which simply represents additional relations and schemata (for example, the storehouse metaphor), from strong restructuring, which involves a change in the core concepts themselves. Shuell (1986) uses the term "tuning" to cover Carey's notion of weak restructuring and further notes that both tuning and restructuring resemble Piaget's concept of accommodation.

We further assert that novices may not have mental models for an unfamiliar domain and will struggle to formulate even rudimentary mental models linking causes to effects. Their learning objective is to employ sensemaking to generate initial mental models of cause/effect stories, whereas experts are revising and adding to current mental models.

Following Posner et al. (1982), we suggest that accommodation itself may be a key learning objective—creating dissatisfaction with an inadequate conception, creating openness to a superior replacement.

Practice

Providing students with practice is necessary for gaining proficiency. But with cognitive skills, practice is not sufficient. For cognitive skills, trainees often may not know what they should be watching and monitoring. They need adequate mental models to direct their attention, but until they get smarter, they may fail to spot the cues that will help them develop better mental models.

VE can help trainees gain this needed practice in a context that allows them to build more robust mental models. Waller, Hunt, and Knapp (1998) found that while short VE training periods were no more effective than paper and pencil exercises, with sufficient exposure to a virtual training environment, VE training actually surpassed real world training. Numerous studies have supported the effectiveness of VEs. Brooks, Fuchs, McMillan, Whitton, and Cannon-Bowers (2006) found that VEs can provide a higher density of experiences and the chance to practice rare and dangerous scenarios safely, and Witmer, Bailey, and Knerr (1995) validated the ability of VE training to transfer to real world settings in a study they conducted with the training of dismounted soldiers in virtual environments.

Managing attention depends on sensemaking. Feedback will not be useful if the trainee does not notice or understand it—and that requires the trainee to know what to attend to and when to shift attention. Barrett, Tugade, and Engle (2004) have suggested that attention management accounts for many of the individual differences in working memory—the ability to focus attention and not be distracted by irrelevancies. For these reasons, we argue that effective practice, whether in actual or in virtual environments, depends on attention management: seeking information—knowing what to seek and when to seek it—and filtering distracting data.

Feedback

Providing students with feedback will not be useful if they do not understand it. For complex cognitive skills, such as leadership, time lags between actions and consequences will create difficulties in sorting out what worked, what did not work, and why. Learners need to engage in sensemaking to discover cause-effect relationships between actions taken at time one and the effects seen at time two. To make things more complicated, learners often have to account for other actions and events that are interspersed between their actions and the consequences. They have to figure out what really caused the consequences versus the coincidental events that had nothing to do with their actions. They have to understand the causes versus the symptoms of deeper causes, and they have to sort out what just happened, the factors in play, the influence of these factors, and the time lags for the effects.

To add to these complications, having an instructor or training tool provide feedback can actually get in the way of transfer of learning (Schmidt & Wulf, 1997) even though it increases the learning curve during acquisition. By placing students in an environment where they are given rapid feedback, the students are not compelled to develop skills for seeking their own feedback. Further, students may become distracted from intrinsic feedback because it is so much easier to rely on the extrinsic feedback. As a result, when they complete what they set out to learn, they are not prepared to seek and interpret their own feedback.

One of the challenges for cognitive learning is to handle time lags between actions and consequences. VE sessions will compress these time lags, which might clarify relationships but will also reduce the opportunity to learn how to interpret delayed feedback. To compensate, VE sessions could add distracters that might have potentially caused the effects as a way to sustain confusion about how to interpret feedback. In addition, VE sessions could be structured to monitor how people interpret the feedback.

For cognitive learning, one of the complications facing instructional designers is that the flawed mental models of the students act as a barrier to learning. Students need to have better mental models in order to understand the feedback that would invalidate their existing mental models. Without a good mental model, students will have trouble making use of feedback, but without useful feedback, students will not be able to develop good mental models. That is why cognitive learning may depend on unlearning as well as learning.

THE PROCESS OF UNLEARNING

For people to develop better mental models they may have to unlearn some of their existing mental models. The reason is that as people gain experience, their understanding of a domain should become more complex and nuanced. The mental models that provided a rough approximation need to be replaced by more sophisticated ones. But people may be reluctant to abandon inadequate mental models, as they may not appreciate the inadequacies. They may attempt to explain away the inconsistencies and anomalies. A number of researchers have described the reluctance to discard outmoded mental models even in the face of contrary evidence. DeKeyser and Woods (1990) have commented on the way decision makers fixate on erroneous beliefs. Feltovich, Spiro, and Coulson (1997) used a garden path paradigm and identified a range of knowledge shields that pediatric cardiologists employed to discount inconvenient data.

Chinn and Brewer (1993) showed that scientists and science students alike deflected inconvenient data. They identified seven reactions to anomalous data that were inconsistent with a mental model: ignoring the data, rejecting the data, finding a way to exclude the data from an evaluation of the theory/model, holding the data in abeyance, reinterpreting the data while retaining the theory/model, reinterpreting the data and making peripheral changes to the theory/model, and accepting the data and revising the theory/model. Only this last reaction changes the core beliefs. The others involve ways to discount the data and preserve the theory.

Klein, Phillips, Rall, and Peluso (2006) described the "spreading corruption" that resulted when people distorted data in order to retain flawed mental models. As people become more experienced, their mental models become more sophisticated, and, therefore, people grow more effective in explaining away inconsistencies. Fixations should become less tractable as cognitive skills improve. Therefore, people may have to unlearn their flawed mental models before they can acquire better ones. Sensemaking here is a deliberate activity to discover what is wrong with one's mental models and to abandon and replace them. Oftentimes, VEs can allow trainees to see the flaws in their mental models by illustrating the potential behavioral outcomes of their current cognitive processes. Being able to understand these flaws is critical for the unlearning process and enabling accommodation.

The process of unlearning that we are presenting resembles the scientific paradigm replacements described by Polanyi (1958) and Kuhn (1962). Another philosopher of science, Lakatos (1976), explained that researchers more readily change their peripheral ideas to accommodate anomalies than their hard-core ideas on which the peripheral ideas are based. As expected, the notion of disconfirmation is central to science education because of the importance and difficulty of changing students' naive theories. And just as scientists resist changing their theories when exposed to disconfirming evidence, so do students. Eylon and Linn (1988) reviewed studies showing that students can be impervious to contradictions. According to Chinn and Brewer (1993), the more a belief is embedded in supporting data and concepts and is used to support other concepts, the greater the resistance. Further, the anomalous data need to be credible, nonambiguous, and presented in concert with additional data in order to have the necessary impact, which presents additional requirements for effective use of VEs.

The term "unlearning" is widely used in the field of organizational learning. Starbuck and Hedberg (2001) stated that "Organizations' resistance to dramatic reorientations creates a need for explicit unlearning . . . Before attempting radical changes, [organizations] must dismantle parts of their current ideological and political structures. Before they will contemplate dramatically different procedures, policies, and strategies, they must lose confidence in their current procedures, policies, strategies, and top managers" (p. 339). We believe that these observations apply to individuals as well as to organizations and that the concept of unlearning needs to become part of a cognitive learning regimen.

Just like organizations, individuals also resist changing their mental models. Chinn and Brewer (1993) refer to Kuhn's (1962) research to suggest that students will be more likely to abandon a flawed set of beliefs if they have an alternative theory/model available. This method may work best when the alternative model is already part of the students' repertoire. For example, Brown and Clement (1989) tried to teach students about the balance of forces in operation when a book is resting on a table. The students initially refused to believe that the table exerts an upward force on the book. So they were asked to imagine that they were supporting a book with their hand. Clearly, their hand was exerting force to keep the book from falling. Next, the students were told to imagine that the book was balanced on a spring. Next, they imagined a book balanced on a pliable wooden plank. Eventually, many of the students came to accept that the solid table must be exerting an upward force on the book. This type of gradual introduction of alternative analogies seems very promising. The alternative explanations make it easier to give up the flawed mental model.

However, in some situations we suspect that the reverse has to happen. People have to lose confidence in their models before they will seriously consider an alternate. Thus, DiBello and her colleagues developed a two-day program that created a VE to help managers think more effectively about their work (DiBello, 2001). The first day was spent in a simulation of their business designed to have the managers fail in the same ways they were failing in real life. This experience helped the managers lose confidence in their current mental models of how to conduct their work. The second day gave the managers a second shot at the simulated exercise and a chance to develop and use new mental models of their work. DiBello and her colleagues have recently ported their program onto *Second Life,* an Internet based virtual world video game, as a more effective means of instruction.

Schmitt (1996) designed similar experiences for the U.S. Marine Corps. His Tactical Decision Games—low fidelity paper and pencil exercises—put individual marines into situations that challenged their thinking and made them lose confidence in their mental models of tactics and leadership. The exercises, like the more technologically advanced VE, provided a safe environment for rethinking some of their closely held beliefs. When the Tactical Decision Games were presented via a VE format, the stress and training impact appear to have been sustained.

Scott, Asoko, and Driver (1991) have described two broad types of strategies for producing conceptual change: creating cognitive conflict and building on existing ideas as analogies. The DiBello and Schmitt approaches fit within the first grouping, to create cognitive conflict. The Brown and Clement work exemplifies the second—introducing analogs as platforms for new ideas.

Chinn and Brewer (1993) have also suggested that asking students to justify their models will facilitate their readiness to change models in the face of anomalous data.

Rouse and Morris (1986) have voiced concerns about invoking the notion of mental models. The concept of a mental model is typically so vague and ambiguous that it has little theoretical or applied value. However, Klein and Hoffman (2008) argue that the term "mental model" is an umbrella that covers a variety of relationships: causal, spatial, organizational, temporal, and so forth. As long as we are clear about which type of relationship we are interested in, much of the murkiness of "mental models" disappears. Doyle and Ford (1998) presented a useful account of mental models of dynamic systems, which they defined as a relatively enduring and accessible, but limited, internal conceptual representation of an external system whose structure maintains the perceived structure of that system. They differentiated their account from the concept of "mental representations," which covers a variety of cognitive structures such as schemas, images, scripts, and so forth.

With regard to cognitive learning, our emphasis is usually on causal relationships. During the learning process, people are engaged in sensemaking to understand and explain how to make things happen. Under the right circumstances, they may also discover better ways to think about causal connections.

People have to diagnose their performance problems, manage their attention, appreciate the implications of feedback, and formulate better mental models by unlearning inadequate models. Learners are not simply accumulating more knowledge into a storehouse. They are changing their perspectives on the world.

That is why we hypothesize that these changes are uneven, rather than smooth and cumulative.

COGNITIVE TRANSFORMATION THEORY

In this section we present an account of the transition process for acquiring cognitive skills. We are primarily interested in how people learn better mental models to achieve a stronger understanding of what has been happening and what to do about it. In contrast to a storehouse metaphor of adding more and more knowledge, we offer the notion of cognitive transformation—that progress in cognitive skills depends on successively shedding outmoded sets of beliefs and adopting new beliefs. We call this account of cognitive learning "Cognitive Transformation Theory" (CTT).

Our central claim is that conceptual learning is discontinuous rather than smooth. We make periodic advances when we replace flawed mental models with better ones. However, during the process of cognitive development our mental models get harder to disconfirm. As we move further up the learning curve or have more expertise, we have to put more and more energy into unlearning—disconfirming mental models—in order to accept better ones.

We do not smoothly acquire knowledge as in a storehouse metaphor. Our comprehension proceeds by qualitative jumps. At each juncture our new mental models direct what we attend to and explain away anomalies. As a result, we have trouble diagnosing the flaws in our thinking. Because of problematic mental models, people often misdiagnose their limitations and discard or misinterpret informative feedback. The previous mental model, by distorting cues and feedback, acts as a barrier to advancement. So progress may involve some backtracking to shed mistaken notions. In addition, flawed beliefs have also influenced the way people encoded experiences in the past. Simply changing one's beliefs will not automatically change the network of implications generated from those beliefs. As a result, people may struggle with inconsistencies based on different mental models that have been used at different times in the past.

Instructional developers have to design interventions that help trainees unlearn their flawed mental models.

We can represent cognitive transformation theory as a set of postulates:

- Mental models are central to cognitive learning. Instruction needs to diagnose limitations in mental models, design interventions to help students appreciate the flaws in their mental models, and provide experiences to enable trainees to discover more useful and accurate mental models.
- Mental models are modular. People have a variety of fragmentary mental models, and they weave these together to account for a novel observation. People are usually not matching events to sophisticated theories they have in memory. They are using fragments and partial beliefs to construct relevant mental models. For most domains, the central mental models describe causal relationships. They describe how events transform into later events. Causal mental models typically take the form of a story.

- Experts have more sophisticated mental models in their domains of practice than novices. Experts have more of the fragmentary beliefs needed to construct a plausible mental model. Therefore, they are starting their construction from a more advanced position. Finally, experts have more accurate causal mental models and have tested and abandoned more inadequate beliefs.
- Experts build their repertoires of fragmentary mental models in a discontinuous fashion. In using their mental models, even experts may distort data, oversimplify, explain away diagnostic information, and misunderstand events. At some point, experts realize the inadequacies of their mental models. They abandon their existing mental models and replace these with a better set of causal beliefs. And the cycle begins again.
- Learning curves are usually smooth because researchers combine data from several subjects. The reason for the smoothness is the averaging of discontinuous curves.
- Experts are fallible. No set of mental models is entirely accurate and complete.
- Knowledge shields are the set of arguments learners can use to explain away data that challenge their mental models (Feltovich et al., 1997). Knowledge shields pose a barrier to developing cognitive skills. People are skilled at holding onto cherished beliefs. The better the mental models, the easier it is to find flaws in disconfirming evidence and anomalous observations. The S-shaped learning curve reflects the increasing difficulty of replacing mental models as people's mental models become more accurate.
- Knowledge shields affect diagnosis. Active learners try to overcome their limitations, but they need to understand what those limitations are. Knowledge shields based on poor mental models can lead learners to the wrong diagnoses of their poor performance.
- Knowledge shields affect feedback. In building mental models about complex situations, people receive a lot of feedback. However, the knowledge shields enable people to discard or neutralize contradictory data.
- Progress depends on unlearning. The better the causal models, the more difficult it is to discover their weaknesses and replace them. In many cases, learners have to encounter a baffling event, an unmistakable anomaly, or an intelligent failure in order to begin doubting their mental models. They have to lose faith in their existing mental models before they can review the pattern of evidence and formulate a better mental model. People can improve their mental models by continually elaborating them, by replacing them with better ones, and/or by unlearning their current mental models. Cognitive development relies on all three processes.
- Individual differences in attitudes toward cognitive conflict will affect success in conceptual change. Dreyfus, Jungwirth, and Eliovitch (1990) noted that bright and successful students responded positively to anomalies, whereas unsuccessful students tended to avoid the conflicts.

Cognitive Transformation Theory generates several testable hypotheses. It asserts that individual learning curves will be discontinuous, as opposed to the smooth curves found when researchers synthesize data across several subjects. CTT suggests a form of state-dependent learning. The material learned with one set of mental models may be inconsistent with material learned with a different mental model. Consequently, learners may be plagued with inconsistencies that reflect their differing beliefs during the learning cycle.

IMPLICATIONS FOR VIRTUAL ENVIRONMENTS

What is difficult about learning cognitive skills in virtual environments? While on the surface, there can appear to be tremendous benefits to taking advantage of virtual environments and the associated technologies to support cognitive skill development, Koschmann, Myers, Feltovich, and Barrows (1994) note that technology in environments often seems to be focused on the capabilities of the technology rather than on the instructional need. In essence, they are often technology focused learning with learning as an afterthought.

Virtual environments are becoming integral to almost all areas of training and educational applications. These virtual environments can include projector based displays, augmented and mixed reality technologies, online structured professional forums, game based learning technologies, and multimodal technologies to name a few. As with the more traditional types of learning discussed in this chapter, Cognitive Transformation Theory can guide the way we develop and use these technologies.

Cognitive Transformation Theory revolves around the principle that mental models are central to cognitive learning. Virtual environments give us the opportunity to examine our mental models and build on them. Simulated environments can allow learners to see how a proposed path of action plays out, thereby allowing them to observe flaws in their mental models and begin the process of improving mental models.

In addition, virtual environments allow for both intrinsic and extrinsic feedback. Many simulations offer scoring or an after action review capability that allows learners to see how they did in comparison to other students or some set standard. More important than the extrinsic feedback, these virtual environments give learners the ability to see how their actions play out and the challenges they may run into based on their mental models, allowing for self-assessment, adjustment, and improvement in cognitive learning.

Because cognitive learning depends heavily on sensemaking, and sensemaking is often complicated by knowledge shields, virtual environment sessions might benefit from designs using garden path scenarios that elicit knowledge shields and give learners a chance to recover from mistaken mindsets and get off the garden path. In a garden path scenario a person is led to accept a proposition that seems obviously true and is then given increasing amounts of contrary evidence gradually leading to the realization that the initial proposition is wrong. The paradigm lets us study how long it takes for participants to doubt and then reject the initial proposition—how long they stay on the garden path.

Virtual environments may also support some of the strategies that Posner et al. (1982) described for facilitating accommodation by helping instructors to diagnose errors and also prepare for the defenses trainees might employ as knowledge shields and by helping instructors track the process of concept change.

CONCLUSIONS

Now we can see what is wrong with the storehouse metaphor of learning described at the beginning of this paper. Learning is more than adding additional information. Learning is about changing the way we understand events, changing the way we see the world, changing what counts as information in the first place. The functions of diagnosis, practice, and feedback are all complex and depend on sensemaking.

To replace the storehouse metaphor we have presented a theory of cognitive transformation. We claim that cognitive skills do not develop as a continual accumulation. Rather, cognitive skills and the mental models underlying them progress unevenly. Flawed mental models are replaced by better ones, but the stronger the mental models the more difficult to dislodge them. As a result, learners explain away anomalies, inconsistencies, inconvenient feedback, and misdiagnose their problems. How we teach cognitive skills, therefore, has to help people unlearn their current mental models before helping them develop better ones. If this unlearning process does not occur, the students will use their current mental models to discount the lessons and the feedback.

Cognitive Transformation Theory may offer a shift in perspective on cognitive learning. It relies on sensemaking as the core function in learning cognitive skills, as opposed to a storehouse metaphor.

These issues pose challenges to the use of VEs for training cognitive skills. The training cannot be treated as a matter of realistically replicating perceptual phenomena. If the technology interferes with diagnosis, distorts cognitive learning objectives, short-cuts the attention management skills needed for practice, and limits the search for and interpretation of feedback, then cognitive learning will be degraded.

Fortunately, a VE can provide a platform for unlearning that can be superior to the natural environment. To be effective for cognitive learning, VE approaches will need to move beyond increasing sensory realism and consider the design of scenarios to promote sensemaking. Cognitive Transformation Theory offers some recommendations for how this might be done. By ensuring that the training environment supports diagnosis, attention management, and feedback, virtual environments can become useful and efficient means of achieving cognitive transformations.

ACKNOWLEDGMENTS

The authors would like to thank Joseph Cohn for his support of this project developed under Contract No. M67854-04-C-8035 (issued by MARCORSYSCOM/PMTRASYS). We would also like to thank Sterling Wiggins, Karol Ross, and Jennifer Phillips for their valuable critiques and inputs.

REFERENCES

Barrett, L. F., Tugade, M. M., & Engle, R. W. (2004). Individual differences in working memory capacity and dual-process theories of the mind. *Psychological Bulletin, 139,* 553–573.

Bloom, B. C. (Ed.). (1956). *Taxonomy of educational objectives: Handbook I. cognitive domain.* New York: David McKay Company, Inc.

Brooks, F., Fuchs, H., McMillan, L., Whitton, M., & Cannon-Bowers, J. (2006). Virtual environment training for *dismounted* teams—Technical challenges. In *Virtual Media for Military Applications* (RTO Meeting Proceedings No. RTO-MP-HFM-136, pp. 22-1–22-10). Neuilly-sur-Seine, France: Research and Technology Organisation.

Brown, D. E., & Clement, J. (1989). Overcoming misconceptions via analogical reasoning: Abstract transfer versus explanatory model construction. *Instructional Science, 18,* 237–261.

Carey, S. (1986). Cognitive science and science education. *American Psychologist, 41,* 1123–1130.

Chi, M., Glaser, R., & Rees, E. (1982). Expertise in problem solving. In R. Sternberg (Ed.), *Advances in the psychology of human intelligence* (Vol. 1, pp. 7–76). Hillsdale, NJ: Erlbaum.

Chinn, C. A., & Brewer, W. F. (1993). The role of anomalous data in knowledge acquisition: A theoretical framework and implications for science instruction. *Review of Educational Research, 63,* 1–49.

Crandall, B., Klein, G., & Hoffman, R. R. (2006). *Working minds: A practitioner's guide to cognitive task analysis.* Cambridge, MA: The MIT Press.

DeKeyser, V., & Woods, D. D. (1990). Fixation errors: Failures to revise situation assessment in dynamic and risky systems. In A. G. Colombo & A. Saiz de Bustamente (Eds.), *Advanced systems in reliability modeling* (pp. 231–252). Norwell, MA: Kluwer Academic.

Dewey, J. (1938). *Experience and education.* New York: MacMillan.

DiBello, L. (2001). Solving the problem of employee resistance to technology by reframing the problem as one of experts and their tools. In E. Salas & G. Klein (Eds.), *Linking expertise and naturalistic decision making* (pp. 71–93). Mahwah, NJ: Erlbaum.

Doyle, J. K., & Ford, D. N. (1998). Mental models concepts for system dynamics research. *System Dynamics Review, 14,* 3–29.

Dreyfus, A., Jungwirth, E., & Eliovitch, R. (1990). Applying the 'cognitive conflict' strategy for conceptual change—some implications, difficulties and problems. *Science Education, 74*(5), 555–569.

Eylon, B.-S., & Linn, M. C. (1988). Learning and instruction: An examination of four research perspectives in science education. *Review of Educational Research, 58,* 251–301.

Feltovich, P. J., Spiro, R. J., & Coulson, R. L. (1997). Issues of expert flexibility in contexts characterized by complexity and change. In P. J. Feltovich, K. M. Ford, & R. R. Hoffman (Eds.), *Expertise in context* (pp. 125–146). Menlo Park, CA: AAAI/MIT Press.

Feyerabend, P. (1962). Explanation, reduction and empiricism. In H. Feigl & G. Maxwell (Eds.), *Minnesota studies in philosophy of science* (Vol. 3, pp. 28–97). Minneapolis: University of Minnesota Press.

Glaser, R., & Chi, M. T. H. (1988). Overview. In M. T. H. Chi, R. Glaser, & M. J. Farr (Eds.), *The nature of expertise.* (pp. xv–xxviii). Mahwah, NJ: Lawrence Erlbaum.

Klein, G., & Hoffman, R. R. (2008). Macrocognition, mental models, and cognitive task analysis methodology. In J. M. Schraagen, L. G. Militello, T. Ormerod, & R. Lipshitz (Eds.), *Naturalistic decision making and macrocognition* (pp. 57–80). Hampshire, England: Ashgate.

Klein, G., Phillips, J. K., Rall, E., & Peluso, D. A. (2006). A data/frame theory of sense-making. In R. R. Hoffman (Ed.), *Expertise out of context: Proceedings of the 6th International Conference on Naturalistic Decision Making* (pp. 113–155). Mahwah, NJ: Lawrence Erlbaum.

Klein, G. A., & Hoffman, R. (1993). Seeing the invisible: Perceptual/cognitive aspects of expertise. In M. Rabinowitz (Ed.), *Cognitive science foundations of instruction* (pp. 203–226). Mahwah, NJ: Lawrence Erlbaum.

Kolb, D. A. (1984). *Experiential learning: Experience as the source of learning and development.* Englewood Cliffs, NJ: Prentice Hall.

Koschmann, T. D., Myers, A. C., Feltovich, P. J., & Barrows, H. S. (1994). Using technology to assist in realizing effective learning and instruction: A principled approach to the use of computers in collaborative learning. *The Journal of the Learning Sciences, 3*(3), 227–264.

Kuhn, T. S. (1962). *The structure of scientific revolutions.* Chicago: University of Chicago Press.

Lakatos, I. (1976). *Proofs and refutations: The logic of mathematical discovery.* Cambridge, England: Cambridge University Press.

Piaget, J. (1929). *The child's conception of the world.* New York: Harcourt Brace.

Polanyi, L. (1958). *Personal knowledge: Towards a post-critical philosophy.* Chicago: University of Chicago Press.

Posner, G. J., Strike, K. A., Hewson, P. W., & Gertzog, W. A. (1982). Accommodation of a scientific conception: Toward a theory of conceptual change. *Science Education, 66*(2), 211–227.

Rouse, W. B., & Morris, N. M. (1986). On looking into the black box: Prospects and limits on the search for mental models. *Psychological Bulletin, 100*(3), 349–363.

Schmidt, R. A., & Wulf, G. (1997). Continuous concurrent feedback degrades skill learning: Implications for training and simulation. *Human Factors, 39*(4), 509–525.

Schmitt, J. F. (1996, May). Designing good TDGs. *Marine Corps Gazette,* 96–98.

Scott, P. H., Asoko, H. M., & Driver, R. H. (1991). Teaching for conceptual change: A review of strategies. In R. Duit, F. Goldberg, & H. Niederer (Eds.), *Research in physics learning: Theoretical issues and empirical studies. Proceedings of an international workshop* (pp. 310–329). University of Kiel, Kiel, Germany: Schmidt & Klannig.

Shuell, T. J. (1986). Cognitive conceptions of learning. *Review of Educational Research, 56,* 411–436.

Starbuck, W. H., & Hedberg, B. (2001). How organizations learn from success and failure. In M. Dierkes, A. B. Antal, J. Child, & I. Nonaka (Eds.), *Handbook of organizational learning and knowledge* (pp. 327–350). Oxford, United Kingdom: Oxford University Press.

Toulmin, S. (1953). *The philosophy of science: An introduction.* London: Hutchinson.

Waller, D., Hunt, E., & Knapp, D. (1998). The transfer of spatial knowledge in virtual environment training. *Presence: Teleoperators and Virtual Environments, 7*(2), 129–143.

Weick, K. E. (1995). *Sensemaking in organizations.* Thousand Oaks, CA: Sage Publications.

Witmer, B. G., Bailey, J. H., & Knerr, B. W. (1995). *Training dismounted soldiers in virtual environments: Route learning and transfer* (ARI Tech. Rep. No. 1022). Alexandria, VA: U.S. Army Research Institute for the Behavioral and Social Sciences. (DTIC No. ADA292900).

Part III: Social Band

Chapter 4

CREATING EXPERTISE WITH TECHNOLOGY BASED TRAINING

Karol Ross, Jennifer Phillips, and Joseph Cohn

Training that is delivered completely or primarily in a technology based format such as desktop simulation has the potential to offer substantial benefits. It can cost less than full-scale simulations or field exercises and can support multiple practice iterations in a short period of time. It is often engaging and motivating (Druckman, 1995; Garris, Ahlers, & Driskell, 2002; Prensky, 2001), as well as easy to distribute. It provides contexts for decision-making practice that are crucial to building expertise in complicated fields of practice. However, guidelines for designing effective training at advanced levels of learning are lacking, leaving potential benefits of such technology unrealized.

The affordances of the technology are only as good as the training design allows them to be. To address this issue, we developed a framework describing performance and the basics of training design for advanced stages of learning. This framework provides insights into how the strengths of technology can best be exploited for learning and lays a foundation for effective training design when technology based training is employed. The framework includes (1) a stage model of cognitive proficiency that describes characteristics of novices, experts, and proficiency levels between the two extremes and (2) guidance for training design to support the advancement of expertise across the stages of performance.

The purpose of training is to move individuals from their current state of skill and knowledge to a higher state. Technology provides a powerful means to accelerate this movement by allowing multiple surrogate experiences and shared insights from real-life operations. But, without a commonly recognized account of the stages of performance, we lack a road map to develop that expertise by tailoring technology to the needs of the training audiences. To build the stage model of cognitive performance, we started with an existing model of proficiency (Dreyfus & Dreyfus, 1980, 1986) and added more recent research findings that delineate knowledge and abilities at the *stages between novice and expert* (Benner, 1984, 2004; Houldsworth, O'Brien, Butler, & Edwards, 1997; McElroy, Greiner, & de Chesnay, 1991).

In this effort we focused on professional expertise in complex, ill-structured knowledge domains and especially on learning at the intermediate stages of development—the most difficult skills to develop and the biggest opportunity for game based and virtual environments.

> An ill-structured knowledge domain is one in which the following two properties hold: (a) Each case or example of knowledge application typically involves the simultaneous interactive involvement of multiple, wide-application conceptual structures (multiple schemas, perspectives, organizational principles, and so on), each of which is individually complex (i.e., the domain involves concept- and case-complexity); and (b) the pattern of conceptual incidence and interaction varies substantially across cases nominally of the same type (i.e., the domain involves across-case irregularity).
>
> (Spiro, Feltovich, Jacobson, & Coulson, 1992, p. 60)

In other words, each individual case is complex, and there is considerable variability across cases. Different situations in these types of professional domains are likely to require application of varying patterns of principles, even in cases of seemingly similar problems or goals. No pat solutions can be employed with regularity. Such domains require professionals to exercise a great deal of judgment to flexibly apply their knowledge. Further, high levels of skill can be acquired only through operational experience with the task. These professions often require that decisions be made under conditions of time pressure and high stakes. Tactical thinking is one such domain, as are many types of medical practice, firefighting, and police work. Surrogate experiences through technology can develop expertise in such complex fields of endeavor when insightful training design takes into account the learning needs at different stages of performance.

THE FIVE-STAGE MODEL OF COMPLEX COGNITIVE SKILL ACQUISITION

The five-stage model describes performance at levels from novice to expert in ill-structured, cognitively complex domains. The Dreyfus and Dreyfus model has been applied to training and instruction within domains such as combat aviation, nursing, industrial accounting, psychotherapy, and chess (Benner, 1984, 2004; Dreyfus & Dreyfus, 1986; Houldsworth et al., 1997; McElroy et al., 1991). Incorporating research findings since 1986 from the applied cognitive psychology and naturalistic decision-making communities into the original model provides the stage model for the framework, summarized in Table 4.1. Each stage is further described below and illustrated by a summary table showing the nature of knowledge and performance, as well as implications for training. We direct the reader to Phillips, Ross, and Cohn (Volume 1, Section 2, Chapter 8) for a discussion of how this model has been applied to the domain of tactical thinking. This application provides a number of examples for all stages of the model.

Stage 1: Novice

Individuals who perform at the novice level have limited or no experience in situations characteristic of the domain in which they seek to gain expertise. They

Table 4.1. Overview of the Stages from Novice to Expert (Reprinted by permission; Lester, 2005)

Stage	Characteristics	How Knowledge Is Treated	Recognition of Relevance	How Context Is Assessed	Decision Making
Novice	–Rigid adherence to taught rules or plans –Little situational perception –No discretionary judgment	Without reference to context	None		
Advanced Beginner	–Guidelines for action based on attributes or aspects[a] –Situational perception is still limited –All attributes and aspects are treated separately and given equal importance			Analytically	
Competent[b]	–Sees actions at least partially in terms of longer-term goals –Conscious, deliberate planning –Standardized and routinized procedures –Plan guides performance as situation evolves[c]				Rational
Proficient[d]	–Sees situation holistically rather than in terms of aspects –Sees what is most important in a situation –Perceives deviations from the normal pattern –Uses maxims, whose meanings vary according to the situation, for guidance –Situational factors guide performance as situation evolves[e]	In context	Present		
Expert[f]	–No longer relies on rules, guidelines, or maxims –Intuitive grasp of situations based on deep tacit understanding –Intuitive recognition of appropriate decision or action[g] –Analytic approaches used only in novel situations or when problems occur			Holistically	Intuitive

[a]Aspects are global characteristics of situations recognizable only after some prior experience.
[b]Original item deleted: "Coping with crowdedness."
[c]Item added.
[d]Original item deleted: "Decision making less labored."
[e]Item added.
[f]Original item deleted: "Vision of what is possible."
[g]Item added.

are typically taught about the situations they will encounter in terms of objective "attributes," such as the number of soldiers in a unit, the range radius of enemy assets, or other measurable quantities that can be recognized without operational or practical exercise experience. Novices are also taught context-free rules, such as the formula for determining how long it will take personnel carriers to get from point A to point B under normal conditions. Because the novice's understanding of the domain is based largely in rules, his or her performance is quite limited and inflexible. As the study of nursing by Benner (1984) points out, rule-guided behavior actually prevents successful performance because a set of rules cannot make clear which tasks are relevant or critical in an actual situation.

A novice under the Dreyfus and Dreyfus (1980) model may have a great deal of textbook or classroom knowledge of the domain, but what places him or her in stage 1 is the shortage of actual lived experience. There is a clear distinction between the level of performance that results when textbook principles and theories are applied and the superior performance achieved when an experience base guides performance. Novices can be dangerous to themselves and others when thrust into operational situations with no "feel for" the nature of that experience. Table 4.2 provides a summary of the novice stage of development and includes training implications for this stage. Experiential training can actually be useful at the introductory stage of performance, because the facts taught at this level are often forgotten before the learner has any relevant experiences. Learning facts in the context of situations can support the retention and later appropriate recall of such declarative knowledge (Bransford et al., 1990).

Stage 2: Advanced Beginner

Advanced beginners have acquired enough domain experience that their performances can be considered marginally acceptable. At this stage, learners can recognize, either on their own or when pointed out to them by an instructor, recurring meaningful "aspects" of the situation. Aspects are global characteristics that are identifiable only through prior experience; the prior experience serves as a comparison case for the current situation. For example, an advanced beginner would be able to grasp that close air support could be helpful in a particular situation after taking part in a previous exercise in which close air support was utilized. A learner at this stage would not know how, when, or where to employ the air assets to the best advantage, but would recognize their potential to help alleviate the situation.

While it is possible to make some of these aspects explicit for an ill-structured domain, it is not possible to form objective rules to govern every situation. Building on the close air support example, it is likely that a different array of factors would determine the applicability of air assets for different situations. A single set of well-defined rules cannot adequately address every instance. With experience, the learner will increasingly pick up on the array of cues that signal opportunities for air support. (For more detailed examples of each stage, see Phillips, Ross, and Cohn, Volume 1, Section 2, Chapter 8.) Technology offers the setting for the practitioner to rapidly build an experience base, understand how similar

Table 4.2. Summary of Stage 1 of Cognitive Skills Acquisition—Novice

Knowledge	Stage 1: Novice Performance	Training Implications
• Objective facts and features of the domain (Dreyfus & Dreyfus, 1986). • Context-free (abstract) rules to guide behavior (Dreyfus & Dreyfus, 1986). • Domain characteristics acquired through textbooks and classroom instruction (Benner, 1984).	• Guided by rules; is limited and inflexible (Benner, 1984). • Shows recognition of elements of the situation without considering the context (Dreyfus & Dreyfus, 1986). • Is variable and awkward (Glaser, 1996). • Focuses on isolated variables (Glaser, 1996). • Consists of a set of individual acts rather than an integrated strategy (Glaser, 1996; McElroy et al., 1991). • Is self-assessed based on how well he or she adheres to learned rules (Benner, 1984; Dreyfus & Dreyfus, 1986). • Reflects a sense of being overwhelmed since all stimuli are perceived to be equally relevant (McElroy et al., 1991).	• Must give learners rules to guide performance (Benner, 1984). • Learners require guidance in the form of instruction or mentoring while developing their experiential knowledge (Houldsworth et al., 1997). • Dialogue with a mentor or instructor enables learner to make sense of his or her experiences and discover he or she learned more than what he or she may have originally thought (Houldsworth et al., 1997). • Structured, situation based learning of facts can increase retention and appropriate recall of declarative information in later stages of learning (Bransford, Sherwood, Hasselbring, Kinzer, & Williams 1990).

cases can vary substantially, and practice understanding different situations and making decisions.

Advanced beginners begin to develop their own "guidelines" that stem from an understanding of the domain attributes and aspects. Guidelines are rules that inform behavior by allowing the practitioner to attach meaning to elements of a situation (Dreyfus & Dreyfus, 1980, 1986). For example, a platoon leader at this level of proficiency may know that the first step in conducting an offensive is to set up a base of fire, and he or she may know that the support position should be a certain distance from the primary objective. However, he or she may not understand that he or she needs to take into account not only distance from the objective, but also angles of fire in order to prevent fratricide. And he or she probably cannot distinguish that the rules and factors critical under one set of circumstances are not necessarily decisive in other operational situations. Spiro et al. (1992) note that in complex domains, the application of different patterns of

Table 4.3. Summary of Stage 2 of Cognitive Skills Acquisition—Advanced Beginner

Knowledge	Stage 2: Advanced Beginner Performance	Training Implications
• Some domain experience (Benner, 1984; Dreyfus & Dreyfus, 1986). • More objective, context-free facts than the novice, and more sophisticated rules (Dreyfus & Dreyfus, 1986). • Situational elements, which are recurring, meaningful elements of a situation based on prior experience (Dreyfus & Dreyfus, 1986). • A set of self-generated guidelines that dictate behavior in the domain (Benner, 1984). • Seeks guidance on task performance from context-rich sources (for example, experienced people, documentation of past situations) rather than rule bases (for example, text-books) (Houldsworth et al., 1997).	• Is marginally acceptable (Benner, 1984). • Combines the use of objective, or context-free, facts with situational elements (Dreyfus & Dreyfus, 1986). • Ignores the differential importance of aspects of the situation; situation is a myriad of competing tasks, all with same priority (Benner, 1984; Dreyfus & Dreyfus, 1986; Shanteau, 1992). • Shows initial signs of being able to perceive meaningful patterns of information in the operational environment (Benner, 1984). • Reflects attitude that answers are to be found from an external source (Houldsworth et al., 1997). • Reflects a lack of commitment or sense of involvement (McElroy et al., 1991).	• It can be beneficial to take off the training wheels and force learners to analyze the situation on their own (Houldsworth et al., 1997). • The learner benefits from having his or her attention directed to certain aspects of the situation by an instructor or mentor. This enables him or her to begin forming principles that can dictate actions (Benner, 1984). • Coaching on cue recognition and discrimination is appropriate and important (Benner, 1984; Benner, 2004). • Coaching on setting priorities is appropriate (Benner, 1984). • Employ strategies to calm learner and decrease anxiety in order to enhance performance capacity (Benner, 2004). • Use diagrams to facilitate the development of accurate mental model development (Scielzo, Fiore, Cuevas, & Salas, 2004).

principles varies from situation to situation, and there is substantial interconnectedness among principles that can be seen only by experiencing one type of situation in a number of varying instantiations. At this stage the practitioner has organized his or her knowledge and experience into principles, but has not built the interconnectedness or developed the ability for flexible application. Table 4.3 provides a summary of the advanced beginner stage of development and includes training implications for this stage.

Stage 3: Competent

Stage 3 is marked by the ability to formulate, prioritize, and manage longer-term goals or objectives. This perspective gives the operator a better sense of the relative importance of the attributes and aspects of the situation. The transition from advanced beginner to competent is highlighted by a shift from highly reactive behaviors, where actions are taken right when a problem surfaces, to planned behaviors, where the learner can see the larger picture and assess what actions must be taken immediately and what can wait until later. While a learner at stage 3 is not as quick or flexible as a stage 4 learner, he or she can typically manage a large set of incoming information and task demands.

The competent performer acts on the situation with a very analytical, hierarchical approach. Dreyfus and Dreyfus (1986) compare this to the problem solving approach described by proponents of information processing. Based on an initial judgment of what part of the situation is most important, the performer generates a plan to organize and thus simplify the situation to improve his or her performance. However, the drawback for competent performers is that their plans drive their behavior to a greater extent than any situational elements that may arise; they tend to hesitate to change their plans midcourse, despite the introduction of new, conflicting information. Simultaneously, competent performers are more emotionally invested in their performances than novices or advanced beginners. Because they actively choose a plan of action for themselves rather than relying on rules offered by a textbook or instructor, they take great pride in success and are distressed by failure (Dreyfus & Dreyfus, 1986). At this point, technology can provide multiple iterations of scenarios to allow the learners to build connections across cases and understand the limits of their plans and the need for adaptability. Table 4.4 provides a summary of the competent stage of development and includes training implications for this stage.

Stage 4: Proficient

Learners at the proficient level have moved away from perceiving situations in terms of independent aspects and attributes and see the situation as an inseparable whole where aspects and attributes are interrelated and woven together. The situation is not deliberately analyzed for its meaning; an assessment occurs automatically and dynamically because the learner has an extensive experience base from which to draw comparisons. However, decisions regarding appropriate actions continue to require some degree of detached analysis and deliberation. With regard to the situation assessment process, proficient individuals experience the event from a specific perspective, with past experiences in mind. Therefore, certain features of the situation stand out as salient, and others fade into the background as noncritical (see, for example, Crandall & Getchell-Reiter, 1993; Hoffman, Crandall, & Shadbolt, 1998; Klein, 1998). Dreyfus and Dreyfus (1980) assert that at this stage, performers are also positively impacted by new information that is obtained as the situation progresses. While competent performers generally cannot change their plans when faced with conflicting information,

Table 4.4. Summary of Stage 3 of Cognitive Skills Acquisition—Competent

Knowledge	Stage 3: Competent Performance	Training Implications
• How to think about the situation in terms of overarching goals or tasks (Benner, 1984). • The relative importance of subtasks depending on situational demands (Benner, 1984; Dreyfus & Dreyfus, 1986). • Particular patterns of cues suggest particular conclusions, decisions, or expectations (Dreyfus & Dreyfus, 1986). • A personalized set of guiding principles based on experience (Houldsworth et al., 1997). • How to anticipate future problems (Houldsworth et al., 1997).	• Is analytic, conscious, and deliberate (Benner, 1984; Dreyfus & Dreyfus, 1986). • Does not rely on a set of rules (Houldsworth et al., 1997). • Is efficient and organized (Benner, 1984; Dreyfus & Dreyfus, 1986). • Is driven by an organizing plan that is generated at the outset of the situation (Dreyfus & Dreyfus, 1986). • Reflects an inability to digress from the plan, even when faced with new, conflicting information (Dreyfus & Dreyfus, 1986). • Reflects an inability to see newly relevant cues due to the organizing plan or structure that directs attention (Benner, 2004). • Reflects an emotionally involved performer who takes ownership of successes and failures (Dreyfus & Dreyfus, 1986). • Focuses on independent features of the situation rather than a synthesis of the whole (Houldsworth et al., 1997).	• Decision-making games and simulations are beneficial at this stage. The scenarios should require the learner to plan and coordinate multiple, complex situational demands (Benner, 1984). • Coaching should encourage learners to follow through on senses that things are not as usual, or on vague feelings of anxiety. They have to learn to decide what is relevant without rules to guide them (Benner, 2004). • Use learner's sense of confusion or questioning (for example, when the plan does not hold up) to improve his or her domain mental models (Benner, 2004).

proficient individuals fluidly adjust their plans, expectations, and judgments as features of the situation change. They have an intuitive ability to recognize meaningful patterns of cues without breaking them down into their component parts for analysis. Dreyfus and Dreyfus (1980) term this ability "holistic similarity recognition." However, the elements that are holistically recognized must still be assessed and combined using sophisticated rules in order to produce a decision or action that meets the individual's goal(s). Technology can provide performers at this advanced stage the opportunity to practice verifying their perceptions in complex situations, practice adaptive behaviors, and gain more fluidity in their performance.

Dreyfus and Dreyfus further describe stage 4 performers as being guided by "maxims" that reflect the nuances of a situation (see also Benner, 1984). These maxims can mean one thing under one set of circumstances, but something else under another set of circumstances. As a simplistic example, consider a building in the midst of an urban combat area whose windows are broken out. This cue could indicate that the building is run down and vacant. It could also indicate that the adversary is occupying the building and has broken out the windows to use it as a base of fire. Other situational cues and factors will need to be considered to determine how to interpret the broken out windows—for example, the adversaries' history of breaking out windows, typical building types that they have utilized in the past, their last known location and projected current location, the presence or absence of undisturbed dust or dirt around the building, and so forth.

Table 4.5 provides a summary of the proficient stage of development and includes training implications for this stage.

Stage 5: Expert

The fifth and final stage of the Dreyfus and Dreyfus model is the expert. At this level the individual no longer relies on analytic rules, guidelines, or maxims; performance becomes intuitive and automatic. The expert immediately understands which aspects of the situation are critical and does not waste time on the less significant aspects. He or she knows implicitly what action to take and can remedy a situation quickly and efficiently. Experts typically need highly realistic situations or actual operational experiences to stimulate development. They benefit from real-life, highly complex scenarios and live exercises in which they can combine the various resources and talents of others around them and conduct challenging discussions with other experts. Table 4.6 provides a summary of the expert stage of development and includes training implications for this stage.

TRAINING DESIGN FOR ADVANCED LEARNING[1]

Surrogate Experiences and Fidelity

Technology based training should be used to build an experience base. That experience base will be activated in every situation encountered in the future to yield more powerful operational performance. Good design of advanced learning experiences can support advancement through stages 2, 3, and 4 of the performance continuum. Design without knowledge of expert cognitive performance and how experiences create advanced learning can waste time and resources, or worse, in some cases negatively affect transfer of training to performance.

[1]For more detailed training principles and examples of application, see Phillips, Ross, and Cohn (Volume 1, Section 2, Chapter 8).

Table 4.5. Summary of Stage 4 of Cognitive Skills Acquisition—Proficient

Knowledge	Stage 4: Proficient Performance	Training Implications
• Typical "scripts" for categories of situations (Klein, 1998). • How to set expectancies and notice when they are violated (Benner, 1984). • How to spot the most salient aspects of the situation (Benner, 1984; Dreyfus & Dreyfus, 1986). • Personalized maxims, or nuances of situations, that require a different approach depending on the specific situation, but not how to apply the maxims correctly (Benner, 1984; Houldsworth et al., 1997).	• Reflects a perception of the situation as a whole rather than its component features (Benner, 1984). • Is quick and flexible (Benner, 1984). • Reflects a focus on long-term goals and objectives for the situation (Benner, 1984). • Utilizes prior experience (or intuition) to assess the situation, but analysis and deliberation to determine a course of action (Dreyfus & Dreyfus, 1986; McElroy et al., 1991). • Reflects a synthesis of the meaning of information over time (Benner, 2004). • Reflects a more refined sense of timing (Benner, 2004).	• Case studies (that is, scenarios) are valuable, where the learner's ability to grasp the situation is solicited and taxed (Benner, 2004). • The learner should be required to cite his own personal experiences and exemplars for perspective on his views of case studies (Benner, 2004). • Teach inductively, where the learner sees the situation and then supplies his or her own way of understanding the situation (Benner, 2004). • Within the scenarios, the facilitation should exhaust the learner's way of understanding and approaching the situation (Benner, 2004). • Scenarios should include irrelevant information and, in some cases, insufficient information to generate a good course of action (Benner, 2004). • Scenarios should contain levels of complexity and ambiguity that mirror real world situations (Benner, 2004). • Do not introduce context-free principles or rules, or decision analysis techniques, within the context of any training or practice (Benner, 1984).

Table 4.6. Summary of Stage 5 of Cognitive Skills Acquisition—Expert

Knowledge	Stage 5: Expert Performance	Training Implications
• How to make fine discriminations between similar environmental cues (Klein & Hoffman, 1993). • How a range of equipment and resources function in the domain (Phillips, Klein, & Sieck, 2004). • How to perceive meaningful patterns in large and complex sets of information (Klein, 1998; Dreyfus & Dreyfus, 1986). • A wide range of routines or tactics for getting things done (Klein, 1998). • A huge library of lived, distinguishable experiences that impact handling of new situations (Dreyfus & Dreyfus, 1986). • How to set expectancies and notice when they are violated (Benner, 1984). • What is typical and atypical for a particular situation (Dreyfus & Dreyfus, 1986; Feltovich, Johnson, Moller, & Swanson, 1984; Klein, 1998).	• Is fluid and seamless, like walking or talking; "integrated rapid response" (Benner, 1984, 2004; Dreyfus & Dreyfus, 1986). • The rationale for actions is often difficult to articulate (Benner, 1984). • Relies heavily and successfully on mental simulation to predict events, diagnose prior occurrences, and assess courses of action (Einhorn, 1980; Klein & Crandall, 1995). • Consists of more time assessing the situation and less time deliberating a course of action (Lipshitz & Ben Shaul, 1997). • Shows an ability to detect problems and spot anomalies early (Feltovich et al., 1984). • Capitalizes on leverage points, or unique ways of utilizing ordinary resources (Klein & Wolf, 1998). • Manages uncertainty with relative ease, by filling gaps with rational assumptions and formulating information-seeking strategies (Klein, 1998; Serfaty, MacMillan, Entin, & Entin, 1997). • Shows efficient information search activities (Shanteau, 1992).	• Scenarios require more complex and detailed contextual information to allow experts to build more sophisticated mental models that contain fine discriminations and more elaborate causal links (for example, full-scale simulations, operational environments) (Feltovich et al., 1984). • Real world planning sessions, which often pull experts from various specialty areas, provide good context within which experts can discuss salient situational cues and preferred actions. • Mentoring others provides an opportunity for experts to "unpack" their thought processes as they model them for others, giving them a better understanding of their own mental models (Ross, Battaglia, Hutton, & Crandall, 2003). • As an alternative to full-scale simulations, tabletop exercises conducted with expert-to-expert peer groups can enable experts to compare the importance of situational factors, cues, and preferred actions.

Expertise is built one challenging experience at a time. A useful experience must be cognitively authentic from the user's point of view. *Cognitive authenticity* (Ross, Halterman, Pierce, & Ross, 1998) is the quality that a surrogate experience has when it can stimulate and allow the learner to begin using the perceptual and recognitional processes of an expert—learning to notice, in the manner of an expert, what is and is not present in different situations and the patterns those cues suggest, and to identify situational factors that inhibit some actions or create leverage points for others. The context must be authentic in relationship to how practitioners experience and act in real-life settings. Building a context to support authentic domain experience is not the same thing as simulating physical fidelity. Reproducing billowing smoke, elegantly drawn leaves on the trees, or precise shadows is artistically rewarding, but irrelevant if those elements are not used in judgments or decisions typical of that situation. Meanwhile, failing to represent a tiny pile of freshly overturned dirt indicating that the nearby entrance of a cave has been disturbed can interfere with an authentic cognitive experience. A good technology based experience is an opportunity from a first-person perspective to learn to perceive like an expert.

Engineering fidelity, the degree to which technology duplicates the physical, functional, and environmental conditions, does not correlate well with psychological fidelity (Fanchini, 1993). Rather than using an engineering approach to fidelity, designs must correspond to the perceptions that practitioners in the domain have of situations. That information comes from the experiences of experts. There is a problem with using subject matter experts (SMEs) to tell us about their cognitive experience. When they are really good, SMEs usually cannot tell us what they are noticing and how they are doing what they do. They say they "just know." They do not go through an analysis process to assess or decide. Their knowledge is not verbally encoded. They do not use words, even internally, to lay out cues, factors, or actions. Their expertise is so ingrained that when asked about a specific critical incident, experts often feel as if they acted on instinct or intuition that cannot be explained, not recalling their mental paths to an assessment or action. Cognitive Task Analysis (CTA; Crandall, Klein, & Hoffman, 2006) helps experts access their strategies, perceptions, and knowledge through in-depth interviews. CTA allows the training designer to see experiences from the point of view of the expert. Expertise is "unpacked" around a specific, critical incident. This information from the SMEs becomes the basis for training design and represents the elements of authenticity that learners must perceive and the cognitive challenges or dilemmas they must work through in order to exercise cognitive processes that will form the basis for expertise.

Advanced training designs must (1) incorporate cognitive challenges found in the field of practice, (2) provide elements of a situation to develop perceptual attunement, and (3) create an immersive experience that stimulates authentic cognitive behavior from the user's point of view and that replicates the cognition of experts in the field.

Complexity, Multiple Representations, and Case Exploration

Inadequate training design for cognitively complex, ill-structured knowledge domains is usually the result of oversimplification and rigid structure. Unfortunately, some training designs that have been used for advanced learning are more suitable for introductory or procedural training, which tends to be linear and hierarchical, and which is frequently task focused, compartmentalized, prepackaged, delivered from one perspective, and comprised of simple analogies. Sometimes training designers simplify and modularize complex knowledge as a means of providing what they believe is easy access to difficult concepts. The designs compartmentalize knowledge, present only clear (and few) cases rather than the many exceptions and variations, and neglect to require application of new knowledge to a variety of situations (Feltovich, Spiro, & Coulson, 1993). The difficulty of representing the complexity of ill-structured domains at the advanced level is often too hard to overcome when using traditional training design processes because these processes encourage oversimplification of information and fail to provide guidance for representing and teaching interrelated concepts where variation among cases is the norm.[2]

Good design at the advanced level provides for extended exploration of operational situations. Exploration is not unstructured discovery learning, but is introduced through prespecified starting points with aids or frameworks for support. One way frameworks can help learners organize their thoughts about a domain as they explore cases is by providing domain specific themes. These themes are drawn from the domain; they are ways that experts express their conceptual understanding such as "know the enemy." Themes provide anchors during exploration, but do not prescribe a preferred method for structuring specific knowledge. They are not a prespecified knowledge schema or a set of mental models that the student should adopt. Themes are also used to guide the design of multiple representations and to support the student in exploring how concepts differ in their meaning and application across situations. "Multiple representations" refers to multiple explanations, multiple analogies, and multiple dimensions of analysis. Representations must be open for exploration without prescribed connections and endpoints—what some advocates of game based training refer to as "free play." However, that free play must be set in well-structured representations of an expert's perceived reality and supported by guided reflection.

Advanced learning is primarily about performance and reflection when grappling with ill-structured problems—those situations that have unclear elements, dynamically evolving goals, and multiple solutions or solution paths. Such problems are emergent dilemmas; they grow from the dynamics of a situational context. Game based or virtual environments are the ideal medium for easy access to multiple explorations of interrelated, dynamic settings needed to build the experience bases that support expertise.

[2]For more information on how the learner progresses through the stages of learning, see Klein and Baxter (Volume 1, Section 1, Chapter 3).

REFERENCES

Benner, P. (1984). *From novice to expert: Excellence and power in clinical nursing practice*. Menlo Park, CA: Addison-Wesley Publishing Company Nursing Division.

Benner, P. (2004). Using the Dreyfus model of skill acquisition to describe and interpret skill acquisition and clinical judgment in nursing practice and education. *Bulletin of Science, Technology & Society, 24*(3), 189–199.

Bransford, J. D., Sherwood, R. D., Hasselbring, T. S., Kinzer, C. K., & Williams, S. M. (1990). Anchored instruction: Why we need it and how technology can help. In D. Nix & R. Spiro (Eds.), *Cognition, education and multimedia* (pp. 115–141). Mahwah, NJ: Lawrence Erlbaum.

Crandall, B., & Getchell-Reiter, K. (1993). Critical decision method: A technique for eliciting concrete assessment indicators from the "intuition" of NICU nurses. *Advances in Nursing Sciences, 16*(1), 42–51.

Crandall, B., Klein, G., & Hoffman, R. R. (2006). *Working minds—A practitioner's guide to cognitive task analysis*. Cambridge, MA: The MIT Press.

Dreyfus, H. L., & Dreyfus, S. E. (1986). *Mind over machine: The power of human intuitive expertise in the era of the computer*. New York: The Free Press.

Dreyfus, S. E., & Dreyfus, H. L. (1980). *A five stage model of the mental activities involved in directed skill acquisition* (Unpublished report supported by the Air Force Office of Scientific Research [AFSC], USAF Contract No. F49620-79-C-0063). Berkeley: University of California at Berkeley.

Druckman, D. (1995). The educational effectiveness of interactive games. In D. Crookall & K. Arai (Eds.), *Simulation and gaming across disciplines and cultures: ISAGA at a watershed* (pp. 178–187). Thousand Oaks, CA: Sage.

Einhorn, H. J. (1980). Learning from experience and suboptimal rules in decision making. In T. S. Wallsten (Ed.), *Cognitive processes in choice and decision behavior* (pp. 1–20). Mahwah, NJ: Lawrence Erlbaum.

Fanchini, H. (1993, September). *Desperately seeking the reality of appearances: The case of sessions on full-scale simulators*. Paper presented at the Fourth International Conference on Human-Machine Interaction and Artificial Intelligence in Aerospace, Toulouse, France.

Feltovich, P. J., Johnson, P. E., Moller, J. H., & Swanson, D. B. (1984). LCS: The role and development of medical knowledge in diagnostic expertise. In W. J. Clancey & E. H. Shortliffe (Eds.), *Readings in medical artificial intelligence: The first decade* (pp. 275–319). Reading, MA: Addison-Wesley.

Feltovich, P. J., Spiro, R. J., & Coulson, R. L. (1993). Learning, teaching, and testing for complex conceptual understanding. In N. Frederiksen, R. J. Milslevy, & I. I. Vehar (Eds.), *Test theory for a new generation of tests* (pp.181–217). Hillsdale, NJ: Lawrence Erlbaum.

Garris, R., Ahlers, R., & Driskell, J. E. (2002). Games, motivation, and learning: A research and practice model. *Simulation & Gaming, 33*(4), 441–467.

Glaser, R. (1996). Changing the agency for learning: Acquiring expert performance. In K. A. Ericsson (Ed.), *The road to excellence* (pp. 303–311). Mahwah, NJ: Lawrence Erlbaum.

Hoffman, R. R., Crandall, B. W., & Shadbolt, N. R. (1998). Use of the critical decision method to elicit expert knowledge: A case study in cognitive task analysis methodology. *Human Factors, 40*(2), 254–276.

Houldsworth, B., O'Brien, J., Butler, J., & Edwards, J. (1997). Learning in the restructured workplace: A case study. *Education and Training, 39*(6), 211–218.

Klein, G. (1998). *Sources of power: How people make decisions*. Cambridge, MA: MIT Press.

Klein, G., & Wolf, S. (1998). The role of leverage points in option generation. *IEEE Transactions on Systems, Man and Cybernetics: Applications and Reviews, 28*(1), 157–160.

Klein, G. A., & Crandall, B. W. (1995). The role of mental simulation in naturalistic decision making. In P. Hancock, J. Flach, J. Caird, & K. Vicente (Eds.), *Local applications of the ecological approach to human-machine systems* (Vol. 2, pp. 324–358). Mahwah, NJ: Lawrence Erlbaum.

Klein, G. A., & Hoffman, R. (1993). Seeing the invisible: Perceptual/cognitive aspects of expertise. In M. Rabinowitz (Ed.), *Cognitive science foundations of instruction* (pp. 203–226). Mahwah, NJ: Lawrence Erlbaum.

Lester, S. (2005). *Novice to expert: The Dreyfus model of skill acquisition*. Retrieved May 1, 2008, from http://www.sld.demon.co.uk/dreyfus.pdf

Lipshitz, R., & Ben Shaul, O. (1997). Schemata and mental models in recognition-primed decision making. In C. Zsambok & G. Klein (Eds.), *Naturalistic decision making* (pp. 293–304). Mahwah, NJ: Lawrence Erlbaum.

McElroy, E., Greiner, D., & de Chesnay, M. (1991). Application of the skill acquisition model to the teaching of psychotherapy. *Archives of Psychiatric Nursing, 5*(2), 113–117.

Phillips, J. K., Klein, G., & Sieck, W. R. (2004). Expertise in judgment and decision making: A case for training intuitive decision skills. In D. J. Koehler & N. Harvey (Eds.), *Blackwell handbook of judgment & decision making* (pp. 297–315). Victoria, Australia: Blackwell Publishing.

Prensky, M. (2001). *Digital game-based learning*. New York: McGraw-Hill.

Ross, K. G., Battaglia, D. A., Hutton, R. J. B., & Crandall, B. (2003). *Development of an instructional model for tutoring tactical thinking* (Final Tech. Rep. for Subcontract No. SHAI-COMM-01; Prime Contract No. DASW01-01-C-0039 submitted to Stottler Henke Associates Inc., San Mateo, CA). Fairborn, OH: Klein Associates.

Ross, K. G., Halterman, J. A., Pierce, L. G., & Ross, W. A. (1998, December). *Preparing for the instructional technology gap: A constructivist approach.* Paper presented at the Interservice/Industry Training, Simulation, and Education Conference, Orlando, FL.

Scielzo, S., Fiore, S. M., Cuevas, H. M., & Salas, E. (2004). Diagnosticity of mental models in cognitive and metacognitive process: Implications for synthetic task environment training. In S. G. Shiflett, L. R. Elliottt, & E. Salas (Eds.), *Scaled worlds: Development, validation, and applications* (pp. 181–199). Burlington, VT: Ashgate.

Serfaty, D., MacMillan, J., Entin, E. E., & Entin, E. B. (1997). The decision-making expertise of battle commanders. In C. Zsambok & G. Klein (Eds.), *Naturalistic decision making* (pp. 233–246). Mahwah, NJ: Lawrence Erlbaum.

Shanteau, J. (1992). Competence in experts: The role of task characteristics. *Organizational Behavior and Human Decision Processes, 53,* 252–266.

Spiro, R. J., Feltovich, P. J., Jacobson, M. J., & Coulson, R. L. (1992). Cognitive flexibility, constructivism, and hypertext: Random access instruction for advanced knowledge acquisition in ill-structured domains. In T. Duffy & D. Jonassen (Eds.), *Constructivism and the technology of instruction: A conversation* (pp. 57–76). Mahwah, NJ: Lawrence Erlbaum.

Chapter 5

CYBERNETICS: REDEFINING INDIVIDUALIZED TRAINING

Elizabeth Biddle, Dennis McBride, and Linda Malone

Adaptive learning has long been discussed and studied with the notion that individualized instruction should be able to optimize learning. However, the application of adaptive learning methods has been primarily through various research and development and prototyping efforts rather than standard practice. This is partly due to the numerous means of tailoring training, such as type of instructional feedback and nature of feedback. It is also partly due to the technologies available to analyze, monitor, and recommend training strategies that optimize a student's unique capabilities and traits along with his or her current state.

Virtual environments (VEs) are rapidly being implemented as primary instructional sources for the training of complex, real world tasks (for example, flying and driving). In addition to providing realistic representations of the operational environments, eliminating safety risks, and reducing costs of operating the actual equipment, VEs allow for the instructor—or automated instructor as will be discussed—to control the environment. This enables the student to obtain experience in responding to catastrophic situations (for example, repeating scenarios to experiment with alternative actions) and the instructor, or automated instructor, to provide the student with feedback in terms of auditory, visual, or other cues that are most appropriate for the situation and the student at that particular moment.

Various approaches (for example, neural nets and blackboard systems) have been implemented for the purpose of providing individualized, automated instructional applications. Due to the complexities, time requirements, and costs involved, there have only been separate, nonrelated implementations. Additionally, these types of systems have, thus far, concentrated on tailoring instruction on the student's observable behavior. However, cognitive (for example, information processing and decision making) as well as noncognitive (for example, motivation and self-efficacy) activities that are not easily observed affect the learning process (for example, Ackerman, Kanfer, & Goff, 1995) and often vary with respect to the contextual domain (Snow, 1989). Additionally, optimal methods of instruction vary depending upon the student's level of expertise.

Noncognitive processes can be detected through evaluation of the student's affective (emotional) reactions (for example, Ackerman, Kanfer, & Goff, 1995; Bandura, 1997). Physiological and neurophysiological monitoring technologies, for cognitive and noncognitive processes, are widely being investigated today as a means of refining the student performance assessment. The future of VE training is to leverage the advances in these technologies to enable a comprehensive assessment of student performance that is used by the training system to adapt the VE to optimize the student's use of the system.

We now elaborate on the above, first by providing a rudimentary treatment of cybernetics and its relationship to learning systems. This section is not intended to transform the reader into a cyberneticist, but rather to provide a top level overview and to suggest what the future may hold for individualized training from a control theoretic perspective. This section will be followed and substantiated with a brief discussion of individual attributes that are important in pedagogy. Finally, we will concentrate on specific physiological phenomenology that, with improving technology, can be measured in real time and exploited further to enhance the learning experience.

CYBERNETICS

Cybernetics is a scientific discipline. It is not a synonym for computers or networks. As we describe below, central to the concept of cybernetic systems is the role of *variable feedback* in guiding *systems* toward *goals*. In the present context, we are focused on the science and technology associated with individually optimized training feedback schema and what they may look like in the future.

The concept and the Greek term "cybernetics" were probably introduced by Plato: *χυβερνήτησ* (kybernētikē). The word actually referred to the art of nautical steermanship, although Plato himself used the term to describe control within the animal, and the roles of what we now call government. The expression was translated in Latin to "governor" and has been used to describe political regulation for centuries. In perhaps the most significant return to the original, technical meaning of the word, James Watt selected the term to describe his 1790 steam engine's mechanism for velocity control. André-Marie Ampère translated to the French *mot juste:* "cybernétique," and from about this time (1834) forward, the concept and its associated terminology have experienced tidal episodes in acceptance and in understanding. The perceived successes or failures of artificial intelligence are probably associated with these vicissitudes.

The popular literature has underwritten an unintended narrowing of the concept by using the diminutive "cyber" in reference to computers, robots, and numerous Internet notions—from cyborgs to cyberspace. The perception that cybernetics is computer specific is not limited to popular opinion. Vibrant (as it should be) within the Department of Homeland Security, for example, is the National Cyber Security Division. Its mission is to protect the nation's computer/communication networks. This unfortunate lexical trajectory is perhaps out of control, even to the point that the field of cybernetics proper might need to rename itself. For the purposes of this chapter, we argue that training specialists benefit from understanding the value of cybernetics in its rigorous scientific application. Fortunately, it happens that the feedback systems that are exploited by training systems today are increasingly computer based. For the purposes of this chapter, *cyber,* popularly implying computer-network oriented, is subsumed by *cybernetics*. As such, cyber systems represent a truly valuable set of tools for training. We are addressing the more inclusive science of cybernetics, and we will finesse the definitional issue in this chapter in the material that follows by introducing the blend, neurocybernetics.

Modern Scientific Cybernetics

The father of modern cybernetics is considered to be Norbert Wiener, who was a professor at the Massachusetts Institute of Technology. Wiener, boy genius, provided invaluable mathematical concepts and solutions to the U.S. military in both world wars during his adulthood. Weiner's (1948) title *Cybernetics: or Control and Communication in the Animal and the Machine* serves both as the seminal contribution for the field and as a suitable technical definition of cybernetics itself. Guilbaud (1959) referred to the emerging discipline as a "crossroads of the science" because cybernetics found itself penetrating several fields, and deriving from several fields. At the heart of the discipline is the phenomenon of feedback for the purpose of control. We will discuss at a top level the important aspects of cybernetics and control theory, but we encourage the interested reader to consult venerable sources such as Wiener (1948), Ashby (1956, 1960), Guilbaud (1959), and Powers (1973).

The science of cybernetics is about systematic state changes or mechanisms, not necessarily (nor even usually) about physical machines. Ashby (1956) begins his introductory treatment with three important interrelated phenomena, and he provides the following everyday example to illustrate. Pale human skin, when exposed to the sun, tends to tan or darken. The skin is thus defined as an *operand* (acted upon), the sun is the *operator,* and the darkened skin is the *transform*. The process thus described is termed a *transition*. For investigators who think in terms of systems of systems, the transition above is only one transition among many others that occur naturally with this particular operator. As Ashby reminds us, other transitions in this solar context include cold soil → warm soil, colored pigment → bleached pigment, and so forth. Multiple, related transitions (particularly those with a common operator) are referred to as a transformation (diurnal global reaction in our solar example). For the purposes of this chapter, we will be concerned with change (transition) in student performance (the operand) as a function of training (the operator). Rudimentary training transitions (for example, maintaining aircraft altitude) are components of the larger transformation (achieving solo status).

The important concept is that of states and state changes. As systems change systematically, subsystem and system states can be described whether the dynamics are discrete or fluid. One notation scheme used by Ashby (1956) is very simple to learn and use, as in Figure 5.1. Here, for transform U, A transitions to D, B to A, and so forth.

U: A B C D E
D A E D D

Figure 5.1. Transformation U, with Five Transitions Internally

For our pilot training example's purposes, let A = reduce power, forward stick pressure, tolerate – 3g (forces); B = discover altitude high; C = discover altitude low; D = trim, fly straight, and level; E = add power, back stick pressure, tolerate + 3g. Now we can make use of the notated transform for achieving straight and level flight on the assigned altitude, or we can make use of a kinematic diagram, which makes more obvious sense out of the transitions as components of the transformation. In the cybernetic context, feedback techniques may very well be different for each transition and may very well change as a function of the number of iterations that the transformation undergoes (in the above case, U, U2, U3, and so forth). This means that feedback for C → E might optimally be primary only, E → D primary and secondary, and so forth (see the Human Development: Learning section in this chapter; see also Figure 5.2).

Virtual environments provide significant opportunity for the design of imaginative and effective, individualized feedback control systems for training optimization. The control theory approach importantly not only allows for, but encourages, the fielding of controlled feedback systems that are dynamic—reinforcement techniques will systematically change in response to measured changes in independent and dependent variables.

Transformations can be much more complex, of course. We provide in Figure 5.3 the following transformation and its associated kinematic graph in order simply to portray the notational relationship between the two.

We can trace the elemental transitions and follow them in the kinematic graph. This exercise shows that the transformation consists of kinematics in which a trajectory progresses either to a stopping point or to a cycle. These are called basins, and they imply stability. The last rudimentary cybernetic concept is that of mechanism. We borrow again from Ashby (1956, who borrowed from Tinbergen, 1951) the following illustration (Figure 5.4) of the three-spined stickleback mating pattern.

The importance here is that this set of transitions represents a machine without input. The behavior is reflexive—it consists of fixed action patterns.

In our application to training, we recognize two important points. First, humans are not dominantly reflexive; we are much more complex behaviorally. We are a highly plasticial (that is, trainable) species. Second, machines (remember, not a

Figure 5.2. Kinematic (Dynamics) Graph for Transformation U, Where the Ultimate Steady State Is Straight and Level Flight

T:	A	B	C	D	E	F	G	H	I	J	K	L	M	N	P	Q
	D	H	D	I	Q	G	Q	H	A	E	E	N	B	A	N	E

The kinematic graph for which is

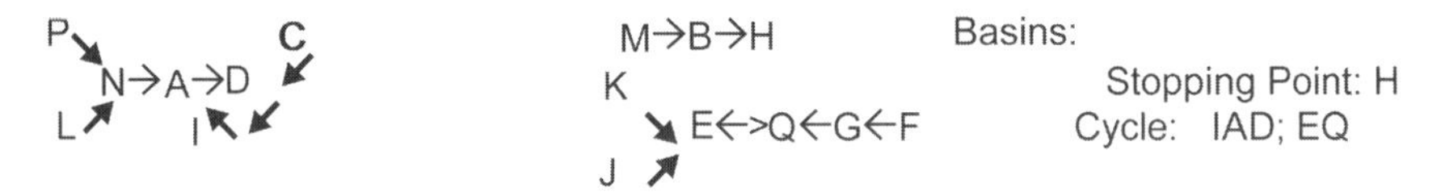

Figure 5.3. Transformation T and Its Kinematic Graph

piece of equipment) for training will be designed to function with input, because the goal of training is not to achieve and maintain homeostasis or equilibrium, but rather to produce state changes in the trainee's performance repertoire.

That humans are highly trainable as indicated above is obviously important. In order to effectively integrate cybernetics and training for the future, we must discuss the product of training—learning, but this important treatment must be provided in the context of learning's bigger picture: human development.

Human Development: Maturation

Developmental psychology focuses on two interacting dynamics. The first is technically delineated as maturation. The natural phenomenon here is that survival skills like walking emerge or "unfold" in the developing individual when (1) physiological critical periods visit the developing organism and when (2) the organism engages the behavior that is being genetically entrained. When this is successful, a developmental milestone is achieved. Thus technically, the infant cannot possibly learn to walk because the developmental window has not made this optional. But the toddler does not learn to walk either. Walking is a survival skill that is encoded in the organism's DNA, such that when the underlying physiology

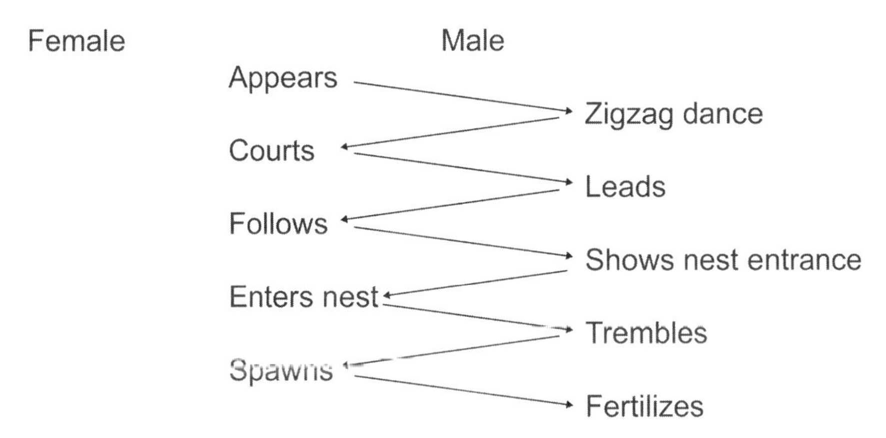

Figure 5.4. The Male-Female Mating Pattern for the Three-Spined Stickleback

is prepared, the child will commence the acquisition process and achieve a developmental milestone for motility. We are reminded of the old adage "When the student is ready, the teacher will come." Disputed in what appears to be more of a political rather than scientific forum, verbal communication, even grammar (not a language per se) may be encoded for maturational debut. Regardless, the underlying physiological processes are becoming clearer: the role of myelin is key.

Myelinization in the nervous system is the process whereby, to use pedestrian terms, gray matter becomes white matter. A deep treatment is beyond the scope of this chapter (for more detail, see, for example, McBride, 2005). The product of myelinization is the formation of a gapped, lipid-protein coating (arising from symbiotic glial and oligodendrocyte cells) around the axons of neurons. This electrical insulation produces an increase in nerve conduction speed by as much as three orders of magnitude (that is, from ca .1 m/s (meters per second) to perhaps 100 m/s). A corresponding decrease in neural refractory time (the time in ms during which the neuron cannot refire) means that effectively, the bandwidth is increased in myelinated cells by as much as a factor of 1,000. For comparison purposes, this transformation is equivalent to upgrading the download of a 4.7 Gb (gigabyte) movie via a 10 Base-T at 10 Mb/s (megabits per second) to a 10 Gb/s OC-192. The former requires 1.5 h; the latter, 5 s.

Based on the high proportion of bandwidth and the resulting control, a human can, for example, run coordinately at maximum sprint speed in one direction, throw a football with deadly accuracy to a target running at maximum speed in an oblique direction, all while being chased by several 300+ pound men with obliteration on their agenda. Humans have significantly more (total and relative) white matter than other species—remarkably more than even its nearest genetic neighbor, the chimpanzee. As a result, humanity has produced what is unthinkable to other species: mathematics, music, trips to the moon, and more. It is quite arguable that myelin is what makes *Homo sapiens* unique.

From conception and probably into adulthood, the process of myelinization continues such that when we achieve bandwidth sufficient to our bladders, we get bladder control; to our extremities, motility and manipulation; to our temporal lobes, speech, arguably language itself; and so forth. Many components of the human nervous system do not even begin the myelinization process until well after (as in many years) parturition. As we learn more about the emergence of myelinization, we see the distinct possibility that a significant number of trained behaviors in the ultimate, highly skilled human repertoire are partially *matured* and then honed through *learning* (the latter will be defined scientifically later). If so, serious cybernetic consideration should be given to this in terms of the design of critical period challenge environments.[1] It is evident that our

[1]Indeed, as of this writing, the Office of Naval Research is exploring this prospect with its Combat Hunter program. One hypothesized notion is that during millions of years of human life prior to modernity, many hunting/fighting skills were matured/learned as children grew from the toddler through the juvenile stage and into adulthood. Since this was probably characteristic of life in the Pleiocene/Pleistocene, critical periods were met with opportunities to acquire instinctive capabilities. However, in our highly synthetic world, where little game hunting-like activity is engaged, perhaps critical periods come and go without producing developmental milestones.

educational systems at some level recognize and exploit the spontaneous, increasing maturity of the growing child and the growing nervous system therein. Mathematics curricula, for example, follow a progressive complex from numbers to algebra to geometry to trigonometry to calculus, and so forth. If McBride and Uscinski (2008) are correct along this line, identification of the progress of the expanding myelin network, measured noninvasively directly, or through corresponding behavioral testing, suggests a significantly new way of thinking about training. That is, by providing abundant exposure to skill learning challenges during the near-completion period of myelinization, in theory, the trainee might absorb the skill as more of an instinct rather than a conditioned response. It should be noted that one can actually acquire a skill even though it did not arise during a developmental critical period, but the behavior will likely be less efficient than it would have been if maturated.

Human Development: Learning

There are several general concepts and technical definitions of learning, but the concepts all converge on the cybernetic notion of directionally transformed behavior. In terms of formal definition, we invoke one that is readily applied at least to perceptual motor skill acquisition. This definition is arguably the most robust for this chapter's purposes because much of cognitive training involves such skill. *Learning is the (rather) permanent change in behavior that comes about as the result of reinforced practice.* That is, progress derives from trial and error learning, wherein the behavior in question (for example, landing an aircraft) is refined through practice and feedback. In this sense, landing the aircraft is not thought to be the product of mere maturation, though essential maturation (for example, eye-hand coordination) must have been achieved in order to begin the learning process. Rather, landing accuracy improves based on the maturationally prepared trainee's exploitation of feedback.

We must comment on elements of the definition of learning in the previous paragraph. First the escape words "(rather) permanent" are partially in parentheses because of a technicality. The learning theory community posits that learning is plastic, that is, it grows with feedback and is permanent. Whereas new behavior can be learned that may override previously learned behavior, the originally acquired behavior is said to be "permanently learned." To provide a shallow argument in support, the so-called tip-of-the-tongue phenomenon is an everyday example of this notion of permanence. That is, finally recalling a name corresponding with a face, days after trying relentlessly, hints at the robustness of a permanently learned association. The words "change in behavior" are fundamentally important. Behavioral psychology respects research into the internal correlates of learning, but the field has classically been focused on observable, molar level behavior. The philosophy is that if *behavior* is to change, behavior must *happen*. This leads to the last component of our definition of the state change, learning. "Reinforced practice" is central to cybernetics. The trainee brings existing skills as joint products of aptitude and experience to the first trial of a new training experience as trials continue. Reinforcement (feedback) guides subsequent performance.

The key then to learning and to cybernetics is the quality and quantity of feedback provided. We contend that if experimental psychology has learned only one fact, it is that feedback—reinforcement—is *sine qui non* to the acquisition of skill. Feedback in training contexts comes in many varieties and utilities. The most fundamental feedback is simply knowledge of results, or KR. In order to improve any skill, the student must know the results of his or her efforts. But there is more to feedback than KR. There are myriad schemes (and a rich literature) for improving performance based on other primary sorts of feedback. And there is an abundance of supplementary (secondary and higher order) feedback methods. The science and technology of optimized feedback for human learning—especially its future—will be discussed, but we must first mathematically characterize learning in order to identify the parameters of the learning process that may be exploitable cybernetically.

Figure 5.5 is an idealized, group averaged, learning curve. It portrays accuracy (landing accuracy) as a function of practice. The curve reveals progress as negatively accelerated and monotonic. Moreover, as detailed theoretically by Noble (1978),[2] a mathematical expression for such is provided in Eq. (1).

$$A = C(1 - e^{-kN}) + T \tag{1}$$

The variables are described in Table 5.1.

Noble indicates that this equation explains 98 percent of the variance for empirical curves for the 10 perceptual motor skills datasets that he analyzed. These skills range from simple ones to more complex ones. Of importance here is the relative contribution of variables N, k, and T. The first, N, represents the quantity of practice engaged by participants. What is clear in the present graph and from interpretation of its mathematical form is that quantity of practice clearly exerts a positive influence on performance. The next independent variable

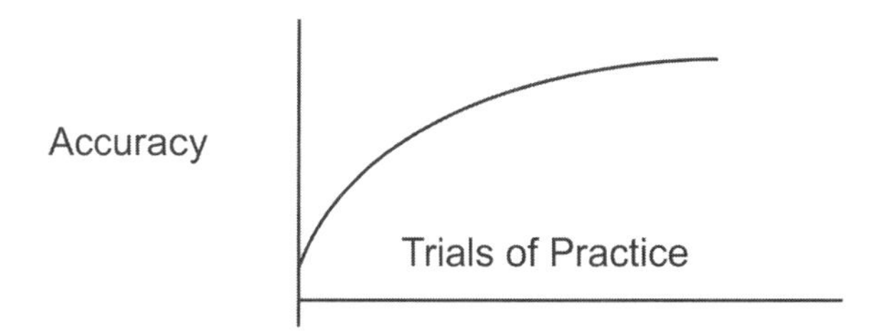

Figure 5.5. An Idealized, Group-Average Learning Curve

[2]A comprehensive treatment of Noble's (1978) theoretical work is beyond the scope of this book. However, the reader is strongly encouraged to pursue an understanding of his work. This approach to the mathematical based study of learning provides significant opportunities for training development beyond our cybernetic approach.

Table 5.1. Variables for Eq. (1)

N = quantity of practice; number of trials, the independent variable
A = accuracy in identifying targets in imagery; positive identifications; dependent variable
C = constant that transforms the accuracy of an ideal participant after N trials to R
T = theoretical joint contribution of experience and hereditary factors (can be negative also)
k = theoretical rate parameter; representative of individual differences in aptitude, and so forth.

of interest is the exponent k. This variable represents the contribution of aptitude factors—these Noble defends empirically and theoretically as the product of genetics. The last variable, T, a joint contribution of experience and heredity, is also very important, though algebraically rather than geometrically (as is the case for k, N). This variable represents a joint contribution of experiential and hereditary factors. Thus N and T represent in whole or part, environmental contributions to the learning curve. T and k, on the other hand, provide partly and fully, respectively, organismic variability that arises largely from heritable sources.

At this point we do not want to fuel an ancient and naive nature versus nurture argument. Clearly both environment and genetics contribute to the acquisition of skilled behavior. The point is that learning, the product of training, can be decomposed into elemental sources of variation with respect to influence. Whether nature-nurture variability is exploited for designing training systems in the future remains to be seen. Our principal point is that training state transformations will comprise multiple, skill-specific transitions, each of which must derive from a technical understanding of how the particular feedback methods succeed or fail for each transition in the transformation.

Human Development: Variability

Discovery of the sources of variation in variables such as N, k, and T is important for the future of training system development. However, as we stressed above, mathematical analysis of grouped, averaged learning data accounts for what it addresses: the data of a pool of trainees, represented typically as simple mean scores. So, how do we proceed from group averages to designing individualized training systems? On the one hand, experimental psychology generally considers variance to be a problem and methodologically treats it this way: for this field, group mean differences are the concern. Differential psychology, on the other hand, thrives on variability. Later in this chapter, we provide a comprehensive look at many individual and group differences, with an eye toward exploiting them. Moreover, we believe that future success in cybernetically conceived training will accrue based on the two communities working together. Former American Psychological Association president Lee Cronbach pleaded for this in 1957.

The (semi-) individualized approach has been sensible in terms of designing training systems for differences among groups—for example, the aging versus youth; male versus female, and so forth. We take this approach to the next step. The optimal training path from *y* intercept to asymptote for one individual is not necessarily the same optimum as for another individual. Fleishman (1953, 1966) showed over a half century ago that different underlying factors (for example, manual dexterity and balance) are called into play in skill acquisition as a function of the type of task engaged (that is, between skilled tasks) and how far up the learning curve subjects have aspired to in training (within skilled tasks). It is reasonable that knowledge of how sub-skills dynamically emerge during training episodes can be used to design differential feedback systems.

As we show in the pages that follow, variation among (and within) people in their respective aptitudes, cognitive styles, emotions, motivations, and so on is very considerable. More importantly, these organismic variables are reliably and validly measurable, consistent with means of exploiting them with sufficient temporal leave. In other words, we are only beginning to foresee the many and rich ways that feedback systems may be designed for individually tailorable, maximal training yield.

Much more is being learned about human learning and, importantly, about the very neural correlates that were ignored or inferred in the past. With continued advances in imagery technology such as functional magnetic resonance imaging (fMRI) and functional near-infrared imaging, it is likely that feedback systems will be organized so that (neural and behavioral) state change sequences can be shepherded most efficiently for desired transforms.

With the above consideration, we endorse the term *neurocybernetics* for what we believe will be an explosive science and technology. In the context of training systems, we suggest that neurocybernetics is the cybernetic exploitation of plasticial, neurobehavioral systems, such that optimized feedback mechanisms operate on an individual's repertory (operand) in order to effect specifically desired training states.

INDIVIDUAL DIFFERENCES IN LEARNING

That people learn differently has been discussed and researched for decades. In the 1970s, computer scientists began focusing their artificial intelligence research toward the development of intelligent tutoring systems to provide instructorless, individualized instruction. Then in the 1980s, Bloom (1984) reported a two-sigma increase in learning outcomes for students who received instruction by a one-on-one human instructor versus students who received traditional classroom instruction. This effect is explained by a human tutor's ability to evaluate student state in combination with task performance and use this information to tailor their interactions. This seminal work initiated a rebirth of individualized training systems research and development to replicate Bloom's findings with a computerized tutor.

Motivation, personality, and perceived autonomy are considered as some of the noncognitive variables that affect student learning (Ackerman, Kanfer, & Goff, 1995; McCombs & Whisler, 1989). Table 5.2 summarizes some of the noncognitive learning variables most commonly discussed and referred to as "affective learning variables" as the effects of these variables on student performance is commonly exhibited through emotional (affective) responses.

We understand, and advocate, that many more factors, such as intelligence and past personal experiences for a start, contribute to individual differences in learning, and the potential interactions and relationships may well be infinite. As likely noticed in these brief descriptions, the variables were described as interacting with each other. For instance, self-efficacy increases when student autonomy, motivation, and self-regulation are stronger—although a specific mathematical representation of this relationship has not been identified, and it is doubtful the relationship is constant between and within individuals. In the next section, we discuss potential methods for obtaining these objective measures.

PHYSIOLOGY OF PERFORMANCE AND EMOTION

The curvilinear relationship between arousal and performance, known as the Yerkes-Dodson law (Yerkes & Dodson, 1908), has long been established. The Yerkes-Dodson law states that performance increases with increased arousal, but only to a certain level. Once the arousal level is too high, performance begins to deteriorate. The maximum optimum arousal level differs among individuals and most likely within individuals depending upon the specific task and current state.

Although there has been debate as to whether physiology precipitates an emotion or is an aftereffect of a change in emotion or mood, it is commonly accepted that physiological processes are associated with emotion. The physiological description of emotion is based on the investigations (for example, Levine, 1986) of the physical activation of arousal (resolved) and stress (unresolved). Therefore, arousal is associated with the "fight or flight" phenomenon in which arousal is increased to either sustain a fight or to enable rapid flight. Regardless of whether the individual fights or flees, he or she resolves the situation and thus reduces his or her arousal. Stress on the other hand, increases in situations in which individuals feel that they cannot control the outcome of the situation, such as reducing or eliminating the stressor. Therefore, arousal is considered a positive response, while stress is a negative response. However, arousal can also be negative if levels become too high or sustained for too long (again exactly how "high" varies depending upon the person and his or her current state).

With respect to physiology, arousal and stress are controlled by sympathetic-adrenal secretions of adrenaline (epinephrine) and cortisol (the interested reader is referred to Levine, 1986). Increased epinephrine secretion leads to an increase in arousal and can occur as a response to both positive and negative stimuli, with the intensity of the stimulus and the person's perception of the stimulus's intensity determining how much the epinephrine secretion rate increases (Frankenhauser, 1986). The secretion of cortisol is related to uncontrollable and

Table 5.2. Noncognitive (Affective) Learning Variables

Affective Learning Variable	Description	Instructional Recommendation
Self-Efficacy	• Student's belief in his or her ability to successfully complete an activity to achieve a goal.	Promote positive student self-efficacy: • Sufficiently challenge the student (Snow, 1989). • Provide positive, timely, and relevant feedback (Bandura, 1997).
Motivation	• Student's desire to succeed that drives him or her to extend an effort to learn.	• Motivated students typically possess positive self-efficacy and self-regulatory skills (Bandura, 1997). • Increased motivation is associated with increased student performance (see Ackerman, Kanfer, & Goff, 1995).
Student Autonomy	• Student control over the instructional process.	• Placing responsibility for learning on student increases student autonomy and motivation (Kember, Wong, & Leung, 1999). • Empowering students to make decisions regarding the learning process and encouraging student initiatives promote student autonomy (Reeve, 1998).
Self-Regulatory Skills	• Coping behaviors the student uses to maintain task focus, maintain confidence in the face of criticism and difficulty, and ensure that sufficient effort is extended on the task.	• Metacognitive, emotion, and social skills improve self-regulation (Hattie, Biggs, & Purdie, 1996; Schunk, 1989).
Personality	• Personality traits influence participation in specific types of activities that are optimized for his or her personality and skills set (Matthews, 1999). • Student prominence in these preferred activities serves the opportunity to skills sets optimized for the preferred environment.	• Instructor cognizance of student's personality traits can be used to guide selection of instructional intervention type.

uncertain situations, which results in feelings of helplessness and distress (Levine, 1986; Frankenhauser, 1986). Therefore, the physiology of emotion can be described as a two-dimensional model (for example, Frankenhauser, 1986) of arousal (adrenaline) versus stress (cortisol).

The traditional physiological measures used to identify emotion are skin conductance (galvanic skin response), electromyogram, electroencephalogram, heart rate, and respiration. Physiological evaluation and identification of emotion are difficult since the interactions of the sympathetic and parasympathetic response systems during arousal and stress are nonlinear. Additionally, physiological responses to external events differ between individuals and within individuals. Neurophysiological measurement technologies such as dense-array electroencephalography, near-infrared spectroscopy, and fMRI allow for evaluation of cognitive activity, in addition to noncognitive activity. Advances in these technologies are lessening the invasiveness and cumbersomeness of the equipment, thus increasing the potential for integrating these technologies within a VE training environment. Description of the various neurophysiological measurement technologies and measurement approaches is beyond the scope of this chapter (the interested reader is referred to Poulsen, Luu, and Tucker, Volume 1, Section 1, Chapter 1).

CLOSED-LOOP TRAINING: BOUNDED ONLY BY IMAGINATION

As described by many of the preceding chapters of this book, data obtained from a virtual environment can be used to evaluate trainee performance and then by an instructor or automated system to modify the virtual environment so as to optimize the student's learning experience in the virtual environment. Further, dynamic monitoring of physiological response to environment interactions or other events can be used to identify changes in the student's noncognitive state. These physiological and neurophysiological measures, in association with task performance measures, can be used to adapt the virtual environment to optimize the training experience as proposed by Sheldon (2001). This approach is to evaluate aspects of student state (for example, stress, frustration, and cognitive activity) and adapt the virtual environment focusing on the manipulation of a couple of instructional intervention types that have been previously demonstrated to be associated with a specific student state response. This capability is rapidly maturing as recent technology advances have decreased the invasiveness and increased the mobility of physiological and neurophysiological measurement devices. In other words, with the continued advances in physiological and neurophysiological monitoring and measurement, we can obtain real time data regarding the *operand*—the student's performance—and use this information to apply a *transform*—instructional feedback—with the goal of achieving a specific *transition*—a quantifiable improvement in student performance.

To realize the above, the maturation of student state assessment and diagnosis capabilities that are able to detect and adapt to changes in student state more similar to a human instructor needs to occur. This will require the integration of learning/artificial intelligence technologies to identify patterns of student state responses and trends with respect to characteristics of the VE and events that occur during the training session. Instead of using a fixed (known) set of interventions, which are in turn provided in response to a known and quantifiable change in student state, student state assessment and diagnosis technologies will enable the provision of varied types of interventions based on a continued monitoring of the student's response to the interventions.

A taxonomy, similar to the taxonomy presented by Klein and Baxter (Volume 1, Section 1, Chapter 3), is needed to provide a basis for automatically adapting the environment to optimize the student's instructional experience. A three-dimensional taxonomy of environment interactions versus physiological response versus instructional intervention type should address various modalities, combinations of modalities, level of stimulation, and connotation of intervention (when applicable). The following list provides a recommendation of potential areas ripe for exploration. Additionally, task-specific interventions, such as injection of scenario events, introduction of a fault or other system change, and so forth also need to be considered.

- Visual: scene/environment, avatars, fidelity, feedback, clutter, color, and so forth;
- Auditory: task-specific and non-task-specific cues, communication with avatars or other humans in the loop, instructional feedback, music, noise, and so forth;
- Tactile: task-specific and non-task-specific cues, environmental, instructional feedback, memory triggers, and so forth;
- Olfactory: task-specific and non-task-specific cues, environment response, memory triggers, and so forth;
- Any combination of the above.

CONCLUSION

As technologies for interpreting student state advance, methods and technologies for responding to student needs in optimal fashion are needed to truly optimize VE technology for training. There is ample evidence that individual differences in composition and their current state affect learning and interactions in a training environment. There has been tremendous research in the investigation of these differences and methods for adapting instruction to the benefit of these variables. There have also been many strides in the physiological and neurophysiological evaluation and understanding of human response. The integration of these fields with performance assessment and instructional methods provides a road to harnessing the power of VE training applications.

REFERENCES

Ackerman, P. L., Kanfer, R., & Goff, M. (1995). Cognitive and noncognitive determinants and consequences of complex skill acquisition. *Journal of Experimental Psychology: Applied, 1*(4), 270–304.

Ashby, W. R. (1956). *An introduction to cybernetics*. London: Chapman & Hall.

Ashby, W. R. (1960). *Design for a brain: The origin of adaptive behavior*. London: Chapman and Hall.

Bandura, A. (1997). *Self-efficacy: The exercise of control*. New York: W. J. Freeman & Co.

Bloom, B. S. (1984). The two sigma problem: The search for methods of group instruction as effective as one-to-one tutoring. *Educational Researcher, 13*(4–6), 4–16.

Cronbach, L. J. (1957). The two disciplines of scientific psychology. *American Psychologist, 12,* 671–684.

Fleishman, E. A. (1953). A factor analysis of intra-task performance on two psychomotor tasks. *Psychometrika, 18,* 45–55.

Fleishman, E. A. (1966). Human abilities and the acquisition of skill. In E. Bilodeau (Ed.), *Acquisition of skill* (pp. 147–167). New York: Academic.

Frankenhaeuser, M. (1986). A psychobiological framework for research on human stress and coping. In M. Appley & R. Trumbull (Eds.), *Dynamics of stress: Physiological, psychological, and social perspectives* (pp. 101–116). New York: Plenum Press.

Guilbaud, G. D. (1959). *What is cybernetics?* New York: Grove.

Hattie, J., Biggs, J., & Purdie, N. (1996). Effects of learning skills interventions on student learning: A meta-analysis. *Review of Educational Research, 66*(2), 99–136.

Kember, D., Wong, A., & Leung, D. (1999). Reconsidering the dimensions of approaches to learning. *British Journal of Educational Psychology, 69,* 323–343.

Levine, P. (1986). Stress. In M. G. Coles, E. Donchin, & S. W. Porges (Eds.), *Psychophysiology: Systems, processes, and application* (pp. 331–353). New York: Guilford Press.

Matthews, G. (1999). Personality and skill: A cognitive-adaptive framework. In P. Ackerman, P. Kyllonen, & R. Roberts (Eds.), *Learning and individual differences: Process, trait, and content determinants*. Washington, DC: American Psychological Association.

McBride, D. K. (2005). The quantification of human information processing. In D. K. McBride & D. Schmorrow (Eds.), *Quantifying human information processing* (pp. 1–41). New York: Rowman & Littlefield.

McBride, D. K., & Uscinski, R. (2008). *The large brain of H. sapiens pays a price*. Manuscript submitted for publication.

McCombs, B. L., & Whisler, J. S. (1989). The role of affective variables in autonomous learning. *Educational Psychologist, 24*(3), 277–306.

Noble, C. E. (1978). Age, race, and sex in the learning and performance of psychomotor skills. In R. T. Osborne, C. E. Noble, & N. Weyl (Eds.), *Human variation: The biopsychology of age, race, and sex* (pp. 51–105). New York: Academic Press.

Powers, W. T. (1973). *Behavior: The control of perception*. Chicago: Aldine.

Reeve, J. (1998). Autonomy support as an interpersonal motivating style: Is it teachable? *Contemporary Educational Psychology, 23*(3), 312–330.

Schunk, D. H. (1989). Self-efficacy and cognitive skill learning. In C. Ames & R. Ames (Eds.), *Research on motivation in education: Vol. 3. Goals and cognitions* (pp. 13–44). New York: Academic Press, Inc.

Sheldon, E. (2001). *Virtual agent interactions*. Unpublished doctoral dissertation, University of Central Florida, Orlando.

Snow, R. E. (1989). Aptitude-treatment interaction as a framework for research on individual differences in learning. In P. Ackerman, R. Sternberg, & R. Glaser (Eds.), *Learning and individual differences: Advances in theory and research* (pp. 13–60). New York: W. H. Freeman & Company.

Tinbergen, N. (1951). *The study of instinct*. Oxford, United Kingdom: Clarendon.

Wiener, N. (1948). *Cybernetics: Or control and communication in the animal and machine*. Cambridge, MA: MIT Press.

Yerkes, R. M., & Dodson, J. D. (1908). The relation of strength of stimulus to rapidity of habit-formation. *Journal of Comparative Neurology and Psychology, 18,* 459–482.

Part IV: Spanning the Bands

Chapter 6

A THEORETICAL FRAMEWORK FOR DEVELOPING SYSTEMATIC INSTRUCTIONAL GUIDANCE FOR VIRTUAL ENVIRONMENT TRAINING

Wendi Van Buskirk, Jessica Cornejo, Randolph Astwood, Steven Russell, David Dorsey, and Joseph Dalton

In recent years the military has placed increased emphasis on advancing training technology because ineffective training can have disastrous consequences. One such area of training technology that has received attention recently is scenario based training (SBT) in virtual environments. In an SBT paradigm, the trainee is presented with scenarios or situations that are representative of the actual task environment (Cannon-Bowers, Burns, Salas, & Pruitt, 1998). SBT has many advantages associated with it, ranging from structured repetition to developing higher order skills (Oser, Cannon-Bowers, Salas, & Dwyer, 1999). One particular advantage of SBT is that it gives trainees opportunities to practice and receive feedback, which are generally considered to be important components of any successful training program. Opportunities for trainees to practice and receive feedback are why SBT is recommended as a medium to train people to handle situations that do not occur frequently in "real life" (for example, nuclear reactor accidents), to handle situations that are life threatening and/or expensive (for example, firefighting, aircraft piloting, and submarine navigation), or to train competencies that "require training in an environment that replicates the critical aspects of the operational setting" (Oser et al., 1999, p. 176).

Despite the advantages of SBT, the science surrounding the implementation of quality instruction within SBT is lacking. Instructional system designers depend heavily on technology to drive the functionality within SBT virtual environments (Oser et al., 1999). However, instructional system designers also need to focus on creating good instruction to support the learning process during training within these environments. In other words, the mere presence of technology does not equate to successful training, but it is easy to overlook this fact when building expensive, impressive virtual environment training systems. It is therefore important for system designers to remember that "in the absence of a sound learning methodology, it is possible that training systems may not fully achieve their intended objectives" (Oser et al., 1999, p. 177). Equal emphasis must be placed on the science of training and the technological medium of training.

We believe this overreliance on technology is due to a lack of guidance regarding which training interventions will result in the most effective training outcomes within SBT environments. For example, an instructor using SBT can give a trainee feedback, but questions remain: Is feedback more effective than another training intervention (for example, deliberate practice)? If so, what type of feedback should be used (for example, outcome, process, velocity, or normative)? When should the instructor introduce the feedback—before, during, or after scenarios? Until guidance is available to address questions such as these, the utility of SBT technology will be limited.

In fact, there is a general consensus among training researchers that more research needs to be carried out in order to determine which training interventions correspond with which learning outcomes such that learning is maximized (Tannenbaum & Yukl, 1992; O'Neil, 2003; Salzman, Dede, Loftin, & Chen, 1999); to date this problem has not received adequate attention (Salas, Bowers, & Rhodenizer, 1998; Salas, Bowers, & Cannon-Bowers, 1995; Salas & Cannon-Bowers, 1997). Even in the face of a proliferation of instructional design theories and models and a wealth of accumulated training knowledge, typical training practice (particularly in simulation based and virtual training environments) lags behind our knowledge of the "science of training" (Salas & Cannon-Bowers, 2001). Therefore, a major challenge facing instructional systems designers is to build systems and training programs based on scientific theories and empirical results rather than on individual preferences and methods of convenience. As summarized by Merrill (1997, p. 51), "which learning strategy to use for a particular instructional goal is not a matter of preference, it is a matter of science."

The goal of this chapter is to address the science of training through the development of a systematic, ontological, and comprehensive organization of learning outcomes and training interventions into a single data structure, which we call the Training Intervention Matrix (TIMx). This matrix will provide an organizational structure that researchers can use to integrate training research results and provide empirically supported guidelines for instructional system designers to follow.

TRAINING INTERVENTION MATRIX

The purpose of the TIMx is to provide instructional system design guidance by linking together two taxonomies—a taxonomy of learning outcomes (LOs) and a taxonomy of training interventions (TIs). Together these two taxonomies will form the foundation architecture of the TIMx. Once the framework is formed, it is our hope that the entries in the intersecting cells, which will contain empirical research results, can be used to derive design guidance. However, the first step to create the TIMx is to develop the two taxonomies.

In order to develop the taxonomies of LOs and TIs, we relied on the five requirements specified by Krathwohl, Bloom, and Masia (1964):

1. A taxonomy must be organized according to a single guiding principle or a set of principles.
2. A taxonomy should be tested by verifying that it is in concurrence with experimental evidence.
3. The order of a taxonomy should correspond to the actual order that occurs among the pertinent phenomena.
4. A taxonomy should be consistent with sound theoretical views.
5. A taxonomy should point "to phenomena yet to be discovered" (Krathwohl et al., 1964, p. 11).

Following these requirements establishes our LOs and TIs as scientifically classified taxonomies rather than simple lists. In the next sections, we focus on how we met the requirement that our LOs and TIs must be organized according to a single guiding principle or set of principles. [To learn how we met Krathwohl et al.'s (1964) four other requirements, see Van Buskirk, Moroge, Kinzer, Dalton, and Astwood (2005)].

X-Axis: Learning Outcomes Taxonomy

We began the development of our taxonomy of LOs by first searching the literature for existing taxonomies. Some of the taxonomies we identified included Fleishman, Quaintance, and Laurie's (1984) taxonomy of human performance, Krathwohl et al.'s (1964) taxonomy of educational objectives, O'Neil's (2003) distance learning guidelines, and Weigman and Rantanen's (2002) human error classification. All the taxonomies provided useful guidance in creating lists of learning outcomes to be included in the final taxonomy. However, many of the taxonomies were created for different purposes and contained items that were not appropriate for the domain in which we were working. Additionally, a few of the taxonomies did not satisfy the five taxonomy requirements listed above. Therefore, we set out to create a new taxonomy while leveraging the work of these researchers. In order to develop our own taxonomy of learning outcomes, we turned to Wei and Salvendy's (2004) human-centered information processing model to serve as our guiding principle (that is, taxonomy requirement #1).

Wei and Salvendy's HCIP Model

Wei and Salvendy's (2004) human-centered information processing (HCIP) model is a worker-oriented model for cognitive task performance. It attempts to identify all of the cognitive aspects of human performance in technical work. We chose this model for several reasons. First, this model is an input-throughput-output model of human task performance. Second, there is an external feedback loop from response to stimuli. This model captures how the trainee perceives, processes, and responds to information. Third, we chose the HCIP model because it breaks memory into its component parts rather than clustering it all into one "memory" box (that is, the model separates long-term, sensory, and working memory). This was important for our purposes because breaking memory into its component parts should facilitate the organization of the learning outcomes. Finally, we selected the HCIP model because it includes teamwork processes, individual differences, and external factors as components. Teamwork processes are important for our purposes because the TIMx addresses the training of both individuals and teams. The inclusion of individual differences and external factors is important because they account for additional sources of variation —variation within and among individuals and variation due to the situation. In future stages of development, these sources of variation will become caveats or moderators in the cells of our matrix.

Modifications to the HCIP Model

Despite the noted advantages of Wei and Salvendy's (2004) model, we adapted it slightly. The main purpose for this was to ensure that our learning outcomes would be mutually exclusive within each module. For example, we combined Wei and Salvendy's original "mental plan and schedule" module and "mental execution" module to form our "cognitive task execution" module. The original two modules had considerable overlap in their definitions and functions, which did not allow us to cleanly fit learning outcomes into just one module. Our revised version is presented in Figure 6.1 and will be referred to as the human-centered information processing-revised model (HCIP-R). Our revised model maintains the advantages of the original HCIP model (that is, input-throughput-output, components of memory, and accounts for teams and individual

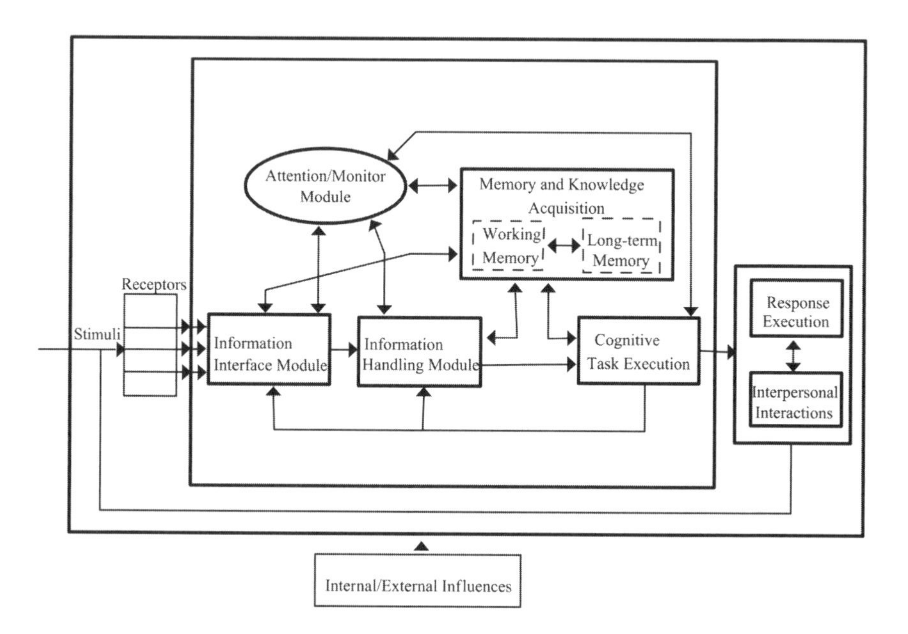

Figure 6.1. Revised HCIP Model

differences) that led us to select it in the first place. The HCIP-R serves as the guiding set of principles for our taxonomy of learning outcomes.

As part of establishing the HCIP-R as our way to organize the LOs in our taxonomy, we created definitions for each module in the model. We also created conceptual definitions for each LO in order to have a common set of terminology. Having this common terminology is important so that the research and empirical findings are associated with the correct LO. For instance, it is possible that what one researcher may call "decision making," another researcher may call "planning." It creates a standard language to discuss and compare research findings. In order to create these definitions, we identified, created, or modified existing definitions for each of our LOs. Once each LO was defined, we assigned it to the appropriate HCIP-R module. Definitions and module classifications are presented in Table 6.1. It is important to note that this may not be an exhaustive list and other learning outcomes we inadvertently overlooked could be included.

Y-Axis: Training Interventions Taxonomy

As with the learning outcome taxonomy, we first searched for existing taxonomies of TIs. Unfortunately, our search of the literature revealed no such taxonomies. We therefore developed our own taxonomy by searching for training strategies in the literature. Also like the LO taxonomy, we met the requirement of selecting an organizing framework for our TIs. The organizing framework we selected is called the Common Distributed Mission Training Station (CDMTS; Walwanis Nelson, Owens, Smith, & Bergondy-Wilhelm, 2003). CDMTS allowed us to organize our TIs into one of four modules, which are defined below (see Table 6.2).

Common Distributed Mission Training Station (CDMTS)

The CDMTS was designed to address simulation and computerized training programs that involve physically distributed teams. It is based on the logic that, regardless of training content, instructors have a common set of needs. The CDMTS is a physical, computerized product that is a training aid for instructors. It was developed based on a framework that presents the interrelationships among several dimensions of military training systems. It is this framework that we use to organize our taxonomy of TIs.

We combined the CDMTS architectural framework dimensions into four general training design modules: student-initiated learning, instructional planning, exercise manipulation, and feedback. Each module corresponds to one of the three general stages of an SBT cycle: pre-exercise, during exercise, and post-exercise. Student-initiated learning was included to encompass TIs that take place without the guidance of an instructor, such as when team members review each other's and the overall team's performance. Instructional planning refers to TIs established before the training exercise occurs. For example, lectures typically would not occur during an exercise. Rather, they typically provide

Table 6.1. Learning Outcomes and Definitions for each HCIP-R Module

Modules and Learning Outcomes	Definitions
Information Interface	**The input of information or data for cognitive processing. The information input can be achieved through perceiving stimuli through physical channels such as visual and auditory.**
Perceptual Judgment	Understanding the information implied by the physical properties of an object (Lederman & Wing, 2003).
Perceptual-Motor Skill	The ability to match voluntary physical motion and perceptional observations (Holden, Flach, & Donchin, 1999).
SA: Perception[a]	The perception of cues in the environment (Endsley, 2000).
Spatial Orientation	One's awareness of oneself in space and the related ability to reason about movement within space (Hunt, 2002).
Visual Scanning	Using vision to search for objects or stimuli in the environment (for example, displays, objects, and rooms) in order to gather information (McCarley et al., 2001; Stein, 1992).
Information Handling	**Captures lower level human information processing. It involves recognizing and translating input from the information interface module into information usable for higher level processing (that is, the cognitive task execution module, the interpersonal interactions module, and so forth).**
Pattern Recognition	The process of identifying or categorizing large perceptual chunks of information in one's environment (Chase & Simon, 1973; Reisberg, 1997).
SA: Comprehension[a]	"Encompasses how people combine, interpret, store, and retain information. It includes the integration of multiple pieces of information and a determination of their relevance to the person's goals" (Endsley, 2000, p. 7).
Cognitive Task Execution	**Higher level cognitive task performance related to examining, evaluating, and reviewing available information in order to determine a course of action (or inaction).**
Organizing/ Planning	The process of calculating a set of actions that will allow one to achieve his or her goal (Garcia-Martinez & Borrajo, 2000).
Problem Solving/ Decision Making	The process of gathering, organizing, and combining information from different sources to make a choice from between two or more alternatives (Lehto, 1997).
Resource Management	Assessing characteristics of resources and needs and determining their appropriate allocation (Gustafsson, Biel, & Garling, 1999).
SA: Projection[a]	"The ability to forecast future situation events and dynamics and their implications" (Endsley, 2000, p. 7).

Interpersonal Interactions	**Behaviors for sharing information and working with others, including both human and synthetic teammates.**
Assertiveness	Both the willingness and the ability to communicate one's own opinions in a manner that will be persuasive to others (Smith-Jentsch, Salas, & Baker, 1996).
Backup Behaviors	Not only offering assistance to others, but also requesting assistance when one knows he or she is overloaded (Smith-Jentsch, Zeisig, Acton, & McPherson, 1998).
Communication	Using proper phraseology, providing complete reports, avoiding excess chatter, and generally using speech to share information and ideas in a manner that others will understand, including listening effectively and fostering open communication (Smith-Jentsch, Johnston, et al., 1998).
Error Correction	Monitoring for errors and taking action to correct these errors when they take place (Smith-Jentsch, Zeisig, et al., 1998).
Explicit Team Coordination	Using planning or communication mechanisms to manage team task dependencies (Espinoza, Lerch, & Kraut, 2002).
Implicit Team Coordination	Coordinating team member actions based on shared cognition and unspoken assumptions about what other team members are likely to do (Espinoza et al., 2002).
Information Exchange	Seeking information from available resources, passing information to appropriate persons, providing accurate "big picture" situation updates, and accurately informing higher commands (Smith-Jentsch, Johnston, & Payne, 1998).
Leadership	Providing guidance or suggestions to others, stating clear team and individual priorities, and appropriately refocusing others in accordance with situational demands (Smith-Jentsch, Johnston, et al., 1998).
Attention	**Captures human cognitive task performance related to attention resource allocation needed for cognitive tasks. The resources are the limited capacity inventory that supplies the attention resources to the other modules for jobs.**
Attention Prioritization	Determining the relative order in which objects or pieces of information are to be focused on or investigated (Shomstein & Yantis, 2004).
Metacognition	Using knowledge, skills, and beliefs concerning one's own cognitive processes and products (Flavell, 1976).
Memory and Knowledge Acquisition	**This module captures human cognitive task performance related to retrieving, storing, retaining, and transferring information needed for cognitive tasks. It is composed of two classes: working memory and long-term memory.**
Declarative Knowledge	Factual knowledge, including recollection of words, definitions, names, dates, and so forth (Weiten, 2001).

Memorization	The rote recall of some material, with no required comprehension and/or ability to integrate that material (Lovett & Pillow, 1995).
Procedural Knowledge	Knowledge of actions, skills, operations, and conditioned responses; knowing how to execute actions (Weiten, 2001).
Strategic Knowledge	"The knowledge that enables the formation of strategies; plans of action determining what kinds of knowledge and tactics should be employed in different problem contexts" (Fentem, Dumas, & McDonnell, 1998).
Influences on Individual Differences[b]	**This module includes state, trait, and environmental factors that moderate the effectiveness of training interventions.**

[a]SA = Situational awareness.

[b]Because of the potentially unlimited number of internal and external influences, we do not attempt to create a list here.

information before an exercise takes place. Therefore, it was included in the pre-exercise module, instructional planning. Exercise manipulation involves TIs that an instructor manipulates in real time, during an exercise, to enhance the training opportunities presented to trainees. Last, feedback refers to TIs that provide trainees information regarding some aspect(s) of an individual's or a team's task and/or team performance. It is important to note that feedback could occur during and/or after an exercise. Therefore, it is its own separate module.

In Table 6.2, our TIs are organized into the appropriate CDMTS module and are accompanied by their conceptual definitions. There are two things to note with regard to the TI taxonomy. First, we realize that this may not be an exhaustive list, and other training interventions may have been overlooked inadvertently. And, second, the TIs are broken down into individual components. We realize that there are other training strategies that incorporate several of the individual components. For example, adaptive guidance incorporates TIs such as feedback, sequencing, practice, and so forth (Bell & Kozlowski, 2002).

FUTURE RESEARCH NEEDS FOR TIMx DEVELOPMENT

With the framework developed (see Figure 6.2), the next iteration of the TIMx development will encompass filling in the cells of the matrix. The most logical approach for filling in the cells is to focus on a LO column. For example, literature describing empirically validated training interventions that can be used to train that specific learning objective would be identified and reviewed. As the cells are filled, the pattern of research results will indicate which TIs are best suited for training the LO. Additionally, the pattern of research results will indicate other areas that need new or continued research attention. For example, a review of the research could show that no empirical research has been conducted on the effectiveness of using sequencing to train pattern recognition within a virtual environment.

Table 6.2. Training Interventions and Definitions for Each CDMTS Module

Module and Training Intervention	Definition of Training Intervention
Instructional Planning (Pre-Exercise)	
Action Learning	A team engages in a specific work-related problem solving task in which there is an emphasis on learning and problem solving through identification of the root problem and development and implementation of an action plan (Goldstein & Ford, 2002).
Advanced Organizers	An adjunct aid that presents a structure of what is to be learned so that new content can be more easily organized and interpreted (Ausubel, 1960; Langan-Fox, Waycott, & Alber, 2000; Chalmers, 2003).
Anchored Instruction	Integrating examples or real world experiences into instructional programs to provide a common frame of reference for learners in order to help them connect to concepts being taught (adapted from Blackhurst & Morse, 1996).
Cross-Training	Team members learn the skills of one or several additional jobs in order to foster mental representations of task interdependency, team role structure, and how the two interact (Noe, 1999; Salas & Cannon-Bowers, 1997).
Error Based Training	Exposing trainees to both positive and negative examples of the behavior being trained (adapted from Baldwin, 1992).
Event Based Approach to Training	Planned events are introduced into simulated scenarios to evaluate specific skills that would be required in these real-life situations (Salas & Cannon-Bowers, 1997).
Exploration Based Training	Training that builds situations in which the learner can make an error and then explore in a trial-and-error way what the cause of the error was and explore alternative strategies to avoid or fix the error (Frese et al., 1991).
Guided Reflection	Instructor prompts the trainee to mentally review the affective, cognitive, motivational, and behavioral aspects of the trainee's performance (Cleary & Zimmerman, 2004).
Heuristic Strategies	Instructions that focus on teaching general rules of thumb to find an acceptable approximate solution to a problem when an exact problem solving method is unavailable or too time consuming (Tversky & Kahneman, 2002).
Intelligent Tutoring Systems	Computer based programs that deliver instruction by first diagnosing a trainee's current level of understanding, then comparing it to an expert model of performance, and finally selecting the appropriate intervention that will advance the trainee's level of understanding (Goldstein & Ford, 2002; Corbett, Koedinger, & Anderson, 1997).

Lecture	Method in which an instructor verbally delivers training material to multiple students simultaneously.
Mental Rehearsal	A trainee is instructed to visualize himself or herself reenacting the target behavior or to visualize himself or herself performing the target behavior perfectly (Davis & Yi, 2004).
Modeling/ Demonstration	Trainees are presented desired learning processes and/or outcomes to mimic (Noe, 1999).
Part Task Training	Training that breaks down a skill or task into components that are practiced separately (Goldstein & Ford, 2002).
Role-Play	A training method in which trainees are given information about a situation and act out characters assigned to them (Noe, 1999).
Sequencing	Practice is provided to learners in a particular order, such as proceeding from easy to difficult or from general to specific. Types of sequencing could include massed/blocked practice, spaced practice, and random practice (Shea & Morgan, 1979; Schmidt & Bjork, 1992).
Student Initiated Learning (Pre-Exercise)	
Guided Team Self-Correction	Group reflection in which trainees self-reflect and gain insights into the nature of their own taskwork, teamwork, and the relation of taskwork and teamwork to the team's overall performance (Smith-Jentsch, Zeisig, Acton, & McPherson, 1998).
Peer Instruction	The students perform tasks individually and then they discuss their responses and their underlying reasoning (Mazur, 1997).
Reflection	The trainee takes the initiative to mentally review the affective, cognitive, motivational, and behavioral aspects of his or her own performance (Cleary & Zimmerman, 2004).
Self-Directed Learning	Learning that takes place outside of a formal training program without the guidance of an actual trainer (Manz & Manz, 1991).
Exercise Manipulation (During Exercise)	
Cueing/Hinting	Techniques that prompt the trainee to access information already known or that prompt the trainee to carry out the next steps required to reach a correct answer or to make connections between the training task and the larger context (Goldstein & Ford, 2002; Hume, Michael, Rovick, & Evens, 1996).

Deliberate Practice	A technique used to provide learning guidance by prompting the learner to perform as needed during skill acquisition or to make connections between the training task and the larger context (Ericsson, Krampe, & Tesch-Romer, 1993).
Didactic Questioning	Features a teacher leading students to concepts through a series of ordered questions that tend to have a single answer (Vlastos, 1983).
Facilitative Questioning	Involves posing open-ended questions to trainees that encourage the trainees to generate their own solutions and ideas without direct input from the teacher (PlasmaLink Web Services, 2006).
Highlighting	Technique used to draw a trainee's attention to a stimulus in order to increase the likelihood of perception of the stimulus (Wickens, Alexander, Ambinder, & Martens, 2004).
Overlearning	The immediate continuation of practice beyond achievement of the criterion (Rohrer, Taylor, Pashler, Wixted, & Cepeda, 2005; Noe, 1999).
Feedback (During and Post-Exercise)	
After Action Review	An interactive process in which trainees discuss task planning and execution under the guidance of an instructor (Scott, 1983).
Environmental Feedback	Provides information about the actual/true relationship between the cues in the environment and their associated outcomes (Balzer et al., 1994).
Normative Feedback	Provides an individual with information about his or her standing relative to others, but is not specific performance-related feedback (Smithers, Wohlers, & London, 1995).
Outcome Feedback	Provides knowledge of the results of one's actions (Ericsson, Krampe, & Tesch-Romer, 1993; Kluger & DeNisi, 1996; Balzer, Doherty, & O'Connor, 1989).
Process Feedback	Conveys information about how one performs the task (not necessarily how well; Kluger & DeNisi, 1996).
Progress/Velocity Feedback	The trainee's performance is compared only with his or her own prior performance on the task. The trainee can gauge the rate of progress at which a performance goal is being reached (Kozlowski et al., 2001).
Scaffolding/Faded Feedback	Support or guidance is given on every trial early in practice and then is gradually withdrawn across practice (Schmidt & Bjork, 1992).

To illustrate this further, consider two research studies that investigate the effectiveness of training decision making using scenario based training simulations. Buff and Campbell (2002) examined the appropriate information or content to include in effective feedback. Specifically, they investigated the effectiveness of process and outcome feedback (as compared to a no feedback practice group) on participant's decision-making performance on a simulated radar display task. Their results showed that participants who received process feedback significantly improved their performance (measured as decision accuracy) from pre- to post-feedback sessions. However, neither participants in the outcome feedback condition nor the no feedback condition improved their performances. Likewise, Cannon-Bowers, Salas, Blickensderfer, and Bowers (1998) also investigated how to train decision making on a simulated radar display task. However, these researchers examined the use of cross-training to train decision making in teams. Their results showed that cross-trained groups made more correct decisions about contacts and did so more quickly than teams assigned to a no training condition. Based on the results there is some indication that process feedback and cross-training may be effective interventions to use to train decision making. Therefore, we could use these results to fill in the cell intersections of decision making and process feedback as well as decision making and cross-training (see Figure 6.2).

Within each cell we anticipate that the effectiveness of the training intervention will depend upon certain caveats, including both internal and external factors. Internal factors include individual difference variables such as goal

Training Intervention Matrix

Learning Outcomes

			Information Interface					Information Handling		Cognitive Task Execution				Attention		Interpersonal Interactions								Memory&Know-ledge Acquisition			
			Perceptual Judgment	Perceptual-Motor	SA: Perception	Spatial Orientation	Visual Scanning	Pattern Recognition	SA: Comprehension	Decision Making \Problem Solving	Organizing/Planning	Resource Management	SA: Projection	Attention prioritization	Meta-Cognition	Assertiveness	Backup Behaviors	Communication	Error Correction	Explicit Team Coordination	Implicit Team Coordination	Information Exchange	Leadership	Declarative Knowledge	Memorization	Procedural Knowledge	Strategic Knowledge
Training Interventions	Instructional Planning (Pre-Exercise)	Action Learning																									
		Activating Prior Knowledge/ Anchored Instruction																									
		Advanced Organizers																									
		Cross Training								✓																	
		Error-Based Training																									
		Event-Based Approach to Training (EBAT)																									
		Exploration (Exploration-Based training)																									
		Guided Reflection																									
		Heuristic Strategies																									
		Intelligent Tutoring Systems																									
		Lectures																									
		Mental Rehearsal																									
		Modeling/Demonstration																									
		Part Task Training																									
		Role Play																									
		Sequencing																									
	Student Initiated Learning (Pre-Exer.)	Guided Team Self-Correction																									
		Peer Instruction																									
		Reflection																									
		Self-Directed Learning																									
	Exercise Manipulation (During Exercise)	Cueing/Hinting																									
		Deliberate Practice																									
		Didactic Questioning (Socratic Method)																									
		Facilitative Questioning																									
		Highlighting																									
		Overlearning																									
		Scaffolding/Faded Feedback																									
	Feedback (During and Post Excercise)	After Action Reviews																									
		Environmental Feedback																									
		Normative Feedback																									
		Outcome Feedback																									
		Process Feedback								✓																	
		Progress/Velocity Feedback																									

Figure 6.2. Training Intervention Matrix

orientation, self-efficacy, and fatigue. External factors include environmental factors such as weather, ambient temperature, and having the required equipment, materials, and organizational support. These variables will ultimately serve as caveats that will fit into the Internal and External Influences module.

Additionally, there is a debate within the literature regarding the distinction between education and training. For example, Kline (1985) suggests that training tends to concentrate on psychomotor skills, while education concentrates on cognitive skills. However, other researchers have found evidence that cognitive skills, such as decision making, can be trained beyond declarative knowledge (Buff & Campbell, 2002; Cannon-Bowers, Salas, et al., 1998). However, what remains to be addressed is whether or not these decision-making strategies can be transferred to other domains. Within the TIMx, we have included a number of traditional educational interventions (for example, lecture and the Socratic method) in addition to traditional training interventions (for example, cueing/hinting and part task training). Therefore, it may be possible to use the TIMX as a framework to address this debate empirically.

Finally, future research also needs to take the use of technology into consideration. The impact of using different types of technology should be studied, as should the impact of the fidelity of simulations. In fact, Muller et al. (2006) argue that research is needed to determine how three types of fidelity (functional, psychological, and physical) interact to provide the most effective training environment. Research should focus on a blended fidelity training solution to "support the examination of training transfer, across low fidelity, high fidelity, and live training environments" that is capable of training both "technical and higher order skill sets" (Muller et al., 2006, p. 10).

SUMMARY

A major challenge facing virtual environment and instructional systems designers is choosing the best way to train certain learning outcomes. Typically, this decision has been based on technology requirements and not on empirical research or the "science of training" (Salas & Cannon-Bowers, 2001). To address this challenge, we created two taxonomies: a learning outcome taxonomy and a training intervention taxonomy. We then linked the taxonomies to create a framework within which we could start scientifically answering questions such as "what is the best way to train decision making?" It is our hope that other researchers will find this framework to be a useful tool and will use it to guide their own research on instructional interventions. We also hope that as the TIMx begins to fill, we can provide empirically supported guidance to instructional system designers about the most appropriate training strategies to use in virtual environments so that their decisions are based on science and not on individual preferences.

ACKNOWLEDGMENTS

We gratefully acknowledge CDR Dylan Schmorrow at the Office of Naval Research who sponsored this work (Contract No. N0001407WX20102).

Additionally, we would like to thank Dr. Ami Bolton, Ms. Melissa Walwanis Nelson, and Ms. Beth Atkinson for their insightful comments during the development of the TIMx.

REFERENCES

Ausubel, D. P. (1960). The use of advance organizers in the learning and retention of meaningful verbal material. *Journal of Educational Psychology, 51,* 267–272.

Baldwin, T. T. (1992). Effects of alternative modeling strategies on outcomes of interpersonal skills training. *Journal of Applied Psychology, 76,* 759–769.

Balzer, W. K., Doherty, M. E., & O'Connor, R. (1989). Effects of cognitive feedback on performance. *Psychological Bulletin, 106,* 410–433.

Balzer, W. K., Hammer, L. B., Sumner, K. E., Birchenough, T. R., Martens, S. P., & Raymark, P. H. (1994). Effects of cognitive feedback components, display format, and elaboration on performance. *Organizational Behavior and Human Decision Processes, 58,* 369–385.

Bell, B. S., & Kozlowski, S. W. J. (2002). Adaptive guidance: Enhancing self-regulation, knowledge, and performance in technology-based training. *Personnel Psychology, 55,* 267–306.

Blackhurst, A. E., & Morse, T. E. (1996). Using anchored instruction to teach about assistive technology. *Focus on Autism and Other Developmental Disabilities, 11,* 131–141.

Buff, W. L., & Campbell, G. E. (2002). What to do or what not to do? Identifying the content of effective feedback. *Proceedings of the 46th Annual Meeting of the Human Factors and Ergonomics Society* (pp. 2074–2078). Santa Monica, CA: Human Factors and Ergonomics Society.

Cannon-Bowers, J. A., Burns, J. J., Salas, E., & Pruitt, J. S. (1998). Advanced technology in scenario-based training. In J. A. Cannon-Bowers & E. Salas (Eds.), *Making decisions under stress: Implication for individual and team training* (pp. 365–374). Washington, DC: American Psychological Association.

Cannon-Bowers, J. A., Salas, E., Blickensderfer, E., & Bowers, C. A. (1998). The impact of cross-training and workload on team functioning: A replication and extension of initial findings. *Human Factors, 40,* 92–101.

Chalmers, P. A. (2003). The role of cognitive theory in human-computer interface. *Computers in Human Behavior, 19,* 593–607.

Chase, W. G., & Simon, H. A. (1973). Perception in chess. *Cognitive Psychology, 4,* 55–81.

Cleary, T. J., & Zimmerman, B. J. (2004). Self-regulation empowerment program: A school-based program to enhance self-regulated and self-motivated cycles of student learning. *Psychology in the Schools, 41,* 537–550.

Corbett, A. T., Koedinger, K. R., & Anderson, J. R. (1997). Intelligent tutoring systems. In M. G. Helander, T. K. Landauer, & P. V. Prabhu (Eds.), *Handbook of human-computer interaction* (2nd ed., pp. 849–874). Amsterdam: Elsevier.

Davis, F. D., & Yi, M. Y. (2004). Improving computer skill training: Behavior modeling, symbolic mental rehearsal, and the role of knowledge. *Journal of Applied Psychology, 89,* 509–523.

Endsley, M. R. (2000). Direct measurement of situation awareness: Validity and use of SAGAT. In M. R. Endsley & D. J. Garland (Eds.), *Situation awareness analysis and measurement* (pp. 147–173). Mahwah, NJ: Lawrence Erlbaum.

Ericsson, K. A., Krampe, R. T., & Tesch-Romer, C. (1993). The role of deliberate practice in the acquisition of expert performance. *Psychological Review, 100,* 363–406.

Espinosa, A., Lerch, J., & Kraut, R. (2002). Explicit vs. implicit coordination mechanisms and task dependencies: One size does not fit all. In E. Salas, S. M. Fiore, & J. Cannon-Bowers (Eds.), *Team cognition: Process and performance at the inter- and intra-individual level.* Washington, DC: American Psychological Association.

Fentem, A. C., Dumas, A., & McDonnell, J. (1998). Evolving spatial representations to support innovation and the communication of strategic knowledge. *Knowledge-Based Systems, 11,* 417–428.

Flavell, J. H. (1976). Metacognitive aspects of problem solving. In L. B. Resnick (Ed.), *The nature of intelligence* (pp. 231–236). Hillsdale, NJ: Erlbaum.

Fleishman, E. A., Quaintance, M. K., & Laurie, A. (1984). *Taxonomies of human performance: The description of human tasks.* San Diego, CA: Academic Press, Inc.

Frese, M., Brodbeck, F., Heinbokel, T., Mooser, C., Schleiffenbaum, E., & Thiemann, P. (1991). Errors in training computer skills: On the positive function of errors. *Human-Computer Interaction, 6,* 77–93.

Garcia-Martinez, R., & Borrajo, D. (2000). An integrated approach to learning, planning, and execution. *Journal of Intelligent & Robotic Systems, 29,* 47–78.

Goldstein, I. L., & Ford, F. J. (2002). *Training in organizations: Needs assessment, development, and evaluation* (4th ed.). Belmont, CA: Wadsworth/Thomson Learning.

Gustafsson, M., Biel, A., & Garling, T. (1999). Outcome-desirability bias in resource management problems. *Thinking & Reasoning, 5,* 327–337.

Holden, J. G., Flach, J. M., & Donchin, Y. (1999). Perceptual-motor coordination in an endoscopic surgery simulation. *Surgical Endoscopy, 13,* 127–132.

Hume, G., Michael, J., Rovick, A., & Evens, M. (1996). Hinting as a tactic in one-on-one tutoring. *The Journal of the Learning Sciences, 5,* 23–47.

Hunt, E. (2002). *Precis of thoughts on thought.* Mahwah, NJ: Erlbaum.

Kline, J. A. (1985, January–February). Education and training: Some differences. *Air University Review.* Retrieved April 15, 2008, from http://www.airpower.maxwell.af.mil/airchronicles/aureview/1985/jan-feb/kline.html

Kluger, A. N., & DeNisi, A. (1996). The effects of feedback interventions on performance: A historical review, a meta-analysis, and a preliminary feedback intervention theory. *Psychological Bulletin, 119,* 254–284.

Kozlowski, S. W. J., Toney, R. J., Mullins, M. E., Weissbein, D. A., Brown, K. G., & Bell, B. S. (2001). Developing adaptability: A theory for the design of integrated-embedded training systems. In E. Salas (Ed.), *Advances in human performance and cognitive engineering research* (Vol. 1, pp. 59–123). Amsterdam: JAI/Elsevier Science.

Krathwohl, D. R., Bloom, B. S., & Masia, B. B. (1964). *Taxonomy of educational objectives: The classification of educational goals.* New York: David McKay Co., Inc.

Langan-Fox, J., Waycott, J. L., & Alber, K. (2000). Linear and graphic advanced organizers: Properties and processing. *International Journal of Cognitive Ergonomics, 4,* 19–34.

Lederman, S. J., & Wing, A. M. (2003). Perceptual judgment, grasp point selection and object symmetry. *Experimental Brain Research, 152,* 156–165.

Lehto, M. (1997). Decision making. In G. Salvendy (Ed.), *Handbook of human factors and ergonomics* (2nd ed., pp. 1201–1248). New York: John Wiley & Sons.

Lovett, S. B., & Pillow, B. H. (1995). Development of the ability to distinguish between comprehension and memory: Evidence from strategy-selection tasks. *Journal of Educational Psychology, 87,* 523–536.

Manz, C. C., & Manz, K. (1991). Strategies for facilitating self-directed learning: A process for enhancing human resource development. *Human Resource Development Quarterly, 2,* 3–12.

Mazur, E. (1997). *Peer instruction: A user's manual.* Upper Saddle River, NJ: Prentice Hall Series in Educational Innovation.

McCarley, J. S., Vais, M., Pringle, H., Kramer, A. F., Irwin, D. E., & Strayer, D. L. (2001, August). *Conversation disrupts visual scanning of traffic scenes.* Paper presented at the Vision in Vehicles conference, Brisbane, Australia.

Merrill, M. D. (1997). Learning-oriented instructional development tools. *Performance Improvement, 36*(3), 51–55.

Muller, P., Cohn, J., Schmorrow, D., Stripling, R., Stanney, K., Milham, L., et al. (2006). The fidelity matrix: Mapping system fidelity to training outcome. *Proceedings of the Interservice/Industry Training, Simulation and Education Conference* [CD-ROM]. Arlington, VA: National Training Systems Association.

Noe, R. (1999). *Employee training and development.* Boston: Irwin McGraw-Hill.

O'Neil, H. F. (2003). *What works in distance learning.* Greenwich, CT: Information Age Publishing.

Oser, R. L., Cannon-Bowers, J. A., Salas, E., & Dwyer, D. J. (1999). Enhancing human performance in technology-rich environments: Guidelines for scenario-based training. In E. Salas (Ed.), *Human/technology interaction in complex systems* (Vol. 9, pp. 175–202). Stanford, CT: JAI Press.

PlasmaLink Web Services. (2006, September 4). *Glossary of instructional strategies.* Retrieved April 16, 2007, from http://glossary.plasmalink.com/glossary.html#F

Reisberg, D. (1997). *Cognition: Exploring the science of the mind.* New York: W. W. Norton.

Rohrer, D., Taylor, K., Pashler, H., Wixted, J. T., & Cepeda, N. J. (2005). The effect of overlearning on long-term retention. *Applied Cognitive Psychology, 19,* 361–374.

Salas, E., Bowers, C. A., & Cannon-Bowers, J. A. (1995). Military team research: 10 years of progress. *Military Psychology, 7*(2), 55–75.

Salas, E., Bowers, C. A., & Rhodenizer, L. (1998). It is not how much you have but how you use it: Toward a rational use of simulation to support aviation training. *International Journal of Aviation Psychology, 8,* 197–208.

Salas, E., & Cannon-Bowers, J. A. (1997). Methods, tools, and strategies for team training. In M. A. Quinones & A. Ehrenstein (Eds.), *Training for a rapidly changing workplace: Applications of psychological research* (pp. 249–279). Washington, DC: American Psychological Association.

Salas, E., & Cannon-Bowers, J. A. (2001). The science of training: A decade of progress. *Annual Review of Psychology, 52,* 471–499.

Salzman, M. C., Dede, C., Loftin, R. B., & Chen, J. (1999). A model for understanding how virtual reality aids complex conceptual learning. *Presence: Teleoperators and Virtual Environments, 8,* 293–316.

Schmidt, R. A., & Bjork, R. A. (1992). New conceptualizations of practice: Common principles in three paradigms suggest new concepts for training. *Psychological Science, 3,* 207–217.

Scott, T. D. (1983). *Tactical engagement simulation after action review guidebook* (Research Rep. No. 83-13). Alexandria, VA: U.S. Army Research Institute for the Behavioral and Social Sciences.

Shea, J. B., & Morgan, R. L. (1979). Contextual interference effects on the acquisition, retention, and transfer of a motor skill. *Journal of Experimental Psychology: Human Learning & Memory, 5,* 179–187.

Shomstein, S., & Yantis, S. (2004). Configural and contextual prioritization in object-based attention. *Psychonomic Bulletin & Review, 11,* 247–253.

Smith-Jentsch, K. A., Johnston, J. H., & Payne, S. C. (1998). Measuring team-related expertise in complex environments. In J. A. Cannon-Bowers & E. Salas (Eds.), *Making decisions under stress: Implications for individual and team training* (pp. 61–87). Washington, DC: American Psychological Association.

Smith-Jentsch, K. A., Salas, E., & Baker, D. (1996). Training team performance-related assertiveness. *Personnel Psychology, 49,* 909–936.

Smith-Jentsch, K. A., Zeisig, R. L., Acton, B., & McPherson, J. A. (1998). Team dimensional training: A strategy for guided team self-correction. In J. A. Cannon-Bowers & E. Salas (Eds.), *Making decisions under stress: Implications for individual and team training* (pp. 271–298). Washington, DC: American Psychological Association.

Smithers, J. W., Wohlers, A. J., & London, M. (1995). A field study of reactions to normative versus individualized upward feedback. *Group & Organization Management, 20,* 61–89.

Stein, E. S. (1992). *Air traffic control visual scanning* (Rep. No. DOT/FAA/CT-TN92/16). Atlantic City International Airport: Federal Aviation Administration Technical Center.

Tannenbaum, S. I., & Yukl, G. (1992). Training and development in work organizations. *Annual Review of Psychology, 43,* 399–441.

Tversky, A., & Kahneman, D. (2002). Judgment under uncertainty: Heuristics and biases. In D. J. Levitin (Ed.), *Foundations of cognitive psychology: Core readings* (pp. 585–600). Cambridge, MA: MIT Press.

Van Buskirk, W. L., Moroge, J. L., Kinzer, J. E., Dalton, J. M., & Astwood, R. S. (2005). Optimizing training interventions for specific learning objectives within virtual environments: A Training Intervention Matrix (TIMx). *Proceedings of the 11th International Conference of Human Computer Interaction* [CD-ROM].

Vlastos, G. (1983). The Socratic Elenchus. In A. Price (Ed.), *Oxford Studies in Ancient Philosophy* (Vol. 1, pp. 27–58). Oxford, England: Blackwell Publishing LTD.

Walwanis Nelson, M. M., Owens, J., Smith, D. G., & Bergondy-Wilhelm, M. L. (2003). A common instructor operator station framework: Enhanced usability and instructional capabilities. *Proceedings of the 18th Interservice/Industry Training Systems and Education Conference* (pp. 333–340). Arlington, VA: American Defense Preparedness Association.

Wei, J., & Salvendy, G. (2004). The cognitive task analysis methods for job and task design: Review and reappraisal. *Behaviour and Information Technology, 23,* 273–299.

Weigman, D. A., & Rantanen, E. (2002). *Defining the relationship between human error classes and technology intervention strategies* (Tech. Rep. No. 1 ARL-02-1/NASA 02-1). Urbana-Champaign: University of Illinois, Aviation Research Lab Institute of Aviation.

Weiten, W. (2001). *Psychology: Themes and variations.* Stamford, CT: Thomson-Wadsworth.

Wickens, C. D., Alexander, A. L., Ambinder, M. S., & Martens, M. (2004). The role of highlighting in visual search through maps. *Spatial Vision, 17,* 373–388.

SECTION 2
REQUIREMENTS ANALYSIS

SECTION PERSPECTIVE

Kay Stanney

The biggest mistakes in any large system design are usually made on the first day.

—Dr. Robert Spinrad (1988)
Vice President, Xerox Corporation
(as cited in Hooks & Farry, 2001)

It is often suggested that the most important step in the training system development lifecycle is comprehensive requirements specification. Yet, it is not uncommon to encounter system solutions that do not meet their intended objectives due to insufficiently articulated requirements (Young, 2001). In fact, it has been estimated that 80 percent of product defects originate during requirements definition (Hooks & Farry, 2001). Requirements should be specified such that training systems can be designed to overcome deficiencies in knowledge, skills, or abilities (KSAs) required to perform a given job (see Milham, Carroll, Stanney, and Becker, Chapter 9). To develop such systems, designers must understand instructional goals so that the training/learning curriculum, content, layout, and delivery mode are designed to prepare personnel to operate, maintain, and support all job components in the required operational environment. To achieve these objectives, it is beneficial to adopt a systems approach to training systems requirement specification (Young, 2004). A systems approach seeks to identify training requirements based on an analysis of job performance requirements data. Training objectives are then formulated from the collected data, which can be used to assess a trainee's progress toward meeting targeted training objectives. By following a systematic requirements specification process, training systems can be designed to ensure that a desired level of readiness is achieved.

LIFECYCLE APPROACHES

Several system development lifecycle (that is, system) approaches can be used to guide the training systems requirements engineering process (Sharp, Rogers, & Preece, 2007; Young, 2004). The most appropriate approach depends on the project type and available resources. An early approach was the waterfall systems development lifecycle, which follows a linear-sequential model that focuses on completion of one phase before proceeding to the subsequent phase. It is easy to use and manage due to its rigidity and would work well for training systems that have very well-defined requirements at the outset or when costs and schedule need to be predetermined, as the progress of system development is readily measurable with this approach. It would not be an appropriate approach for complex training systems, where requirements are at a moderate to high risk of changing, as it is inflexible, slow, costly, and cumbersome due to its rigid structure. In an effort to be more flexible and accommodate changes in requirements more effectively than the waterfall approach, use of multiple iterations or spirals of requirements planning, gathering, analysis, design, engineering, and evaluation have been adopted (compare the incremental approach, the evolutionary lifecycle, and the spiral approach; Boehm, 1988). The Rapid Applications Development (RAD) approach (Millington & Stapleton, 1995) is an iterative approach that attempts to quickly produce high quality systems through the use of iterative prototyping, active user involvement, and risk reduction through "time boxing" (that is, time-limited cycles of ~6 months to achieve system/partial system builds) and Joint Application Development (JAD) workshops in which users and developers convene to reach consensus on system requirements. With time boxing and JAD, RAD aims to produce systems quickly and with a tight fit between user requirements and system specifications; thus, it can lead to dramatic savings in time, cost, and development hours. However, the time-limited cycles can lead to a sacrifice in quality and there can be a tendency to push difficult problems to future builds in an effort to demonstrate early success. RAD would be appropriate for training system development efforts that have a high level of user community support, but would not be fitting for large, complex projects with distributed teams and limited user community involvement. Another alternative is the agile system development lifecycle, which involves tight iterations (one to three weeks) through four phases (that is, warm-up, development, release/endgame, and production), with a product being delivered for user feedback at the end of each iteration. During each iteration, developers work closely with users to understand their needs and implement and test solutions that address user feedback. The agile approach would be appropriate for training systems that have emergent and rapidly changing requirements, but would be less fitting for projects that cannot handle the increased system complexity/flexibility and cost that are often associated with the agile approach (for example, training systems with high reliability or safety requirements). While there are several lifecycle approaches, for training systems development it is likely that the suitable choice will be a flexible approach, such as RAD or agile system development.

REQUIREMENTS ENGINEERING

Regardless of the lifecycle approach followed during a systems development effort, each approach has a requirements engineering phase in which four central activities are conducted: (1) elicitation, (2) modeling and analysis, (3) specification, and (4) verification and validation (see Figure SP2.1; Young, 2001, 2004). Elicitation focuses on gathering requirements from stakeholders. The collected data are then modeled and analyzed to ensure each requirement is necessary, feasible, correct, concise, unambiguous, complete, consistent, verifiable, traceable, nonredundant, implementation free, and usable (Hooks & Farry, 2001; Young, 2004). Next, the requirements are documented in a specification document. Documented requirements are then verified and validated with stakeholders to ensure that the system being developed meets the needs of the intended stakeholders.

Elicitation

Requirements elicitation is the process by which the requirements for a training system are discovered. It comprises a set of activities that enables the understanding of the goals for, objectives of, and motivations for building a proposed training system. It seeks to garner an understanding of the target operational domain and associated tasks such that requirements can be elicited that result in a training system that satisfies its intended goals. It involves identifying target

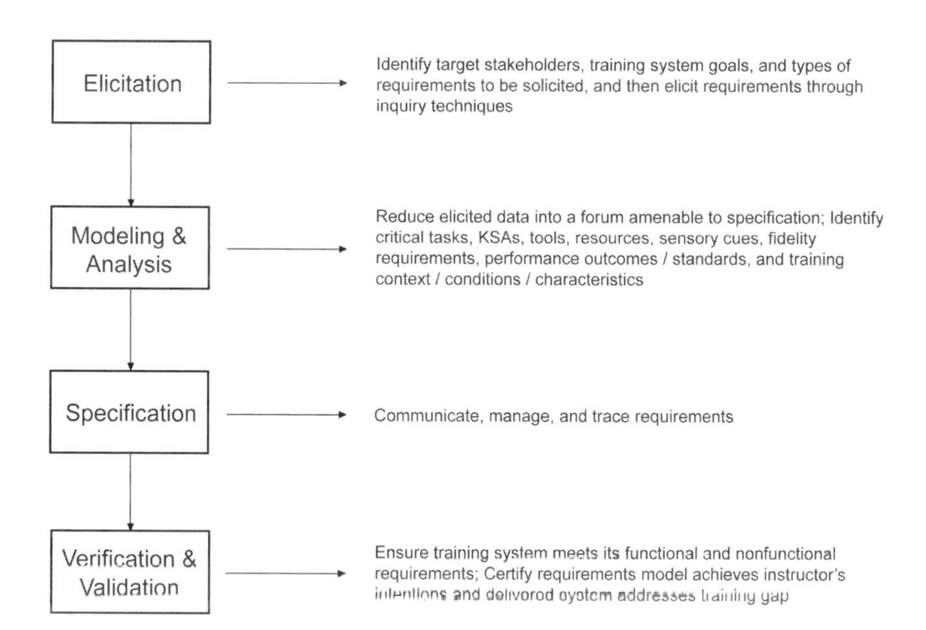

Figure SP2.1. Four Central Activities Involved in Training Systems Requirements Engineering

stakeholders, identifying training system goals, identifying the types of requirements to be elicited, and then eliciting the requirements through a set of inquiry techniques.

IDENTIFYING STAKEHOLDERS

Requirements should address the viewpoints of a variety of stakeholders (that is, alternative viewpoints), including users (for example, trainees and instructors), customers (for example, acquisition community), system developers, quality assurance teams, and requirements analysts. It is thus important to take a multiperspective approach to eliciting requirements to ensure all viewpoints are incorporated (Sharp et al., 2007).

IDENTIFYING TRAINING SYSTEM GOALS

Training system goals can be identified through training needs analyses (TNAs), changes to operational requirements, or changes to equipment or systems. TNA is a process of gathering and interpreting data from the target training community and operational environment in an effort to identify performance gaps and formulate training solutions (see Cohn, Stanney, Milham, Bell Carroll, Jones, Sullivan, and Darken, Volume 3, Section 2, Chapter 17). A comprehensive TNA provides the basis from which to design training solutions that realize substantial improvements in human performance by closing the identified performance gap in a manner that is compatible with current training practices.

Once the training gap is identified, a task analysis (TA) should be conducted in order to gain a complete understanding of the target operational context. The TA will identify specific training goals, including the specific tasks and associated knowledge (that is, foundational information required to accomplish target training tasks), skills (that is, an enduring attribute that facilitates learning on target training tasks), and abilities (that is, an enduring attribute that influences performance on target training tasks) to be trained. The TA will also uncover the supporting tools, resources (that is, reference used to perform target training tasks), operational context that is, platform, system, and environment), sensory modality and fidelity requirements, and use scenarios (see Phillips, Ross, and Cohn, Chapter 8) required for such training (Integrated Learning Environment [ILE], 2006). These goals can then be translated into learning/training and system requirements that drive the training systems development process.

TYPES OF REQUIREMENTS

A requirement is a statement of the capabilities, physical characteristics, or quality factors that a training system must possess to achieve its intended training objectives (Young, 2004). There are several types of requirements, which fall into two broad categories: functional requirements (that is, what the system must do) and nonfunctional requirements (that is, the constraints on the types of solutions that will meet the functional requirements, including user, usability, performance, interface, operational requirements, environment, facility, resource, verification, acceptance, documentation, security, portability, quality, reliability, maintainability, and safety requirements). For training systems (ILE, 2006), additional requirements include trainee requirements, instructor requirements (defines the instructor prerequisites and qualifications [for example, education, military rank, civilian grade, and experience] and the number of instructors required per training session [that is, instructor/student ratio]), and curriculum requirements (for example, mission tasks to be trained, content, layout, delivery mode, course throughput, and so forth).

ELICITATION TECHNIQUES

There are several techniques for eliciting requirements (Zhang, 2007; Zowghi & Coulin, 2005). Traditional techniques include introspection, review of existing doctrine, analysis of existing data, interviews (both open-ended and structured), surveys, and questionnaires. Collaborative techniques include conducting focus groups and/or JAD/RAD workshops, prototyping, brainstorming, and participatory design (see Nguyen and Cybulski, Chapter 11). Cognitive techniques include task analysis, protocol analysis, and knowledge acquisition techniques (for example, card sorting, laddering, repertory grids, and proximity scaling techniques) (see Mayfield and Boehm-Davis, Chapter 7). Contextual approaches include ethnographic techniques (for example, participant observation and enthnomethodology), discourse analysis (for example, conversation analysis and speech act analysis), and sociotechnical methods (for example, soft systems analysis). In general, structured interviews are one of the most effective elicitation techniques (Davis, Dieste, Hickey, Juristo, & Moreno, 2006). Observational techniques (that is, task analysis, protocol analysis, and contextual approaches) provide means by which to develop a rich understanding of the target domain and are effective in eliciting tacit requirements that are difficult to verbalize during interviews (Zhang, 2007). Each technique has strengths and weaknesses, and thus it is best to use a combination of approaches (Goguen & Linde, 1993) or a synthetic approach that systematically combines several of the techniques (Zhang, 2007). Table SP2.1 provides a comparative summary of requirements elicitation techniques.

Modeling and Analysis

Once requirements data have been elicited, the data must be modeled and analyzed to reduce them into a form amenable to specification. In terms of modeling, an effort is made to express the requirements in terms of one or more models that support further analysis and specification. Through delineation of precise models, details missed during elicitation can be identified; the resulting models can also be used to communicate the requirements to developers (Cheng & Atlee, 2007). There are several approaches to modeling functional requirements, and these can focus on the organization (for example, Enterprise Models), the data (for

Table SP2.1. Comparative Summary of Requirements Elicitation Techniques

Elicitation Type	Elicitation Technique	Strengths	Shortcomings
Traditional	Introspection	Easy to administer	Cannot introspect contextual data; when experts are used as source for intro-spection, they may not reflect behaviors of naive trainees
	Reviewing existing doctrine	Provides foundational knowledge from which to reason about tasks, environment, and other data sources; may provide detailed requirements for current system	Doctrine does not always match reality; often verbose and replete with irrelevant detail
	Interviews	Rich collection of data —both objective and subjective perceptions; represent multiple stakeholders' opinions; flexible, can probe in depth, and adapt on the fly	Data volume can be cumbersome to structure and analyze; potential for large variability among respondents; difficult to capture contextual data
	Questionnaires	Quickly collect data from large number of respondents; can administer remotely; can collect subjective perceptions (for example, attitudes and beliefs) and target trainee characteristics	Difficult to capture contextual data; must be careful to avoid bias in respondent selection; difficult to analyze open-ended questions
Collaborative	Focus groups; brainstorming; JAD/RAD workshops	Fosters natural interaction between people; gauge reaction to early product concepts; foster stakeholder consensus and buy-in; team dynamics can lead to rich understanding of trainee needs	Ad hoc groups may be uncomfortable for participants; few may dominate discussion; possibility of "groupthink"

	Prototyping	Provides an early view of what is feasible; develop understanding of desires and possibilities; good when there is a high level of uncertainty about requirements; good for obtaining early feedback from stakeholders; stimulates discussion	Need system build to apply; can be costly; may assume trainees accept prototype design when, in fact, they accept only its behaviors
Cognitive	Task analysis	Obtain detailed task and rich contextual data	If not well structured, can be time consuming and difficult to analyze rich data
	Protocol analysis	Tap tacit knowledge; obtain contextual data if embedded in work context; reveal shortcomings of existing systems	Based on introspection and thus may be biased or unreliable; cannot capture social dimension
	Knowledge acquisition	Can elicit tacit knowledge, classification knowledge, hierarchical knowledge, and mental models; elicited knowledge is represented in standardized format	Does not model performance and contextual data; assumes knowledge is structured
Contextual	Ethnographic	Obtain rich contextual data; uncover existing work patterns and technology usage; identify division into and goals of social groups	Extremely labor intensive; can be difficult to analyze rich data; difficult to assess proposed system changes

	Discourse analysis	Identify division into and goals of social groups; consider effect of new system on existing social structure; uncover value system of organization; tap tacit knowledge	Only applicable to situations with substantial social interaction and verbal data; labor intensive; cannot readily apply prior to prototype stage

example, Entity-Relationship-Attribute models), the functional behaviors of stakeholders and systems (for example, object-oriented models), the stakeholders (that is, viewpoint models), or the domain itself (that is, a model of the context into which the system will operate) (Kovitz, 1999; Nuseibeh & Easterbrook, 2000). Nonfunctional requirements are generally more difficult to model, but it is still essential to operationalize them such that they are measurable (see Cohn, Chapter 10). In terms of training system design, the modeling process should allow for learning theory and research to be built into the model such that it guides training system design.

Once the requirements data have been modeled, they can then be more readily analyzed. Requirements analysis focuses on evaluating the quality of captured requirements and identifying areas where further elicitation and modeling may be required. As with modeling, there are several approaches to requirements analysis, including requirements animation (for example, simulate operational models or goal-oriented models), automated reasoning (for example, goal-structured analysis, analogical and case based reasoning, and knowledge based critiquing), and consistency checking (for example, model checking in terms of syntax, data value and type correctness, circularity, and so forth).

Whichever analysis approach is used, when developing requirements for training systems, the analysis process must first identify the training gap (see Figure SP2.2). Then the analysis should identify triggers that serve as learning or training requirements that fill the gap (ILE, 2006). For each learning/training requirement, the analysis must identify the critical tasks to be trained, as well as the desired levels of proficiency (see Phillips, Ross, and Cohn, Chapter 8), targeted KSAs, necessary tools, resources, sensory cues, fidelity levels, desired performance outcomes (cognitive, affective, psychomotor, verbal, and social), and training criteria by which to assess acceptable performance (that is, performance standards) (see Milham, Carroll, Stanney, and Becker, Chapter 9). It must also specify the training context (that is, use cases) and desired training conditions (for example, platform, environment, time pressure, and stress level) required to achieve training objectives. The desired training characteristics should also be specified (for example, learning curve, coordination/teaming requirements, chain of command, anticipated performance errors, and remediation strategies), as should the estimated costs of the training and associated return on investment

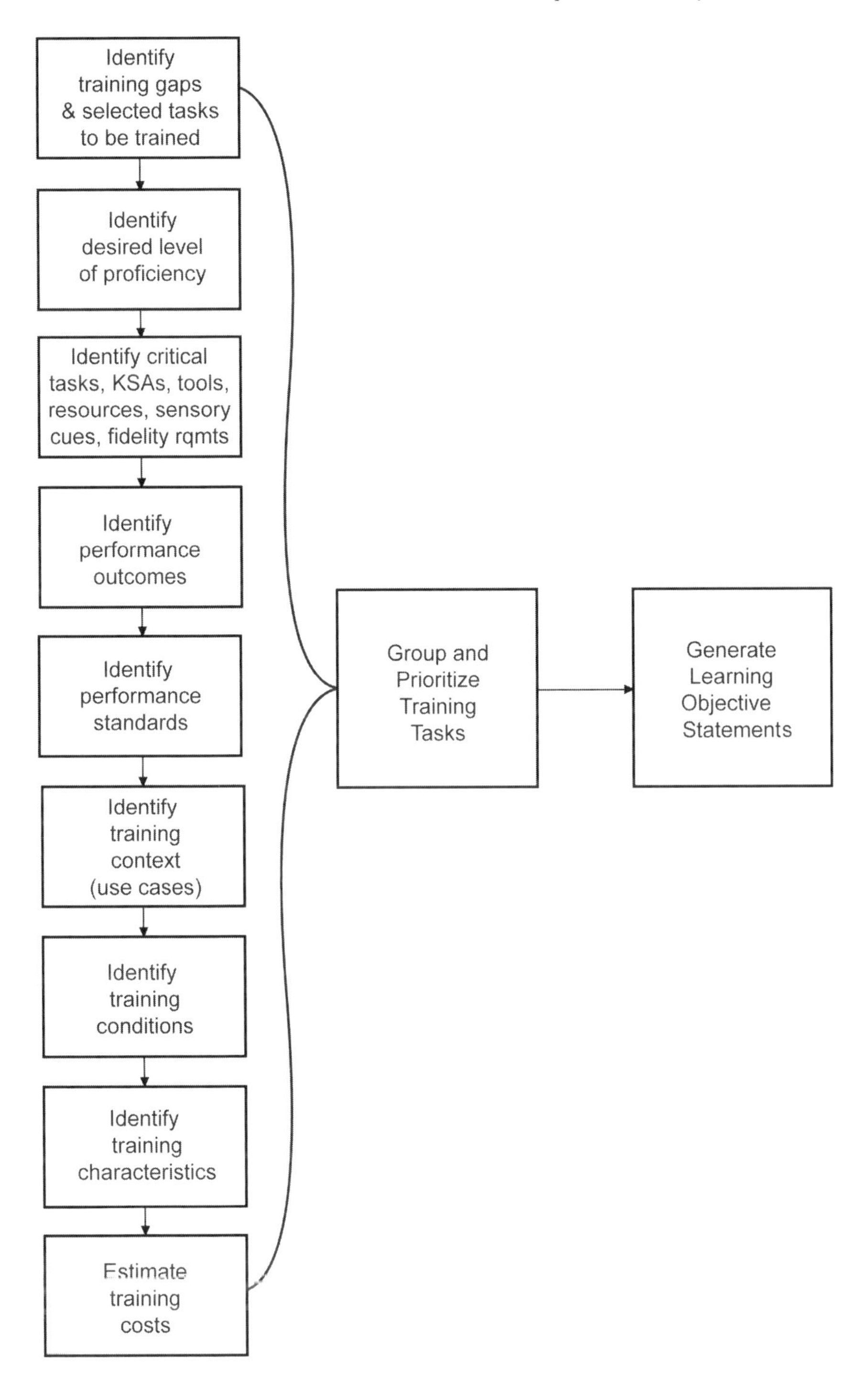

Figure SP2.2. Learning Analysis Process (Adapted from ILE, 2006)

(see Cohn, Chapter 10). Training tasks can then be grouped by skill requirements and prioritized.

The result of the learning analysis process is the generation of a learning objective statement (LOS), which "establishes content (and training technology) linkage with the full spectrum of work proficiency required for mission readiness and professional expertise" (ILE, 2006). The LOS aligns the identified training gaps, training objectives (that is, targeted KSAs), content, sequencing, delivery mode, student assessment, and program evaluation. In generating the LOS, it is important to recognize that the effectiveness of a training solution for addressing each targeted KSA depends on the fidelity of the training solution. Cohn et al. (2007) suggest developing a "blended" training solution, with initial acquisition of declarative knowledge and basic skills being trained via classroom lectures and low fidelity training solutions, basic procedural knowledge and problem solving skills being trained and practiced via medium fidelity training solutions, and consolidation of learned declarative knowledge and basic skills and procedures, practice of acquired knowledge and skills (for example, mission rehearsal), as well as development of more advanced strategic knowledge and tactical skills being trained via high fidelity training solutions. If, during the modeling stage, the resultant models are linked to learning theory, this should facilitate the development of a blended training solution.

Specification

The requirements specification process involves communicating requirements, requirements management, and requirements traceability (Hooks & Farry, 2001; Kovitz, 1999; Nuseibeh & Easterbrook, 2000; Sharp et al., 2007). The manner in which requirements are documented influences how effectively the requirements can be interpreted and realized in the training system design. Each documented requirement should be necessary, feasible, correct, concise, unambiguous, complete, consistent, verifiable, traceable, nonredundant, implementation free, and usable (Young, 2004). There are a number of specification languages, ranging from formal (for example, Z notation, Vienna Development Method) to semiformal (for example, State-Transition Diagram, Unified Modeling Language, and Data Flow Diagrams) to informal (for example, natural language) that can be used to document training system requirements specifications. Companies often adopt specification templates that are specific to their needs. The LOS is a template that can be followed for training system requirements specification. Learning objective statements are generally broken into three main components (Arreola, 1998):

1. An action word that describes the targeted performance (that is, competency) to be demonstrated after training;
2. A statement of the performance criterion (that is, performance standard) that represents acceptable performance;
3. A statement of the conditions under which the trainee is to perform during training.

Kovitz (1999) provides heuristics that can be followed to improve the quality of requirements documentation, regardless of the format in which requirements are documented.

Requirements traceability relates to how readily requirements can be read, navigated, queried, and changed (Gotel, 1995; Nuseibeh & Easterbrook, 2000). It can be achieved by cross-referencing, by using specialized templates that provide links between document versions, or by restructuring the requirements specification according to an underlying network or graph that keeps track of requirements changes. A general-purpose tool (for example, hypertext editor, word processor, and spreadsheet), which supports cross-referencing between documents, or a database management system, which provides tools for documenting, editing, grouping, linking, and organizing requirements, can be used to support traceability.

Verification and Validation

Verification and validation (V&V) is the process of ensuring a training system meets its functional and nonfunctional requirements. Verification focuses on ensuring a training system design satisfies its specified requirements (Young, 2001). Verification seeks to provide evidence through cognitive walkthroughs, functional and performance testing, and demonstrations that the requirements have been met in the delivered training system. Requirements validation involves certifying that the requirements model is correct in terms of the instructor's intentions, thereby ensuring that the right problem (that is, training gap) has been solved in the delivered system. Validation establishes that the requirements model specifies a training system solution that is appropriate for meeting trainees' and instructors' needs and thus often requires the involvement of stakeholders (see Nguyen and Cybulski, Chapter 11). During validation, the focus is on evaluating external consistency (that is, agreement between the requirements model and the problem domain), ambiguity (that is, the requirement can be interpreted only one way), minimality (that is, no overspecification), and completeness (that is, there is no omission of essential information needed to make the requirements model valid). To ensure the effectiveness of training systems, it is essential that the V&V process considers how best to evaluate training effectiveness (see Cohn, Stanney, Milham, Carroll, Jones, Sullivan, and Darken, Volume 3, Section 2, Chapter 17). The training effectiveness evaluation should assess how effectively the training system design facilitates development and maintenance of targeted training objectives and transfer of training back to the targeted operational environment.

Validation can be conducted via a combination of inspection and formal approaches (for example, automated reasoning, such as analogical and case based reasoning, knowledge based critiquing, and consistency checking) and informal techniques (for example, prototyping, running scenarios, animating requirements, and conducting cognitive walkthroughs). The V&V process aims to ensure requirements have been interpreted correctly in the training system solution, are effective and reasonable in filling the training gap, and are recorded in a manner that can be readily verified.

STATE OF THE ART IN REQUIREMENTS SPECIFICATION

Requirements engineering has evolved over the past few decades into a practice supported by various techniques and tools. Over this time, the following aspects have been identified as best practices in requirements specification: (1) conducting requirements modeling and analysis within the organizational and social context in which the training system is to operate, (2) developing requirements specifications that are solution independent, with a focus on modeling stakeholders' goals and scenarios that illustrate how these goals are to be achieved rather than modeling the desired functionality of a new system (that is, information flows and system state), and (3) placing emphasis on analyzing and resolving conflicting requirements and reasoning with models that contain inconsistencies (Nuseibeh & Easterbrook, 2000).

As we look to the future of training systems requirements engineering (see Table SP2.2), several areas are important to address (Cheng & Atlee, 2007; McAlister-Kizzier, 2004; Nuseibeh & Easterbrook, 2000):

- In terms of elicitation, there is a need to develop technologies that improve the precision, accuracy, and variety of requirements elicited, particularly with regard to techniques for identifying stakeholders (that is, instructors, trainees, and so forth) and their requirements with respect to the particular target context/environment.
- In terms of elicitation, there is a need to develop tools that assist in structuring ethnographic observations and instructor interviews so that they are targeted toward the following: (1) identifying learning objectives, associated KSAs, and performance criterion and (2) better characterizing the instructional setting, group dynamics, communication patterns/modes, organizational/political climate, and other cultural factors that influence instruction.
- In terms of modeling, considerable recent research has focused on improving scenario based modeling approaches, and the focus is now on developing techniques for creating, combining, and manipulating models (for example, model synthesis, model composition, model merging) and supporting requirements model reuse.
- In terms of modeling, emphasis should be placed on developing models that build learning theory and research into the modeling process to guide training system design.
- In terms of modeling, there is also a need to develop modeling techniques that deal with inconsistent, incomplete, and evolving requirements models; of particular value would be self-managing systems that accommodate varying, uncertain, incomplete, or evolving requirements at run time. This is particularly important given the dynamic nature of CONOPS (concept of operations).
- In terms of analysis, there is a need to develop techniques that effectively link learning analysis to training system design (that is, processes that enable translation of learning objectives into training system designs). Such techniques should aim to

Table SP2.2. Future Needs in Training Systems Requirements Engineering

Requirements Engineering Activity	Future Needs
Elicitation	• Technologies that improve the precision, accuracy, and variety of requirements elicited • Tools for structuring ethnographic observations and instructor interviews to support requirements elicitation
Modeling	• Techniques for creating, combining, and manipulating models (for example, model synthesis, model composition, and model merging) • Tools that support requirements model reuse • Tools that build learning theory and research into the modeling process to guide training system design • Techniques that deal with inconsistent, incomplete, and evolving requirements models (for example, self-managing systems)
Analysis	• Techniques that effectively link learning analysis to training system design (that is, processes that enable translation of learning objectives into training system designs) • Techniques for supporting the prioritizing of requirements • Techniques for bridging the gap between current contextualized elicitation approaches and their inputs (for example, multimedia inputs—video/audio) and more formal specification and analysis techniques
Specification	• Requirements management techniques that automate traceability, determine the maturity and stability of elicited requirements, support scaling of requirements specification, and support security of requirements databases • Tools that allow instructors to quickly update and modify training system requirements to reflect the changing tactics, techniques, and procedures that target evolving opponent strategies • Techniques for analyzing the impact a particular training system architectural choice (for example, delivery mode) has on the ability to satisfy current and future requirements • Techniques that support outsourcing of downstream development tasks (that is, filling the gap between requirements specification and development teams that are geographically distributed)

Verification and Validation	• Techniques for evaluating how effectively a training system design specification facilitates the development and maintenance of targeted training objectives and transfer of training back to the targeted operational environment • Tools for supporting the use of novelty, value, and surprisingness to explore the creative outcome of requirements engineering • Methods (for example, animations and simulations) for providing information to stakeholders (that is, instructors, trainees, and so forth) to elicit their feedback

systematically specify a blended fidelity training solution, which integrates the optimal mix of classroom instruction, training technologies, and live events throughout a given course of training, to ensure that a desired level of readiness is achieved.

- In terms of analysis, there is also a need to develop improved techniques for analyzing the quality of elicited requirements, particularly with regard to revealing misunderstandings or questions that require further elicitation and techniques for prioritizing requirements such that an optimal combination of requirements can be identified and implemented.
- In terms of modeling and analysis, there is a need to develop new techniques for formally modeling and analyzing contextual properties of the target environment into which the training system is to be immersed and bridging the gap between current contextualized elicitation approaches and their inputs (for example, multimedia inputs—video/audio) and more formal specification and analysis techniques.
- In terms of specification, there is a need to develop requirements management techniques that (1) automate the task of documenting traceability links among requirements and between requirements and downstream artifacts, (2) determine the maturity and stability of elicited requirements, (3) support scaling of requirements specification (that is, techniques that allow for organizing and manipulating large numbers of requirements), and (4) support security of requirements databases.
- In terms of specification, there is also a need to develop requirements specification tools that allow instructors to quickly update and modify training system requirements to reflect the changing tactics, techniques, and procedures that target evolving opponent strategies.
- In terms of specification, there is a need to develop techniques for analyzing the impact a particular training system architectural choice (for example, delivery mode) has on the ability to satisfy current and future requirements.
- In terms of specification, there is also a need to support globalization through the development of requirements specification techniques that support outsourcing of downstream development tasks (that is, filling the gap between requirements specification and development teams that are geographically distributed).
- In terms of verification and validation, there is a need to develop techniques for evaluating how effectively a training system design specification facilitates the

development and maintenance of targeted training objectives and transfer of training back to the targeted operational environment.

- In terms of verification and validation, there is a need to develop tools for supporting the use of novelty, value, and surprisingness to explore the creative outcome of requirements engineering (see Nguyen and Cybulski, Chapter 11).
- In terms of verification and validation, there is a need to develop improved methods (for example, animations and simulations) for providing information to stakeholders (that is, instructors, trainees, and so forth) to elicit their feedback.

CONCLUSIONS

This chapter has provided an overview of the training systems requirements specification process (that is, elicitation, modeling and analysis, specification, and verification and validation), discussed some of the issues associated with the process and the value of good requirements specification, as well as contemplated future directions for the field. In order to avoid the "big mistakes" that often transpire during large training system development efforts, it is essential to adopt rigorous requirements engineering practices that fully characterize the capabilities, physical characteristics, and quality factors that a training system must possess to achieve its intended training objectives.

REFERENCES

Arreola, R. A. (1998). *Writing learning objectives.* Retrieved July 2, 2007, from http://www.utmem.edu/grad/MISCELLANEOUS/Learning_Objectives.pdf

Boehm, B. W. (1988). A spiral model of software development and enhancement. *IEEE Computer, 21*(5), 61–72.

Cheng, B. H. C., & Altee, J. M. (2007). Research directions in requirements engineering. In *Future of Software Engineering (FOSE'07)* (pp. 285–303). Los Alamitos, CA: IEEE Computer Society.

Cohn, J. V., Stanney, K. M., Milham, L. M., Jones, D. L., Hale, K. S., Darken, R. P., & Sullivan, J. A. (2007). Training evaluation of virtual environments. In E. L. Baker, J. Dickieson, W. Wulfeck, & H. O'Neil (Eds.), *Assessment of problem solving using simulations* (pp. 81–105). Mahwah, NJ: Lawrence Erlbaum.

Davis, A., Dieste, O., Hickey, A., Juristo, N., & Moreno, A. M. (2006). Effectiveness of requirements elicitation techniques: Empirical results derived from a systematic review. In *14th IEEE International Requirements Engineering Conference (RE'06)* (pp. 179–188). Los Alamitos, CA: IEEE Computer Society.

Goguen, J. A., & Linde, C. (1993). Techniques for requirements elicitation. In S. Fickas & A. Finkelstein (Eds.), *Proceedings, Requirements Engineering '93* (pp. 152–164). Los Alamitos, CA: IEEE Computer Society.

Gotel, O. (1995). *Contribution structures for requirements traceability.* London: Imperial College, Department of Computing.

Hooks, I. F., & Farry, K. A. (2001). *Customer-centered products: Creating successful products through smart requirements management.* New York: American Management Association.

ILE. (2006). *Navy ILE instructional systems design and instructional design process* (MPT&ECIOSWIT-ILE-GUID-1). Retrieved May 25, 2007, from https://ile-help.nko.navy.mil/ile/contentItems/Navy%20ILE%20ISD%20Process_20070815.pdf

Kovitz, B. L. (1999). *Practical software requirements: A manual of contents and style.* Greenwich, CT: Manning Publications.

McAlister-Kizzier, D. L. (2004, February). *Research agenda to assess the effectiveness of technologically mediated instructional strategies 2004.* Paper presented at the 23rd Annual Organizational Systems Research Association (OSRA) Conference, Pittsburgh, PA.

Millington, D., & Stapleton, J. (1995). Special report: Developing a RAD standard. *IEEE Software, 12*(5), 54–56.

Nuseibeh, B., & Easterbrook, S. (2000). Requirements engineering: A roadmap. *Proceedings of the Future of Software Engineering* (pp. 35–46). New York: ACM Press.

Sharp, H., Rogers, Y., & Preece, J. (2007). *Interaction design: Beyond human-computer interaction* (2nd ed.). Hoboken, NJ: John Wiley & Sons.

Young, R. R. (2001). *Effective requirements practices.* Boston: Addison-Wesley.

Young, R. R. (2004). *The requirements engineering handbook.* Boston: Artech House.

Zhang, Z. (2007, March). *Effective requirements development—A comparison of requirements elicitation techniques.* Paper presented at the System Quality and Maintainability (SQM2007), Amsterdam, The Netherlands. Retrieved June 19, 2007, from http://www.cs.uta.fi/re/rem.pdf

Zowghi, D., & Coulin, C. (2005). Requirements elicitation: A survey of techniques, approaches, and tools. In A. Aurum & C. Wohlin (Eds.), *Engineering and managing software requirements* (pp. 19–46). Heidelberg, Germany: Springer-Verlag.

Part V: Methods

Chapter 7

APPLIED METHODS FOR REQUIREMENTS ENGINEERING

Tom Mayfield and Deborah Boehm-Davis

Virtual environments (VEs) are an exciting and powerful medium that promises much in the way of realistic and accurate operating scenarios, more cost-effective training, and better training transfer (Cohn, Volume 1, Section 2, Chapter 10; Foltz, LaVoie, Oberbreckling, and Rosenstein, Volume 1, Section 3, Chapter 17). Specifically, VEs offer the following potential operating benefits:

- Simpler training and reduced operator error through better system/equipment interfaces;
- Reduced downtime, improved efficiency, and economy in personnel during operations;
- Downstream cost savings due to the reduced likelihood of costly design changes; and training benefits;
- Interaction at the knowledge level that leads to improved understanding of the process being controlled and a more proactive approach to operability;
- Reduced reliance on rule behavior that reduces the risk of mindset incidents and places less emphasis on procedural training to cope with specific emergencies, which rarely occur in the way envisaged.

However, realizing these potential benefits requires careful attention in developing the requirements for the VE. In most respects, developing requirements for VEs is no different from developing requirements for any other complex system. Specifically, developers should follow basic human factors processes, including the use of task analysis techniques to understand the functions that must be accomplished by the system. This chapter will describe methods that can be used to decide on the training requirements and evaluate the VE proposed to meet the tasks underpinning those requirements.

Specifically, deciding which tasks and features of tasks are essential to creating a successful training program can be carried out only if a detailed task analysis breakdown is done. Task analysis (TA) is a universal human factors (HF) methodology for decomposing complex activities into understandable "chunks." TA in its various forms is the only breakdown methodology that will give the VE designer the detail needed to understand the functions that must be accomplished by the system and to decide on the level of fidelity required. Use of TA techniques is important in developing the constraints that will influence performance in VE training environments. Although the traditional approach focuses on a technology-centered approach, successful training systems need to be informed by a human-centered approach.

Requirements specification also needs to be informed by the trade-off between cost and realism. In evaluating cost, it is important to evaluate both the cost of producing the trainer and the cost of not adequately training the users. Although the development of a VE for training may be expensive, the cost of failure to train may be exponentially higher. Such trade-offs may become apparent when major emergencies occur, such as at Three Mile Island, where the inability to simulate the exact emergency conditions during training led to wrong assumptions being made that turned a recoverable incident into a major disaster (President's Commission on the Accident at Three Mile Island, 1979), or the British Airways Flight 009 incident where the captain's unorthodox use of the Boeing-747 training simulator to simulate a four-engine failure helped him to prevent a possible crash into the ocean and loss of life, thanks to his greater understanding of the aircraft's behavior (Diamond, 1986).

Realism covers both the level of fidelity of individual features of the simulation, as well as its overall appearance. The history of simulation suggests that careful attention to the tasks that need to be trained can allow developers to build simulators that represent needed functionality at a reasonable cost. Early aircraft trainers (such as the Link trainers designed by Edward A. Link in the United States in the 1930s; L3 Communications, 2007) were little more than boxes with hoods, with basic aircraft controls and displays, and a few degrees of freedom to simulate flight. However, they were sufficient to train basic familiarity with the controls, and they reduced training time in the actual aircraft. Vehicle manufacturers, whether cars, tanks, ships, or aircraft, have used mock-ups to show layouts, often in wood, with paper drawings to represent the operating panels. The nuclear industry, through regulatory requirements, provides full-scale simulators of power plants that represent every facet of the design from layout to operational verisimilitude.

Those who plan to use VEs should carefully consider what functions and features need to be represented in the training environment. In the case of VEs, this will include the physical tasks associated with access and egress, movement and manipulation of tools and controls, as well as the cognitive tasks associated with operating the virtual world. It is also important to understand the elements of tasks to be simulated in the VE and identify "purposeful behavior" (Cockayne & Darken, 2004). This behavior relates mainly to the physical movement component of the task (motor), but extends to knowing what to do or where to go (cognitive) and when to start and stop (perceptual).

In some cases, features will be incorporated in the VE because they are central to the tasks that are being trained. In other cases, certain features that are not critical to the task being trained need to be incorporated to create a realistic scenario that will be accepted by the user. For example, in training a pilot, simulating access and egress to the vehicle may not be critical for training flight skills, but strapping in and safety checks may need to be simulated to set up realistic scenarios. Similarly, learning to train and fire a tank main armament may not require learning the task of removing shell cases, at least in initial training. However, if shell cases create interference that must be addressed in operating the armament, it may need to be included in the final simulation.

The level of fidelity with which individual features and functions must be represented in the synthetic environment in order to effectively convey needed knowledge to the user must also be considered. For example, one might consider whether an object can be shown as a complete entity, that is, with no further breakdown needed, or whether it must be shown as individual parts so that trainees can carry out specific tasks on each part. However, this may be difficult to determine a priori. Take, for example, a situation where a pilot is trained to fly through a particular region using video rather than interactive world mapping. The impact on cognitive processing and effectiveness of operation may not be clear. In this example, the lack of interactivity means that the training experience will bear only a passing resemblance to what the pilot will experience in the real world; however, the impact on performance is not as clear.

In the following sections, TA methods are outlined that will provide techniques for realizing such requirements (see Table 7.1). The methods described in the first two sections (data collection and description) allow the analyst to identify

Table 7.1. Task Analysis Methods for Requirements Gathering

Method	Issues Identified	Outcomes
Collect Task Requirements Data	Information on normal, standard, and emergency procedures; information on users (individuals and groups); timing	Collection of individual pieces of data needed to identify tasks to be represented in the VE
Describe the Data Collected	Information on timing and appropriate level of analysis for individual tasks	Static or dynamic representation of the tasks required in the VE
Assess Risk	Information on personal and system safety	Modifications to requirements identified for the VE
Assess System Performance	Information on system performance and potential improvements	Modifications to requirements identified for the VE
Assess Interactions across Levels of the Organization	Information on selection, training, and workplace design; communication and allocation of function; mental models and device design	Modifications to the requirements identified for the VE

requirements for which tasks need to be represented in the VE. The following three sections identify information that can be used to modify the initial requirements based on issues of risk, system performance, or interactions among different users of the system.

KEY ELEMENTS TO CONSIDER IN DEVELOPING THE TA

In developing a TA for a virtual environment, several key features should be considered. First is *timing.* As with many systems, it is important to begin the analysis early in the virtual environment development process. The benefits of early TA include user acceptance and less rework.

A second key element is *breadth.* In developing a virtual environment, it is important to cover only those requirements necessary to capture the relevant tasks. This means that decisions will need to be made up front about what will be required to make the virtual environment work.

A third element is selection of the appropriate *level of analysis.* In general, analysis levels range from the macro to the micro level. At the macro level, concerns tend to focus on system-level issues, which mean that they will focus on physical and behavioral attributes of the system and on issues of person-to-person communications. At the micro level, the focus tends to be at the keystroke level, where the user is interacting with the system. At this level of analysis, cognitive attributes, such as the user's goal in executing particular actions, become important and therefore must be represented in the analysis. This level is where issues of perception, analysis, response times, and human error typically are considered. In developing requirements for a virtual environment, both macro and micro tasks will likely be necessary. Thus, it will be important to select techniques that fit with each of these levels of analysis (Preece, 1994).

A fourth element that must be considered is *concurrency;* that is, the analysis should take into account when resources may need to be shared and whether sufficient resources exist for such sharing. This concern for sufficient resources can be applied to a variety of purposes—optimizing performance or safety, reducing risk, understanding or improving operations, or developing training. In addition, TA techniques can provide information to specify the allocation of functions, characteristics of the system users, or level of staffing to improve the ability to share resources. These techniques can also provide information on job organization, task and interface design, and methods for skills and knowledge acquisition.

Finally, the TA process must provide for *design feedback;* that is, the TA process should be iterative, allowing for design changes based on initial testing or evaluation. For example, initial testing may reveal the possibility of simplifying the VE without compromising the training requirement. Suppose that the designer has been trying to render a complex-shaped piece of equipment. However, initial testing reveals that users need to know that their access to another component may be blocked by this component during a maintenance operation. It may be sufficient to program this complex shape as a simple block rather than rendering it faithfully.

KEY ELEMENTS FOR A SUCCESSFUL TASK ANALYSIS

- Start early.
- Ensure that only relevant tasks are represented in the VE.
- Ensure the appropriate level of analysis.
- Ensure that sufficient resources are allocated to the system.
- Ensure an opportunity for design feedback.

COLLECTING TASK REQUIREMENTS DATA

Elicitation of task details can be seen as a three-legged stool. The first leg represents operating procedures that provide a view of the system operation in the way it is expected to be used and that include safety requirements as well as standard and emergency modes of operation. The second leg represents the users' perceptions of the job and task breakdown, which can be obtained from discussion and interviews. Finally, the third leg represents observational data that show how the system is actually used, along with user "fixes" and shortcuts.

Without one of the legs, the stool is unbalanced. In knowledge elicitation terms, the data are incomplete and will not reflect the system goals from all three stakeholding entities—design, operations, and users. The following sections discuss the three elements that make up the legs and show how data might be collected to support the development of requirements.

Procedures Data

In most descriptions of task data gathering, little if anything is said about the use of procedures. There are a number of reasons for this, not least that rarely during the design stage are procedures attempted, possibly due to continuous changes that may make procedures outdated as soon as they are written. Although individual pieces of equipment are often delivered with a manual, system operation may not be fully defined. In fact, it is not uncommon to find procedures still being written after installation and even during commissioning. When developing VEs for training, however, it is likely that the system already has been designed and implemented (see Ricci, Owen, Pharmer, and Vincenzi, Volume 3, Section 1, Chapter 2), so there may well be normal, standard, and emergency operating procedures available. These procedures manuals will provide a valuable first cut at developing the task analysis. They provide basic data on how the designer expected the system to work and are often further validated by the installation engineers once their job is completed. These data often are in an easily assimilated form and may be broken down by person and place, as well as by control and display. With the availability of modern computing systems, the hard copy manuals and procedures will usually be backed up by interactive electronic manuals, which make deriving the task analysis even less time consuming.

The procedural data may need to be transcribed into a more usable form for the data description process, and this process may differ depending on whether the data will be analyzed by the same person who collected the data or someone new. Characterizing the procedural tasks into taxonomies (Fleishman & Quaintance, 1984) is one way of sorting out real world tasks for comparison with their virtual world counterparts to aid the VE designer. Generally, the transcription is done through a spreadsheet or simple database. With some training, direct transcription to a hierarchical task analysis (HTA), task decomposition, or even operational sequence diagram (OSD) can be achieved (Kirwan & Ainsworth, 1993; Shepherd, 2001).

Data Elicited from Operators/Users

In collecting data from operators/users, it is important to recognize that there may be different "types" of users operating the system. The views of each of the user groups are likely to be different, as their perception of the tasks will be different from both the procedures and the observed actions. It is also important to assess the ability of the operator to use the technology and to recognize that all the users must be matched to the task demands of the system. Further, career-training avenues need to be available within the organization for skill advancement and retention of personnel. Thus, it will be important to collect data from all types of users. Data from these user surveys can be used to build a database of operating information that can be structured into guidance for VE designers. Typical user data elicitation techniques include the following (see Volume 1, Section 2 Perspective):

- Questionnaires—sets of predetermined questions in a fixed order containing closed (Yes/No), open-ended (How do you . . . ?), level of agreement, and/or level of preference or attitude items;
- Structured one-on-one interviews—sets of predetermined questions face-to-face so that open-ended questions or indecisive answers can be explored;
- Round table group discussions—may elicit team task behaviors or, with a group of the same users, allow interaction within the group to bring out task behaviors individuals might forget;
- Walkthrough/talkthroughs—one-on-one interviews structured around describing user actions without the benefit of the equipment or system in front of him or her;
- Verbal protocols—one-on-one interviews while using the equipment or system, where the user explains why an action is being taken.

Each method has advantages and disadvantages (Kirwan & Ainsworth, 1993). The first four, with their total reliance on user memory, can all suffer from incomplete information. However, they are helpful as users will provide details of shortcuts, linking activities (not covered in procedures, but found to be necessary to complete a process), and alternative ways of achieving operating goals. Any discrepancies between what the procedures say should be done and what the users say they do will be confirmed by collecting observational data as described below.

Observational Data

The capacity of any human observer is necessarily limited; thus, the collection of observational data should be initially guided by expert opinion or previous data/research in the context. However, the collection of ancillary data should be as broad as possible (for example, videotapes, activity logs, communication transcripts, and so forth) so that new viewpoints or theories can also be explored on an ad hoc basis. Care should be taken to make both the observation and ancillary measures as unobtrusive and nonreactive as possible (Webb, Campbell, Schwartz, & Sechrest, 1966).

Generally, naturalistic observation methods and case studies are used in psychology to gain a detailed account of a system's structure and functioning over time. Methods drawn from different disciplines tend to focus on different levels as the basis for observation and analysis. For example, protocol analysis has been used by cognitive psychologists to gain a view of process at a level on the order of seconds to minutes (Ericsson & Simon, 1980, 1984). Cognitive task analysis techniques, also used by cognitive scientists, create a more detailed analysis of behavior at the level of milliseconds (Schraagen, Chipman, & Shalin, 2000). These methods can be supplemented by interviews, questionnaires, or systematic surveys targeted at specific processes or events (Dillman, 1978). At the other end of the spectrum, ethnographic and other contextual methods (see Section 2 Perspective) used in anthropology (Brislin, Walter, & Thorndike, 1973) can give a similarly rich account of the thinking and actions of one or more actors in a cultural environment across longer time spans, such as days, weeks, or months. The use of these methods for preliminary observation will allow the requirements engineer to explore natural human system function and performance based on his or her current understanding while also extracting the maximum possible amount of information to support the formulation and evaluation of additional insights.

Another method for collecting task requirements data is by observation of operators as they are engaged in task completion. A good observational technique "capture[s] all the significant visual and audible events, in a form which makes subsequent analysis reasonably easy, and which does not influence the performance of the task" (Kirwan & Ainsworth, 1993, p. 54). Observation can be done through a strict recording of the activities being observed (such as with the use of some form of video recorder), or it may be done through the use of individuals not engaged in the task who record their observations. These observers may be subject matter experts, but this is not a requirement. Data can be collected continuously throughout the observational period or intermittently throughout the observation period.

The most common form of intermittent observation is application of the activity sampling technique (Hoffman, Tasota, Scharfenberg, Zullo, & Donahoe, 2003). This technique measures the user's behavior at predetermined times. This technique starts with categorization of the activities that might take place over the course of the task. Then, a sampling schedule is developed. The analyst records the targeted behaviors at intervals, either noting by tally the number of times that a specific activity is completed in a window of time or by capturing the sequence of activities completed within a time window. Critical incident technique is a specific form of activity sampling that looks at the key events in an operation rather than preset or random samples. It is often used where the process has specific safety requirements, such as nuclear or chemical operation and processing.

COLLECTING TASK REQUIREMENTS DATA

Keys to Success

- Represent operating procedures that provide a view of the system operation in the way it is expected to be used and that include safety requirements, as well as standard and emergency modes of operation.
- Represent the users' perceptions of the job and task breakdown.
- Represent observational data that show how the system is actually used, along with user fixes and shortcuts.

Outcomes

- Normal, standard, and emergency procedures
- Information on users (individuals or groups, whether supervised or unsupervised)
- Information on activities conducted and timing of activities

---

METHODS FOR DESCRIBING DATA COLLECTED

Having collected the task data, the analyst needs to decide how the tasks should be broken down to provide the human factors input to the requirements analysis. That is, defining the requirements goals for the analysis is an important step in choosing the method to describe the data. In most cases, one specific method will not provide all the answers needed. For instance, HTA is a great method for breaking down complex procedures, but does not provide a timeline, and concurrent tasks are difficult to show. Link analysis will show interconnecting tasks and provide a spatial representation, but tasks have to be identified separately and again there is no timeline. Choosing the right method will partly depend on the type of data collection method, but mainly on the expected results from the task analysis. Using TA during design will generally mean that data are less reliable, and more fragmented, than when a TA is done on a mature equipment or system.

For VE training systems, the analyst will be more concerned with ensuring that the VE designer has the tasks broken down to a sufficient level of detail such that they can program the virtual world with the most efficient use of memory and provide accurate response times. The following sections outline some of the more common task analysis methods used to provide task description methods suitable for gathering the requirements for the VE.

Task analysis can be seen as a way of taking complex relationships—systems and jobs—and breaking them down into progressively simpler elements—tasks and subtasks, which can then be used as requirements for the operational and training systems. For instance, many of the early methods were concerned with the more physical aspects of system operation and were aimed at breaking tasks down to a level that enabled the controls and displays that would be needed to successfully carry out routine tasks to be defined. Typical of these task decomposition methods are HTA (Shepherd, 2001), OSDs, link analysis, and the task decomposition method itself.

These methods are highly structured and demanding in time and resources, but provide a wealth of task-level descriptors that can be used for requirements gathering. Operational sequence diagrams are regularly used in military contexts, or wherever operations are tightly controlled and regulated. Link analysis has been used in a variety of commercial and military applications, with particular success in control room design and air traffic control operations. Hierarchical task analysis has been used to analyze nuclear power plant operations, to provide a basis for training manuals, and to reorganize organizations. It has been shown to work at the multisystem, system, equipment, and component levels. It is particularly noteworthy that cognitive task analysis methods designers have used the more physically based hierarchical task analysis as a descriptive basis for the goals, operators, methods, and selection rules used in GOMS (goals, operators, methods, and selection rules), natural GOMS language, and cognitive, perceptual, and motor GOMS (John & Kieras, 1996a, 1996b). This link between physical and cognitive TA is extremely useful when deciding on a level of analysis, particularly with reference to understanding both the physical and cognitive aspects of a system or equipment operation. It also serves as an information collection tool to systematically expand the basic description of the task element activities.

Static Descriptions (For Example, Charting/Networking)

A conventional decomposition method is the "tree" diagram as, for instance, in an organizational chart or a fault tree. Predominantly a static representation, an event is broken down into features that reflect the specific area under analysis. This might be the cause and effect of an accident or how one organizational entity interfaces with another. Generally the data will not describe individual tasks, but a collection of events or, as in the case of petri nets, interactive nodes that model the behavior of a system (Peterson, 1977). The descriptions provided by these methods might be more useful to the VE designer than some of the other task analysis methods, as they may reflect traditional engineering conventions.

Dynamic Descriptions (For Example, Simulations and Computational Models)

Another way of describing data that have been collected is through the use of simulations or models. Simulations reproduce the behavior of a system (or part of a system) and typically allow for interaction with the system. Models represent some portion of a system that is of interest. Some include representations of user cognition, while others focus more on behavioral outcomes (Barnes & Beevis, 2003; Bisantz & Roth, 2008; Gugerty, 1993). Further, some models provide static predictions of performance (John & Kieras, 1996a, 1996b), while others are computational and stochastic, such as adaptive control of thought–rational (Anderson & Lebiere, 1998), executive process/interactive control (EPIC; Kieras & Meyer, 1997), Soar (Rosenbloom, Laird, & Newell, 1993), and constraint based optimal reasoning engine (Eng et al., 2006).

Most modeling techniques require expertise both in the application domain (for example, supervisory control) and in interface design. They also each make assumptions about the primitive operations present in the system and about their users. This requires extensive experience with the model on the part of the designer. Fortunately, these models are starting to make the transition from academia to industry. Work is beginning to appear that exercises the models and tests their limits (Gray & Boehm-Davis, 2000; John & Vera, 1992). Gugerty (1993) has described some steps that might be taken to allow these models to be successfully applied in industry; in addition, work is being done to develop tools that can allow designers, even those without a cognitive psychology background, to apply these techniques (John, Prevas, Salvucci, & Koedinger, 2004; John & Salvucci, 2005).

DATA COLLECTION

Keys to Success

- Select analysis method
- Represent the task
  a. Statically
  b. Dynamically

Outcome

- Static or dynamic representation of the tasks required to be represented in the VE

METHODS FOR ASSESSING POTENTIAL SOURCES OF RISK

In developing training, it is critical to know what information needs to be conveyed to users or what experiences users need to have in order to remain safe when using the new system. In developing requirements for a VE to be used for training, it is important to identify the risk characteristics of the system so as to provide an accurate simulation in the VE of the problems that might be encountered. A number of methods are available to expert HF appraisers to assess the safety of existing or proposed systems. Barrier analysis and work safety analysis (Kirwan & Ainsworth, 1993) both examine the extent to which protective measures exist within a system. Event trees (Gertman & Blackman, 1994; Kirwan & Ainsworth, 1993; Kirwan & James, 1989; Park, 1987) and failure modes and effects analyses (Crow, 2002; Kirwan & Ainsworth, 1993), both of which have their origins in systems reliability metrics, examine human reliability and the consequences that derive from human failure. Hazard and operability analyses attempt to identify system design issues based on input from expert personnel (Benedyk & Minister, 1998; Kirwan & Ainsworth, 1993), while influence diagrams (Howard, 1990) provide a graphic representation of the factors that are identified as contributory to safety problems. Finally, the management oversight risk tree technique (Johnson, 1973, 1980) allows for an examination of the influence that the management structure of an organization has on safety.

Another approach to evaluating the human interaction with a system is the use of anthropometric models, such as JACK (Badler, 1989; Phillips & Badler, 1988; UGS, 2004), which allow system designers to construct a computer based representation of the product and animate its use based on three-dimensional models of the human. These models may use data from standards or empirical studies (Badler, 1989; Phillips & Badler, 1988; You & Ryu, 2004). These models have proven useful in developing methods for training (requirements for how to train). For example, JACK has been used to develop safe procedures for manual handling operations, such as lifting or using objects in awkward postures, where the actual conditions are likely to be hazardous or a full-scale training rig is impracticable.

ASSESSING RISK

Keys to Success

- Apply event trees, failure mode and effects analyses, and anthroprometric models to evaluate risk potential

Outcome

- Assessment of risk that can be used to
  c. Identify the risk characteristics of the system that must be represented in the VE simulation
  d. Modify the design of the system to reduce training requirements

---

METHODS FOR ASSESSING THE SYSTEM (FOR EXAMPLE, CHECKLISTS AND SURVEYS)

If the ultimate aim of task analysis is to improve the interface between humans and machines by a more detailed understanding of the tasks involved in system operations, then there needs to be a way of determining whether an improvement has been made. Methods that can help do this are typically checklists and surveys and absolute judgment. The first two are fairly straightforward, and a number of guidance documents exist to help carry out such assessments, such as NUREG-0700 (U.S. Nuclear Regulatory Commission, 1981) for nuclear power systems, MIL-STD 1472 (U.S. Department of Defense, 1999) for the military, and the Questionnaire for User Interface Satisfaction (Chin, Diehl, & Norman, 1987) for general interface concerns. Such documents provide a wealth of information, including what to look for, specific design requirements, and even what questions to ask. If part of the original TA data collection was done by questionnaire, then a basis for comparison is available to help determine if an improvement has been made. Within the VE, an assessment of the training value can assess the extent to which training transfer is perceived as effective and efficient (see Cohn, Stanney, Milham, Carroll, Jones, Sullivan, and Darken, Volume 3, Section 2, Chapter 17).

Absolute judgment is assessment by a group of experts who reach agreement on a range of criteria applied to the system or equipment. Generally, this technique does not produce the level of accuracy that one might like (Miller, 1956; Nielsen & Phillips, 1993), especially if used instead of normal task analysis data gathering techniques, where much more consistent results can be found. Expert assessments also can be limited in identifying usability and safety problems. For example, it has been shown (Rooden, Green, & Kanis, 1999) that any given expert is likely to identify a relatively unique subset of problems and that the problems identified will be influenced by the materials available for review (for example, a real product, a set of drawings, a mock-up of the product, or a video of user trials).

ASSESSING SYSTEM PERFORMANCE

Keys to Success

- Apply checklists, surveys, and expert appraisals to evaluate system (human and machine) performance

Outcome

- Assessment of performance that can be used to
  e. Modify the requirements for the VE
  f. Modify the design of the system to reduce training requirements

---

METHODS FOR ASSESSING INTERACTIONS ACROSS LEVELS OF THE ORGANIZATION

The individual analyses that have been described should identify key constraints that influence performance and that need to be represented in requirements. Aspects such as workplace tools and job design, for example, may set boundaries on the performance level that can be achieved in a given situation. However, to get the greatest benefit from this approach to developing requirements, the central framework of the analyses should be extended by a consideration of the levels of analysis represented within a work organization. For example, this process should determine whether an identified key social cognitive process has plausible links to lower level cognitive psychological or physiological processes or links to upper level industrial/organizational and sociological/anthropological processes. This extension across levels is difficult because theories from individual scientific domains tend to focus on one level to the exclusion of the others. Nevertheless, the vertical integration across levels can give important connections that combine the theoretical views abstracted from different domains into a coherent whole.

Consideration of these issues can also be undertaken while the user is undergoing training in the VE. That is, although these issues should be considered when developing requirements for the VE, observation of human performance while immersed in the VE may also provide insights into redesign of the system being modeled.

Organizational Issues—Selection and Training and Workplace Design

Many industries today are opting for "lean" organizations, even while expecting no loss in quality or worsening error rates (see Hendrick, 2008, for a review of organizational- and system-level issues). These reductions in staffing have implications for the skills required by operators and for the training they require. For example, staffing studies will require complex scenarios so as to fully understand the ramifications of fewer people. However, there will still be a danger in this approach as these scenarios may be sufficient to understand normal operations, but not the problems that more usually arise in emergency situations and when automated systems have broken down.

Understanding individual versus organizational roles and responsibilities and changes caused by reorganization are also important. Overlaying the organizational structure on top of the TA breakdown for the individual may present the VE designer with an opportunity to show where gaps may lie. Very often, it is the unscheduled tasks that staff members carry out to fill gaps in the process that cause operational problems when roles are removed or subsumed. Using a VE for training may make those gaps more visible. It might also allow for feedback to the organization on more efficient workplace layouts. VE may also allow staff members to train for circumstances when the technology does not work perfectly or for emergency situations.

Small Group Issues—Communication and Function Allocation Issues

An advantage of VE over traditional training environments is the interactive aspect of the experience. The VE allows not only for allocation of function between the user and the system as it will operate in the real environment, but it may also allow for adaptive allocation based on measures of performance. VE also can allow small groups to rehearse communications as they may exist in the actual environment.

Observation of performance in the training VE may also be useful in making recommendations for changes to either function allocation or small group communications. For example, observation of the training environment may make problems with communication among users and/or operators or issues with function allocation more visible.

Individual Issues—Mental Model and Device Design Issues

The goal of all training programs is to impart information or knowledge to the user. Thus, it is important to verify what knowledge the VE training environment has conveyed to the user. Carroll and Olson (1987) propose three basic representations to characterize what users know: (1) simple sequences, (2) methods, and (3) mental models. Simple sequences refer to the sequence of actions that must be taken to perform a given task. These sequences are steps that allow users to get things done. They do not require that the user understand why the steps are being performed. Methods refer to the knowledge of which techniques or steps are necessary to achieve a specific goal. This characterization of knowledge, unlike simple sequences, incorporates the notion that people have general goals and subgoals and can apply methods purposefully to achieve them. Mental models refer to a more general knowledge of the workings of a system. Specifically, mental models are defined as "a rich and elaborate structure, reflecting the user's understanding of what the system contains, how it works, and why it works that way" (Carroll & Olson, 1987, p. 6). However, mental models are assumed to be incomplete (Norman, 1983); thus, the possession of a mental model for a system trained in a VE does not necessarily mean that the user has a technically accurate or complete representation of the system's functioning. Thus, it is critical to assess user knowledge of the system after training.

Any misunderstandings that are common to a number of users can be considered a technical requirement to change the design of the system or the training interface to help more accurately convey system functioning.

ASSESS INTERACTIONS ACROSS LEVELS OF THE ORGANIZATION

Keys to Success

- Evaluate organizational policies on selection and workplace design
- Evaluate communication patterns and allocation of functions within small groups
- Evaluate individual mental models that result from device design

Outcomes

- Assessment of organizational constraints that can be used to
  - g. Modify the requirements for the VE
  - h. Modify the design of the system to reduce training requirements

---

SUMMARY

VEs promise much in the way of realistic and accurate training scenarios. The question is whether this increased realism will lead to more cost-effective training or better transfer of training. This chapter argues that these benefits can be achieved only through the application of the task analysis techniques seen in human factors to support an operator (user)-centered approach to training requirements gathering rather than the more conventional technology-oriented approach.

REFERENCES

Anderson, J. R., & Lebiere, C. (1998). *The atomic components of thought.* Mahwah, NJ: Erlbaum.

Badler, N. I. (1989, April). *Task-driven human figure animation.* Paper presented at the National Computer Graphics Association 89, Philadelphia, PA.

Barnes, M., & Beevis, D. (2003). Human system measurements and trade-offs in system design. In H. R. Booher (Ed.), *Handbook of human systems integration* (pp. 233–263). New York: John Wiley & Sons.

Benedyk, R., & Minister, S. (1998). Evaluation of product safety using the BeSafe method. In N. Stanton (Ed.), *Human factors in consumer products* (pp. 55–74). London: Taylor & Francis, Ltd.

Bisantz, A., & Roth, E. (2008). Analysis of cognitive work. In D. A. Boehm-Davis (Ed.), *Reviews of human factors and ergonomics* (Vol. 3, pp. 1–43). Santa Monica, CA: Human Factors and Ergonomics Society.

Brislin, R. W., Walter, J. L., & Thorndike, R. M. (1973). *Cross-cultural research methods.* New York: John Wiley & Sons.

Carroll, J., & Olson, J. R. (Eds.). (1987). *Mental models in human-computer interactions: Research issues about what the user of software knows.* Washington, DC: National Academy Press.

Chin, J., Diehl, V., & Norman, K. (1987, September). *Development of an instrument measuring user satisfaction of the human-computer interface.* Paper presented at the ACM CHI 88, Washington, DC.

Cockayne, W., & Darken, R. P. (2004). The application of human ability requirements to virtual environment interface design and evaluation. In D. Diaper & N. A. Stanton (Eds.), *The handbook of task analysis for human-computer interaction* (pp. 401–422). Mahwah, NJ: Lawrence Erlbaum.

Crow, K. (2002). *Failure modes and effects analysis.* Palos Verdes, CA: DRM Associates.

Diamond, J. (1986). *Down to a sunless sea: The anatomy of an incident.* Retrieved July 19, 2007, from www.ericmoody.com

Dillman, D. A. (1978). *Mail and telephone surveys: The total design method.* New York: John Wiley & Sons.

Eng, K., Lewis, R. L., Tollinger, I., Chu, A., Howes, A., & Vera, A. H. (2006, April). *Generating automated predictions of behavior strategically adapted to specific performance objectives.* Paper presented at the Human Factors in Computing Systems, Montreal, Quebec, Canada.

Ericsson, K. A., & Simon, H. A. (1980). Verbal reports as data. *Psychological Review, 87,* 215–251.

Ericsson, K. A., & Simon, H. A. (1984). *Protocol analysis.* Cambridge, MA: MIT Press.

Fleishman, E. A., & Quaintance, M. K. (1984). *Taxonomies of human performance: The description of human tasks.* Orlando, FL: Academic Press.

Gertman, D. I., & Blackman, H. S. (1994). *Human reliability & safety analysis data handbook.* New York: John Wiley & Sons, Inc.

Gray, W. D., & Boehm-Davis, D. A. (2000). Milliseconds matter: An introduction to microstrategies and to their use in describing and predicting interactive behavior. *Journal of Experimental Psychology: Applied, 6,* 322–335.

Gugerty, L. (1993). The use of analytical models in human-computer-interface design. *International Journal of Man-Machine Studies, 38,* 625–660.

Hendrick, H. (2008). Macroergonomics: The analysis and design of work systems. In D. A. Boehm-Davis (Ed.), *Reviews of human factors and ergonomics,* (Vol. 3, pp. 44–78). Santa Monica, CA: Human Factors and Ergonomics Society.

Hoffman, L. A., Tasota, F. J., Scharfenberg, C., Zullo, T. G., & Donahoe, M. P. (2003). Management of patients in the Intensive Care Unit: Comparison via work sampling analysis of an acute care nurse and physicians in training. *American Journal of Critical Care, 12*(5), 436–443.

Howard, R. A. E. (Ed.). (1990). *Influence diagrams.* New York: John Wiley & Sons Ltd.

John, B. E., & Kieras, D. E. (1996a). The GOMS family of user interface analysis techniques: Comparison and contrast. *ACM Transactions on Computer-Human Interaction, 3*(4), 320–351.

John, B. E., & Kieras, D. E. (1996b). Using GOMS for user interface design and evaluation: Which technique? *ACM Transactions on Computer-Human Interaction, 3*(4), 287–319.

John, B. E., Prevas, K., Salvucci, D., & Koedinger, K. (2004, April). *Predictive human performance modeling made easy.* Paper presented at the CHI, Vienna, Austria.

John, B. E., & Salvucci, D. D. (2005, October–December). Multipurpose prototypes for assessing user interfaces in pervasive computing systems. *IEEE Pervasive Computing, 4,* 27–34.

John, B. E., & Vera, A. H. (1992, May). *A GOMS analysis of a graphic, machine-paced, highly interactive task.* Paper presented at the Human Factors in Computing Systems, Monterey, CA.

Johnson, W. G. (1973). *MORT oversight and risk tree* (Rep. No. SAN 821-2). Washington, DC: U.S. Atomic Energy Commission.

Johnson, W. G. (1980). *MORT safety assurance system.* New York: Marcel Dekker.

Kieras, D., & Meyer, D. E. (1997). An overview of the EPIC architecture for cognition and performance with application to human-computer interaction. *Human-Computer Interaction, 12,* 391–438.

Kirwan, B., & Ainsworth, L. K. (Eds.). (1993). *A guide to task analysis.* London: Taylor & Francis, Ltd.

Kirwan, B., & James, N. J. (1989, June). *The development of a human reliability assessment system for the management of human error in complex systems.* Paper presented at Reliability 89, Brighton Metropole, England.

L3 Communications. (2007). *Link simulation & training: Setting the standard for over 75 years.* Retrieved July 16, 2007, from http://www.link.com/history.html

Miller, G. A. (1956). The magical number seven, plus or minus two: Some limits on our capacity for processing information. *Psychological Review, 63,* 81–97.

Nielsen, J., & Phillips, V. L. (1993, April). *Estimating the relative usability of two interfaces: Heuristic, formal, and empirical methods compared.* Paper presented at the ACM INTERCHI'93 Conference on Human Factors in Computing Systems, Amsterdam, The Netherlands.

Norman, D. (1983). Some observations on mental models. In D. Gentner & A. L. Stevens (Eds.), *Mental models* (pp. 7–14). Hillsdale, NJ: Lawrence Erlbaum.

Park, K. S. (1987). *Human reliability: Analysis, prediction, and prevention of human error.* Amsterdam: Elsevier.

Peterson, J. L. (1977, September). Petri nets. *ACM Computing Surveys (CSUR), 9,* 223–252.

Phillips, C., & Badler, N. I. (1988, October). *Jack: A toolkit for manipulating articulated figures.* Paper presented at the ACM/SIGGRAPH Symposium on User Interface Software, Banff, Canada.

Preece, J. (1994). *Human-computer interaction.* New York: Addison-Wesley Publishing Company.

President's Commission on the Accident at Three Mile Island. (1979). *The need for change, the legacy of TMI: Report of the President's Commission on the accident at Three Mile Island (aka "Kemeny Commission report").* Washington, DC: U.S. Government Printing Office.

Rooden, M. J., Green, W. S., & Kanis, H. (1999, September). *Difficulties in usage of a coffeemaker predicted on the basis of design models.* Paper presented at the Human Factors and Ergonomics Society 43rd Annual Meeting, Houston, TX.

Rosenbloom, P., Laird, J., & Newell, A. (Eds.). (1993). *The soar papers: Research on integrated intelligence.* Cambridge, MA: MIT Press.

Schraagen, J. M., Chipman, S. F., & Shalin, V. L. (Eds.). (2000). *Cognitive task analysis.* Mahwah, NJ: Lawrence Erlbaum.

Shepherd, A. (2001). *Hierarchical task analysis.* New York: Taylor & Francis.

U.S. Department of Defense. (1999). *Department of Defense design criteria standard (MIL-STD 1472F).* Washington, DC: Author.

U.S. Nuclear Regulatory Commission. (1981). *Guidelines for control room design reviews (NUREG-0700).* Washington, DC: Author.

UGS. (2004). *Jack.* Retrieved October 18, 2007, from http://www.ugs.com/products/efactory/jack/

Webb, E. J., Campbell, D. T., Schwartz, R. D., & Sechrest, L. (1966). *Unobtrusive measures: Nonreactive research in the social sciences.* Chicago: Rand McNally.

You, H., & Ryu, T. (2004, September). *Development of a hierarchical estimation method for anthropometric variables.* Paper presented at the Human Factors and Ergonomics Society 48th Annual Meeting, New Orleans, LA.

Chapter 8

CREATING TACTICAL EXPERTISE: GUIDANCE FOR SCENARIO DEVELOPERS AND INSTRUCTORS

Jennifer Phillips, Karol Ross, and Joseph Cohn

In a previous chapter (see Ross, Phillips, and Cohn, Volume 1, Section 1, Chapter 4), a five-stage model of learning in complex cognitive domains was presented based on the work of Dreyfus and Dreyfus (1986). Characteristics of a learner's knowledge and ability were described at each of these stages: novice, advanced beginner, competent, proficient, and expert. With regard to the differential skill sets, a notional strategy was provided for enhancing a learner's progression from one stage to the next. Cognitive Transformation Theory was introduced in Klein and Baxter (Volume 1, Section 1, Chapter 3), postulating learning as a sensemaking activity and describing a learner's progression to higher levels of proficiency as a function of replacing and improving domain-specific mental models. This chapter extends the research and theory discussed in Chapters 3 and 4 to the question of how virtual environments (VEs) can be optimally employed to hasten the movement of learners along the continuum from novice to expert in the complex cognitive domain of tactical thinking.

The premise of this chapter is that the design requirements for training scenarios and instructional strategies differentially depend on the learner's level of proficiency. Though VEs can be excellent settings for learning when properly developed, the instructional implementation of these training systems today is often far removed from what is known to be effective for intermediate and advanced learning (for example, the advanced beginner, competent, and proficient stages). A road map for successful design and employment of training in VEs does not exist. Simulation developers have spent much time and money reproducing the most faithfully realistic experiences they could, trusting that experience alone in these environments will create expertise. Now, as a result of efforts to integrate research findings and describe the process by which complex cognitive skills develop in naturalistic domains, it is possible to shed more light on learning stage-specific requirements for VE training systems (Ross, Phillips, Klein, & Cohn, 2005). This chapter provides a brief overview of a Cognitive Skill Acquisition framework and then applies the framework to the tactical thinking domain in order to provide initial actionable guidance to scenario developers and instructors who utilize VE training to improve complex tactical decision and judgment skills.

THE COGNITIVE SKILL ACQUISITION FRAMEWORK

Ross and her colleagues (2005) presented a Cognitive Skill Acquisition framework to describe the learning process in ill-structured, cognitively complex knowledge domains and subsequent training implications for each of the five stages of learning. Readers are referred to Ross, Phillips, and Cohn (Volume 1, Section 1, Chapter 4) for a description of the distinctions among the five levels of performance, which are summarized in Table 8.1. The Cognitive Skill Acquisition framework views learning as the process of moving from one stage to the next. Therefore, its training implications reflect in part the goal of getting learners to exhibit the characteristics of the next stage along the continuum.

The framework can be applied to specify training requirements in range of domains, as it is applied in this chapter to tactical thinking, so long as

- The boundaries of the target domain are clear,
- The presence of cognitively complex challenges in performance are evident (that is, it is a cognitively complex domain rather than a rule-driven, procedural domain), and
- An analysis of cognitive performance has been conducted and the nature of expertise development is understood, such as through a cognitive task analysis.

The process for applying the Cognitive Skill Acquisition framework to a particular domain requires that the general characteristics of performance for each of the five stages be customized to the domain using cognitive task analysis or similar data. It may be most useful to generate these domain-specific characteristics along a set of themes, as is illustrated in the context of the tactical thinking domain by the example in Table 8.2.

Principles for Learning Progression

Individuals do not develop general, context-independent cognitive skills such as decision making, sensemaking, and problem detection. They get better at these activities when they develop their mental models in a specific domain (Glaser & Baxter, 2000; Spiro, Feltovich, Jacobson, & Coulson, 1992). Their mental models support understanding, reasoning, prediction, and action (Genter, 2002; Ross et al., 2005). Five principles regarding the learning process for ill-structured, cognitively complex domains have been derived from an extensive review of the research literature addressing expertise and the nature of learning in complex cognitive (or ill-structured) domains (see Klein and Baxter, Volume 1, Section 1, Chapter 3; Ross, Phillips, and Cohn, Volume 1, Section 1, Chapter 4).

Principle 1. The nature of training befitting of novices is qualitatively different from training that is effective for advanced learners. Novices respond well to introductory learning that provides rigid rules, structure within the domain, and

Table 8.1. Overview of the Stage Model of Cognitive Skill Acquisition (Reprinted by Permission; Lester, 2005)

Stage	Characteristics	How Knowledge Is Treated	Recognition of Relevance	How Context Is Assessed	Decision Making
Novice	Rigid adherence to taught rules or plans Little situational perception No discretionary judgment	Without reference to context	None	Analytically	Rational
Advanced Beginner	Guidelines for action based on attributes or aspects Situational perception is still limited All attributes and aspects are treated separately and given equal importance	In context			
Competent	Sees action at least partially in terms of longer-term goals Conscious, deliberate planning Standardized and routinized procedures		Present		
Proficient	Plan guides performance as situation evolves Sees situation holistically rather than in terms of aspects Sees what is most important in a situation Perceives deviations from the normal pattern Uses maxims, whose meanings vary according to the situation, for guidance Situational factors guide performance as situation evolves			Holistically	
Expert	No longer relies on rules, guidelines, or maxims Intuitive grasp of situations based on deep tacit understanding Intuitive recognition of appropriate decision or action Analytic approaches used only in novel situations or when problems occur				Intuitive

Table 3.2. General and Domain-Specific Characteristics for the Advanced Beginner Stage

STAGE 2: ADVANCED BEGINNER			
General Characteristics		Characteristics in Tactical Thinking Domain	
Knowledge	Performance	Tactical Thinking Profile	Example
• Some domain experience (Benner, 1984; Dreyfus & Dreyfus, 1986) • More objective, context-free facts than the novice, and more sophisticated rules (Dreyfus & Dreyfus, 1986) • Situational elements, which are recurring meaningful elements of a situation based on prior experience (Dreyfus & Dreyfus, 1986) • A set of self-generated guidelines that dictate behavior in the domain (Benner, 1984) • Seeks guidance on task performance from context-rich sources (for example, experienced people and documentation of past situations) rather than rule bases (for example, textbooks) (Houldsworth, O'Brien, Butler, & Edwards, 1997)	• Is marginally acceptable (Benner, 1984) • Combines the use of objective, or context-free, facts with situational elements (Dreyfus & Dreyfus, 1986) • Ignores the differential importance of aspects of the situation; situation is a myriad of competing tasks, all with same priority (Benner, 1984; Dreyfus & Dreyfus, 1986; Shanteau, 1992) • Shows initial signs of being able to perceive meaningful patterns of information in the operational environment (Benner, 1984) • Reflects attitude that answers are to be found from an external source (Houldsworth et al., 1997) • Reflects a lack of commitment or sense of involvement (McElroy, Greiner, & de Chesnay, 1991)	• *Mission*. Understands that own mission must support intent, but is unable to operationalize intent • *Enemy*. Understands the impact of the enemy on own mission, but regards enemy as static being • *Terrain*. Recognizes important terrain features and avoids nonsubtle problem areas such as chokepoints, but remains unable to leverage terrain to own advantage • *Assets*. Understands how to apply organic asset capabilities to particular mission requirements • *Timing*. Acknowledges that timing and sequencing are important • *Big Picture*. Fails to understand how own mission and activities function as a part of the larger organization • *Contingencies*. Does not consider contingencies • *Visualization*. Is unable to visualize battlefield	• Advanced beginners will show some signs of experiential knowledge, but will still struggle. In urban combat, for example, they are likely to use only the latest intelligence (rather than situational cues) to estimate the enemy's current strength and location. They will not conceptualize the enemy as a force that could move or take action, with the exception of engaging friendlies from the current position. They will look at buildings and take note of their sizes and locations, but not recognize the implications for the mission—for example, brick buildings will not be interpreted to be better strongholds than buildings constructed with Sheetrock. They will match subordinate units to particular tasks, but are unlikely to mix assets across units (for example, attach a rifle team to an engineer unit for security).

modularized facts and knowledge. Advanced beginner, competent, and proficient individuals are more likely to improve their performances as a result of experiential training where they can practice decisions, assessments, and actions and then reflect on their experiences. The following mistakes are often made in the design of training events for advanced learners:

- Simplifying the interrelationships among topics and principles in order to make them easier to understand and employing a single analogy, prototype example, organizational scheme, perspective, or line of argument (Spiro et al., 1992). Training for advanced learners should not seek to simplify concepts that are complex. Simulations should introduce several cases of a single complex principle in order to demonstrate its applicability across a range of circumstances.
- Overemphasizing memory by overloading the learner with the need to retrieve previous knowledge. Training should require learners to apply principles rather than carry the heavy baggage of detailed rules and content (Spiro et al., 1992).

Principle 2. People can improve their mental models by continually elaborating them or by replacing them with better ones. However, at each juncture the existing mental models direct what learners attend to and how they interpret environmental cues. This makes it difficult for learners to diagnose what is lacking in their beliefs and take advantage of feedback. Knowledge shields (Feltovich, Johnson, Moller, & Swanson, 1984) work against learners' attempts to get better and smarter by permitting them to explain away inconvenient or contradictory data. Training scenarios can support the elaboration of mental models by prohibiting common ways of achieving an outcome so that learners must find viable alternatives. Further, scenarios can break learners out of their knowledge shields by purposefully presenting inconvenient data that turn out to be central to an accurate assessment of the situation.

Principle 3. The most dramatic performance improvements occur when learners abandon previous beliefs and move on to new ones. Some call this "unlearning." Old mental models need to be disconfirmed and abandoned in order to adopt new and better ones, and this path to expert mental models is discontinuous. Development does not occur as a smooth progression. It often requires some backtracking to shed mistaken notions (see Klein and Baxter, Volume 1, Section 1, Chapter 3). Training scenarios designed with unmistakable anomalies or baffling events can serve to break learners out of their current perceptions.

Principle 4. Learners who can assess their own performance will improve their mental models more quickly than their peers. The knowledge needed to self-assess is built into domain mental models, so by refining this skill learners are also enhancing their understanding of the dynamics of the domain itself. Also, self-assessment is more efficient than assessments provided by a second party; they occur continually, with immediacy, and within each and every experience. Skilled mentors can help learners develop self-assessment skills by helping them diagnose their weaknesses and discover where their mental models are too simplistic.

Principle 5. Experiences are a necessary, but not sufficient, component for the creation of expertise. Training experiences can be a waste of time or even be harmful when they do not allow adequate opportunity for domain-appropriate mental model building, target the right kind of challenges, support performance, and provide insights. It has been noted that "a key feature of ill-structured domains is that they embody knowledge that will have to be used in many different ways, ways that cannot all be anticipated in advance" (Spiro et al., 1992, p. 66). To develop mental models for such complex performance, the learner must be immersed in multiple iterations of experiences from different vantage points to make numerous connections.

THE COGNITIVE SKILL ACQUISITION FRAMEWORK APPLIED TO TACTICAL THINKING

Eight themes of tactical thinking performance delineated by Lussier and his colleagues as a result of a cognitive task analysis (Lussier, Shadrick, & Prevou, 2003) provide the structure for the Cognitive Skill Acquisition framework for tactical thinking:

- *Focus on the Mission and Higher Headquarters' Intent*
- *Model a Thinking Enemy*
- *Consider Effects of Terrain*
- *Know and Use All Assets Available*
- *Consider Timing*
- *See the Big Picture*
- *Consider Contingencies and Remain Flexible*
- *Visualize the Battlefield*

Developmental sequences vary from domain to domain, and training interventions must match these naturally occurring sequences. In the case of tactical thinking, the first four themes—*Mission, Enemy, Terrain,* and *Assets*—are hypothesized to represent mental models and develop before the last four themes—*Timing, Big Picture, Contingencies,* and *Visualization,* which are higher order cognitive processes or mental manipulations of the first four mental models (Ross, Battaglia, Hutton, & Crandall, 2003; Ross, Battaglia, Phillips, Domeshek, & Lussier, 2003). The training implications described below consider this hypothesized developmental sequence in conjunction with the Cognitive Skill Acquisition framework.

Implications for Tactical Thinking Training in Virtual Environments

At the novice stage of development of tactical thinking skills the training value of VEs is quite different than at the advanced stages. For novices, VE training should provide learners with support in operationalizing the facts, rules, and processes they learn through other forms of training. In other words, VE training must enable them to practically apply the knowledge they are gaining in order to establish their own experience based mental models about "how things work" on the battlefield and when it is appropriate to use specific procedures and tactics. Advanced stage performers, however, have already developed these basic mental models about friendly assets, mission tasks, terrain features, and the enemy. At these stages, the role of VE training is to facilitate development of a rich base of varied experiences resulting in highly elaborated mental models. Decision making, sensemaking, and other naturalistic cognitive activities can be practiced in complex environments with varied goals, situational constraints, and mission types to produce tactical thinkers who can respond flexibly and effectively in most any situation.

This section is organized by stage of learning. Within each stage, indicators of that proficiency level are described. The indicators should be used by instructors or scenario developers to anticipate how the learner will think through tactical problems and also as means of comparing characteristics from one stage to the next, with the goal of producing performance associated with the next stage up. In addition, specific scenario design and instructional requirements are presented.

Novices

Indicators of Proficiency Level

A novice is likely to show the following behaviors:

- When asked about his mission, regurgitates the mission order, but fails to reference the commander's intent,
- When asked about assets at his disposal, provides a textbook or standardized characterization of their capabilities,
- When asked about the enemy, does not know typical tactics or capabilities of the particular enemy in question. The novice may provide a theoretical set of capabilities based on a class of enemy (for example, "A Soviet commander would . . . " or "An insurgent would . . ." or "A Middle Eastern adversary would . . .").
- When asked about terrain, goes through a classroom-taught checklist, such as observation, cover and concealment, obstacles, key terrain, and avenues of approach.

Scenario Design Components

Existing military training is very strong for individuals at the novice level. Novices require standard rules to anchor their thinking and knowledge about how to execute procedures. However, introduction of VE simulations into the novices' training program can assist them in developing an understanding of when and how the rules and procedures apply operationally. Simulations for novices should utilize a ground based (rather than a bird's eye) perspective to immediately familiarize them with cue sets such as they will find in the real world. Further, scenarios should do the following:

- Focus on utilization of assets and requirements for mission accomplishment. The content of training scenarios should enable novices to practice executing procedures and tactics (for example, establishing a blocking position, executing 5 meter and 25 meter searches) *in context*. Learners should practice on a range of scenarios that illustrate how tactics must be implemented somewhat differently depending on the particulars of the situation. Simulations should also require learners to allocate assets to various mission tasks and receive embedded feedback about the effective range of weapons in context and the time it takes to traverse between points given situational factors such as road conditions and weather. They should illustrate that a unit is not a single fused entity as it appears on a tactical map, but rather consists of moving pieces and parts (such as people and vehicles). This enables learners to begin forming mental models of assets that can be split up or attached to other units and to begin conceptualizing groupings that occupy more than a single static grid coordinate on a map.

- Incorporate simple aspects of a dynamic enemy. Training scenarios should exhibit that the enemy is not static. For example, design scenarios in which the enemy moves or splits up his forces. For novices, this is sufficient introduction of the enemy.
- Incorporate simple but meaningful terrain features. Simple terrain may include features found in rural settings, such as hills or berms and paved or dirt roads. In urban settings, simple terrain might include one- and two-story buildings and intersections. These features are in contrast to more complex terrain, which may include wooded areas in rural settings that could be sparse or dense and therefore have different mobility affordances. In urban settings, highly complex terrain would include underground sewer systems or densely populated areas with several roads and buildings. For novices, scenarios should include such terrain features as hills or other elevated areas that impact line of sight. They should introduce features that will make clear the difference between cover and concealment, such as buildings (cover and concealment, depending on the construction) or automobiles (concealment, but not cover). They should present dirt versus paved roads that differentially affect rates of movement.

Instructional Strategies

Novices will be best served by practicing on a range of scenarios with different assets available and different mission requirements. The goal is to support basic mental model development across a wide range of asset and mission types in order to produce an understanding of asset capabilities and mission tasks in context.

At the novice level, instructors or coaches are necessary to guide and direct the learning process more so than at the later stages. Following VE training sessions, an instructor-led after action review should focus on the lines of questioning below and probing regarding the learners' experiences. For every topic addressed in the after action review, the instructor should ask *why* learners made particular decisions or situation assessments. Illuminate the thought process of the learner in each case.

- Asset capabilities. What was learned about how to use the assets' capabilities in the context of the situation?
- Mission. What actions were taken and why? Was the mission accomplished? Why or why not?

- Enemy. What was learned about the enemy? What was surprising about the enemy? How might learners think about the enemy differently next time?
- Terrain. How did terrain features impact the mission?

Advanced Beginners

Indicators of Proficiency Level

An advanced beginner is likely to show the following behaviors:

- When asked about the mission, describes the mission and the commander's intent, but is unable to operationalize that intent within the context of the mission and the battlefield environment,
- Is unable to differentiate mission priorities,
- Makes straight matches of assets to mission tasks,
- Articulates the enemy's capabilities, but does not consider situational factors that impact the enemy's probable goals or capabilities,
- Identifies basic terrain features that will impact the mission, and
- May experience generalized anxiety about performing well without making mistakes, because he or she has no sense of what part of the mission is most important to perform well (Benner, 2004).

Scenario Design Components

Advanced beginners are ready to make meaning out of the experiences they glean from simulations. At this level, VEs can supplement existing training by enabling learners to practice implementing tactics that have been newly introduced and employing assets whose capabilities they are learning. In addition, VE simulations for advanced beginners should incorporate enemy and terrain models that are more complex than those in the novice scenarios. Specifically,

- Scenarios should reflect an intelligent, dynamic adversary. The enemy should *not* follow the templates that have been taught in classroom instruction or case study analysis. Enemy forces should move and take action while the learner deliberates about a course of action. The goal is to break the learner out of the mindset that the enemy will be predictable and static.
- Scenarios should incorporate terrain that has significant impact on the workability of potential courses of action. For example, movement along a straight, flat road should result in being spotted and engaged by the enemy. Furthermore, enemy courses of action should leverage terrain features (for example, pin friendly forces in a choke point) to illustrate the role of terrain on the battlefield.
- Asset capabilities should continue to be exercised. Scenarios should incorporate units that are not full strength or units that have had assets attached or detached. At platoon echelon and below, friendly assets must be depicted as individual moving pieces (soldiers/marines and vehicles) rather than as unit icons that move as a whole. Further, some scenarios should reward learners for thinking ahead about what other assets might be needed or keeping a reserve to deal with future events. Other scenarios should reward the decisive employment of the learner's full force. Learners need to

develop an understanding of the trade-offs of keeping a reserve element or not and begin to project ahead to assess what might happen in the future that will require preparation and readiness. Finally, scenarios can illustrate how assets can be used to acquire information to reduce levels of uncertainty. Learners should receive useful information (for example, about the enemy's activities or other important battlefield features) from assets that are positioned to see a wider view than the learners themselves; in this way, advanced beginners can develop mental models about how to proactively acquire information.

- The mission required by the scenario should be relatively simple and straightforward and should correspond to tactics and missions that have been taught in classroom or analogous instructional settings. However, some advanced beginner scenarios should incorporate mission tasks that must be prioritized such that learners fail if they do not address the higher priority task first.

Instructional Strategies

Advanced beginners would benefit from practicing with the same scenario several times, with performance feedback following each trial and detailed explanations of how performance has improved or denigrated from one trial to the next. Multiple iterations allow learners at this level to understand how different uses of assets and various courses of action impact the outcome. Instructors should encourage experimentation, even with courses of action that are judged to be non-optimal. It is important for learners to internalize the specific reasons that some courses of action produce better results than others. In addition, learners may find unexpected positive outcomes from a particular course of action. If possible, instructors should be able to introduce small alterations in the environmental conditions to illustrate how variations in situational factors influence the workability and "goodness" of available courses of action.

Like novices, advanced beginners still require an instructor to guide and direct their learning process. After action reviews should be instructor led and can address the following lines of questioning:

- Utilization of assets. What worked, what did not work, and what factors need to be considered when deciding how to employ assets (for example, morale? readiness?)?
- Mission tasks. How were the tasks approached, and why? Which approaches were beneficial, and which were not? Why and why not?
- Enemy. What did the enemy do, and why? How did learners know what he was doing? What information led to their assessments? Were their assessments accurate, and why or why not?
- Terrain. What features were noticed during planning? How did terrain impact the mission during execution, and why? How would the learners approach the terrain layout differently next time?

Competent Individuals

Indicators of Proficiency Level

Heightened planning is the hallmark of the competent stage. A competent performer is likely to show the following behaviors:

- Is able to predict immediate futures and therefore takes a planful approach (Benner, 2004),
- Experiences confusion when the plan does not progress as predicted (Benner, 2004),
- Experiences anxiety that is specific to the situation as opposed to generalized anxiety (for example, am I doing this right with regard to this part of the situation?) (Benner, 2004),
- Differentiates mission priorities,
- Deliberately analyzes what has to occur in order for intent to be achieved,
- Considers trade-offs of using assets for various purposes and for keeping a reserve,
- Projects forward about what other assets might be needed as the mission progresses,
- Generates ideas about what the enemy might be thinking and what the objective might be, but does not have a specific assessment that drives decisions,
- Considers the enemy's capabilities in the context of the terrain and other situational factors, and
- Incorporates terrain features into the plan and considers the effects of the terrain on assets employed or needed.

Scenario Design Components

Scenarios for competent performers should enable continued development of *Asset, Mission, Enemy,* and *Terrain* mental models, but in the context of the *Consider Timing* and *Consider Contingencies* cognitive processes. That is, scenarios should present situations where success relies on the timing and sequencing of the operation, planning for contingencies and adapting contingency plans as the mission progresses.

- Scenarios should introduce surprises during the execution of missions to provide practice in rapidly responding to changing situations. For example, friendly units could become unable to perform (for example, because they cannot reach their intended position or because a weapon system breaks down), the enemy could move in a nontraditional way or bring a larger force than was expected, key roads could be too muddy to traverse or blocked by locals, or higher headquarters (HQ) could deliver a new fragmentary order based on an opportunistic target.
- Scenarios should present conflicts that require prioritization of mission tasks. Learners need to be forced to determine which part of the mission order is most important to higher headquarters based on the commander's intent. Success should be contingent on taking actions that support intent.
- Mission orders should incorporate strict time requirements or the need to synchronize assets or effects, and the scenarios should build in realistic timing of force movement and engagement with the enemy. When success relies on appropriate timing of actions, learners will be forced to make judgments about how long the prerequisite tasks or movements will take. These cases will enable learners to strengthen their mental models about the timing of certain tasks and set up opportunities to learn how to adjust when events do not happen in the planned sequence.
- Scenarios should require proper sequencing of tasks in order for the learner to accomplish the mission. That is, learners should be able to see how the mission breaks down

when certain tasks, such as thorough route reconnaissance, are not accomplished prior to other tasks, such as moving forces along a route.

- Scenarios should introduce the utility of nonorganic and nonmilitary assets. Learners can be encouraged to request assets from higher headquarters or another unit by realizing that the mission can be accomplished only by accessing those assets. Also, scenarios can present civilian resources such as host nation police, village elders, or relief workers who can provide information or serve important roles (such as communicating with the local populace).

Instructional Strategies

At the competent level, instructors play a key role in mental model development, but their participation at the competent, proficient, and expert levels is not required as persistently as it is for novices and advanced beginners. In lieu of an instructor, feedback can be delivered by developing expert responses against which learners can compare their own performances. The unit leader could function as facilitator and elicit peer feedback from participants. Alternatively, feedback can be generated within the VE system by illuminating situational cues, factors, or demands that should have prompted learners to change their approaches or move to a contingency plan. Regardless of the instructional medium, the following issues should be addressed with individuals at the competent level:

- Prior to execution, contingencies. What are the different ways the plan could play out, and how would the learner know if that were happening?
- Prior to execution, the enemy. What might the enemy be attempting to do, and why? How might the learner assess the enemy's objectives as the situation plays out? What information should the learner be seeking?
- Prior to execution, terrain. What are the critical terrain features on the battlefield? How might they impact both friendly and enemy courses of action? How might terrain be leveraged and used against the enemy? How might the enemy leverage terrain features and use them against friendlies?
- Mission plan. Why did the plan break down? What should have been the early indicators that the plan would not play out as intended?
- Situation. What were the cues and factors available? How might they have been interpreted?
- Timing and sequencing. What issues regarding timing and sequencing needed to be considered, and why?
- The Big Picture. What was higher HQ trying to accomplish? What was the learner's role in accomplishing the larger mission? Did the learner contribute in useful ways to the larger mission?

Proficient Individuals

Indicators of Proficiency Level

The proficient stage is marked by a qualitative leap from being guided by the formal plan to being guided by the evolving situation. Proficient individuals intuitively recognize changes to the situation, but still deliberate about how to handle the new circumstances (that is, determine what course of action will meet the objectives). Proficient performers are likely to show the following behaviors:

- Describe that they changed their perspective or situation assessment during the course of a situation or notice that the situation is not actually what they anticipated it to be (Benner, 2004),
- Recognize changes in the situation (for example, due to new information) that will impact or interfere with achieving intent,
- Deliberately analyze courses of action to determine the best one for the situation,
- Recognize the utility and importance of nonmilitary assets, such as civilian officials or village elders,
- Consider their organic assets as parts of a larger team of friendly assets working to achieve a common goal,
- Articulate timing and sequencing issues,
- Assess the enemy's objectives and intent based on situational factors, and
- Describe key aspects of terrain for both friendly and enemy courses of action.

Scenario Design Components

Scenarios for proficient individuals should incorporate high levels of complexity, ambiguity and uncertainty, sophisticated coordination requirements, and situations that evolve and change rapidly into tough dilemmas. More specifically,

- Scenarios should present situations where accomplishing the commander's intent requires a different approach than accomplishing the explicit mission tasks.
- Scenarios should incorporate an enemy who uses nonconventional forces and techniques. For example, the enemy could use civilian vehicles, dress deceptively, or otherwise mislead.
- Scenarios should incorporate substantial situational changes during execution to force the learner to revise the existing course of action or develop a new one on the fly. Proficient performers should be skilled at recognizing how the situation has changed, but they require multiple repetitions in order to develop and refine the action scripts within their mental models.
- Scenarios should incorporate feedback on secondary and tertiary consequences of action. For example, in a counterinsurgency mission, an emotion-driven decision to provide assistance to desperate locals rather than to continue with the original mission may have consequences for mission accomplishment and domino into a larger impact on the operation. Depending on the situation, an action like this could prompt locals to set unwarranted expectations about how relief is provided, bog down relief efforts for a greater need elsewhere, or have political ramifications.
- Scenarios should require timing, sequencing, and coordination between and across units rather than only within the learner's own organic assets. This enables learners to form mental models of friendly forces as a larger team effort and to understand the capabilities and limitations of other dissimilar units (for example, air or artillery).

Instructional Strategies

The facilitation, in whatever form it takes, should exhaust the learner's way of understanding and approaching the situation. Learners should be required to cite their own personal experiences for perspective on their views of the situation depicted in the scenario (Benner, 2004). Benner recommends that instructors teach inductively, where the learners see the situation and then supply their own way of understanding the situation.

When an instructor is available, semistructured time-outs during execution of the scenarios are beneficial. These periods of inquiry and reflection encourage learners to discuss their current interpretations of the situation, their mental simulations of how the situation may play out, and their ideas about what courses of action can produce the desired results. Discussion among the learners is nearly as valuable for proficient performers as the probes and dialogue with the instructor. Likewise, after action reviews should encourage dialogue and questioning between the instructor and learners about their interpretations of the situation, their mental simulations and visualizations of the battlefield, and especially their consideration of how various courses of action supported or failed to support the mission goals.

When an instructor is unavailable, alternate approaches can provide adequate substitutions. First, individuals at this stage of development can learn quite a bit from their peers. Semistructured after action reviews can be provided to groups of learners to guide their discussion of the exercise. The reviews should focus on the same questions used when an instructor is present—how the situation was assessed, how learners projected into the future, and the rationale for the courses of action employed or adjustments made. In addition, learners should be encouraged to share past experiences that have influenced their thinking about the scenario.

Just as competent learners are likely to benefit from expert responses, proficient learners can also use information generated from experts as another instructor-free approach. For proficient performers, expert responses should include very detailed information about how experts thought about the scenario at multiple intervals within the scenario. This information can be generated by conducting in-depth, cognitive task analysis-like interviews with a few expert tacticians. Learners should be provided with experts' interpretations of the situation, including the cues and factors they recognized pertaining to the enemy objective, the friendly status, and other aspects of the battlefield (for example, terrain or noncombatants). They should see the experts' projections (that is, mental simulations) about how the situation would play out and the rationale for those projections as well as visualizations of first-, second-, and third-order consequences. There should be a discussion about the courses of action taken by the experts along with a detailed rationale regarding asset allocation, prioritization and primary goal(s), and aspects of timing and/or sequencing.

Other topics to review following VE training sessions, with or without instructor leadership, include the following:

- The larger picture. What is the larger organization trying to accomplish? How can the learner develop opportunities for the larger organization or otherwise feed the overall objective over and above his or her own mission tasks?
- Enemy intent. What is his likely intent? What aspects of the situation could have revealed clues about his intent? How can intent be denied by friendly forces?
- Contingencies. In what other ways could the situation have played out? What situational cues would suggest those particular outcomes? What responses (that is, courses of action) would be appropriate for the various contingencies?
- Actions. What courses of action could be taken in response to changes in the dynamics of the situation? What are the relative advantages and disadvantages of each?

Experts

Indicators of Proficiency Level

When an individual moves from proficient to expert status, the main change is his or her ability to intuitively recognize a good course of action. Experts typically show the following behaviors:

- Exhibit good metacognition (Kraiger, Ford, & Salas, 1993), meaning they can accurately gauge their own abilities to deal with the situation at hand;
- Intuitively generate a plan and take actions that will achieve the commander's intent;
- Eliminate obstacles to higher headquarters' intent or present opportunities to other units to support achievement of intent;
- Fluidly leverage and coordinate organic, nonorganic, and nonmilitary assets;
- Understand how to deny the enemy intent;
- Use the terrain to create advantages for friendlies and disadvantages for the enemy.

Scenario Design Components

Experts involved in VE training sessions may reap the greatest benefit as a mentor to less experienced tacticians. By coaching and being forced to communicate what they know to others, they reflect on and thus strengthen their existing mental models. It may also be possible to develop garden path scenarios (Feltovich et al., 1984) in VE scenarios to challenge the fine discriminations within experts' mental models.

Instructional Strategies

Experts benefit from peer discussions reflecting on shared real world experiences, full-scale exercises or simulations, or operational planning sessions (for example, plan development to address a potential crisis situation in a real world "hot spot"). Discussions could be structured to address the following:

- Enemy intent. What was (is) the enemy's objective, and why? What situational cues and factors led to that assessment? At what point in the mission did the enemy's intent and course of action become clear? What were the key indicators?

- Big picture. How did (could) individual units, or joint/coalition forces, work together to meet the overarching mission? Were assets shared in ways that supported mission accomplishment? What other configurations of assets could have addressed the larger mission intent rather than unit-specific orders?
- Contingencies. Did the mission play out in unexpected ways that were not imagined in contingency planning sessions? When was the change noticed? Were there early indicators that could have revealed the new direction to commanders sooner?
- Visualization. What were (are) the friendly and enemy leverage points on the battlefield? How did (could) friendly forces deny enemy intent by using the terrain, nonconventional assets (for example, civilians), and other resources or strategies?

SUMMARY

This chapter, Klein and Baxter's chapter (Volume 1, Section 1, Chapter 3), and Ross, Phillips, and Cohn's chapter (Volume 1, Section 1, Chapter 4) set out to integrate a broad range of research and applications to form a coherent framework for improving VE training for tactical decision making. The framework is grounded in empirical research, but it is by no means complete. Several questions remain, especially with regard to application of the principles to specific training development efforts. These chapters provide a starting point from which the training community can research and evaluate the assertions and refine and evolve the framework to make it more useful as a guide for VE training that will effectively prepare tactical decision makers for current and future challenges.

REFERENCES

Benner, P. (1984). *From novice to expert: Excellence and power in clinical nursing practice.* Menlo Park, CA: Addison-Wesley Publishing Company Nursing Division.

Benner, P. (2004). Using the Dreyfus model of skill acquisition to describe and interpret skill acquisition and clinical judgment in nursing practice and education. *Bulletin of Science, Technology & Society, 24*(3), 189–199.

Dreyfus, H. L., & Dreyfus, S. E. (1986). *Mind over machine: The power of human intuitive expertise in the era of the computer.* New York: The Free Press.

Feltovich, P. J., Johnson, P. E., Moller, J. H., & Swanson, D. B. (1984). LCS: The role and development of medical knowledge in diagnostic expertise. In W. J. Clancey & E. H. Shortliffe (Eds.), *Readings in medical artificial intelligence: The first decade* (pp. 275–319). Reading, MA: Addison-Wesley.

Gentner, D. (2002). Mental models, psychology of. In N. J. Smelser & P. B. Bates (Eds.), *International Encyclopedia of the Social and Behavioral Sciences* (pp. 9683–9687). Amsterdam, The Netherlands: Elsevier Science.

Glaser, R., & Baxter, G. P. (2000). *Assessing active knowledge* (Tech. Rep. for Center for the Study of Evaluation). Los Angeles, CA: University of California.

Houldsworth, B., O'Brien, J., Butler, J., & Edwards, J. (1997). Learning in the restructured workplace: A case study. *Education and Training, 39*(6), 211–218.

Kraiger, K., Ford, K., & Salas, E. (1993). Application of cognitive, skill-based, and affective theories of learning outcomes to new methods of training evaluation. *Journal of Applied Psychology, 78,* 311–328.

Lester, S. (2005). Novice to expert: The Dreyfus model of skill acquisition. Retrieved May 1, 2008, from http://www.sld.demon.co.uk/dreyfus.pdf

Lussier, J. W., Shadrick, S. B., & Prevou, M. I. (2003). *Think like a Commander prototype: Instructor's guide to adaptive thinking* (ARI Research Product No. 2003-01). Alexandria, VA: U.S. Army Research Institute for the Behavioral and Social Sciences.

McElroy, E., Greiner, D., & de Chesnay, M. (1991). Application of the skill acquisition model to the teaching of psychotherapy. *Archives of Psychiatric Nursing, 5*(2), 113–117.

Ross, K. G., Battaglia, D. A., Hutton, R. J. B., & Crandall, B. (2003). *Development of an instructional model for tutoring tactical thinking* (Final Tech. Rep. for Subcontract No. SHAI-COMM-01; Prime Contract DASW01-01-C-0039 submitted to SHAI, San Mateo, CA). Fairborn, OH: Klein Associates.

Ross, K. G., Battaglia, D. A., Phillips, J. K., Domeshek, E. A., & Lussier, J. W. (2003). Mental models underlying tactical thinking skills. *Proceedings of the Interservice/ Industry Training, Simulation, and Education Conference* [CD-ROM]. Arlington, VA: National Training Systems Association.

Ross, K. G., Phillips, J. K., Klein, G., & Cohn, J. (2005). *Creating expertise: A framework to guide technology-based training* (Final Tech. Rep. for Contract No. M67854-04-C-8035 submitted to MARCORSYSCOM/PMTRASYS). Fairborn, OH: Klein Associates.

Shanteau, J. (1992). Competence in experts: The role of task characteristics. *Organizational Behavior and Human Decision Processes, 53,* 252–266.

Spiro, R. J., Feltovich, P. J., Jacobson, M. J., & Coulson, R. L. (1992). Cognitive flexibility, constructivism, and hypertext: Random access instruction for advanced knowledge acquisition in ill-structured domains. In T. Duffy & D. Jonassen (Eds.), *Constructivism and the technology of instruction: A conversation* (pp. 57–75). Mahwah, NJ: Lawrence Erlbaum.

Part VI: Requirements Analysis

Chapter 9

TRAINING SYSTEMS REQUIREMENTS ANALYSIS

Laura Milham, Meredith Bell Carroll, Kay Stanney, and William Becker

Recent and upcoming advances in virtual environment (VE) technology provide the infrastructure upon which to build state-of-the-art, interactive training systems (Schmorrow, Solhan, Templeman, Worcester, & Patrey, 2003; Stanney & Zyda, 2002). An important advantage afforded by the use of VE systems is the ability to instruct and practice targeted training objectives otherwise restricted by the resource costs, potential dangers, and/or limited availability of live training. VEs afford instructors the ability to increase training efficiency and effectiveness by capitalizing on several advances in simulator technology (for example, appropriate levels of cue fidelity, immersion, portability, practice iteration, and so forth). However, not all VE training systems realize this potential. Developing operationally, theoretically, and empirically driven requirements for VE training systems is a key component to ensure training effectiveness; this chapter discusses two methods for achieving such requirements specifications: operational requirements analysis and human performance requirements analysis.

Operational requirements analysis (ORA) focuses on identifying training goals at contextually appropriate task and expertise levels. Through this process, systems developers can take into account the training goals and gaps to be addressed by the system and allow critical contextual considerations to drive system requirements. Systems developed without regard to the operational context can result in suboptimal training solutions, which are ineffective, expensive, and limited in future utility. The ORA process is similar to long-standing system engineering practices that follow a logical sequence of activities to transform an operational need into a specification of the preferred system configuration and associated performance requirements (Goode & Machol, 1957), the difference being context. In ORA, system requirements evolve through a deep understanding of the mission context associated with the operational environment. The outcome from ORA is then fed forward to human performance requirements analysis (HPRA), which translates ORA data into metrics that can be used to assess basic skilled performance and higher order skill sets, such as the development of situation awareness and the conduct of decision making. HPRA seeks to ensure the training system is meeting its overall intended training goal of providing the skill sets needed to support mission outcomes.

TRAINING SYSTEM REQUIREMENTS ANALYSIS

The overall goal of training systems requirements analysis is to design a system that trains to standards. The process illustrated in Figure 9.1 commences at identification of these standards and cycles through the four stages necessary to ensure that a training system affords attainment of performance that meets these standards. These stages include the following: (1) identification of training goals, (2) development of design specifications that facilitate training goals being met, (3) development of metrics to evaluate if training goals are being met, and (4) development of training management methods that promote attainment of training goals (see Figure 9.2).

ORA is concerned with the domain based side of the training system requirements analysis process, which compels consideration of the target operational environment and mission context. HPRA aims to integrate what is known about human performance and learning to measure learning of targeted training objectives. These two distinct, yet interdependent drivers work in cooperation throughout the process to ensure that effective requirements are specified. As such, it is

Figure 9.1. Training System Design Lifecycle (Adapted from Milham, Cuevas, Stanney, Clark, & Compton, 2004)

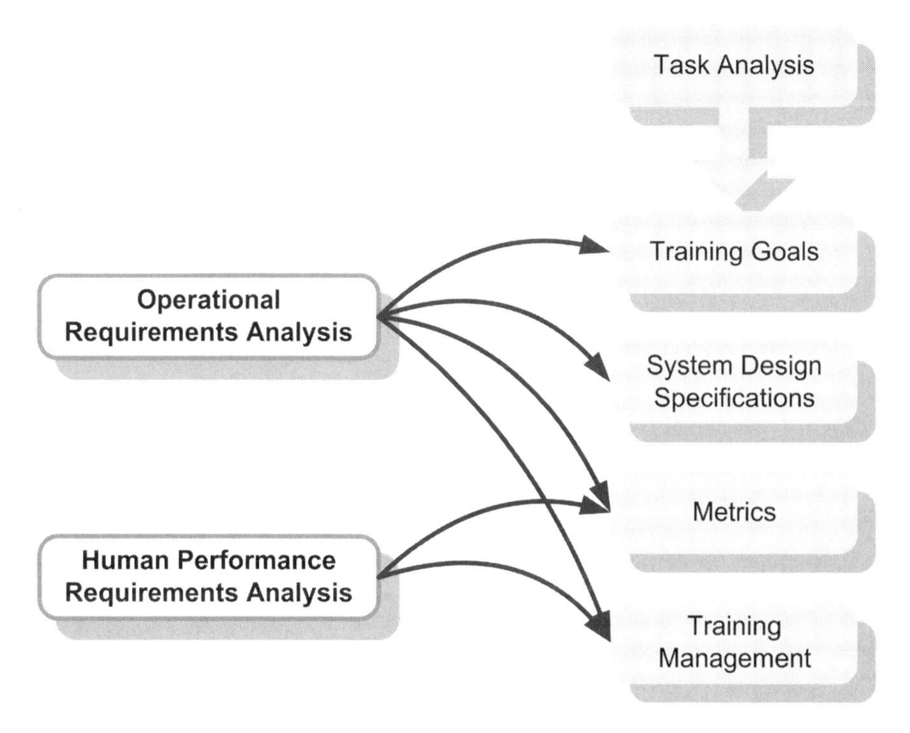

Figure 9.2. Training System Design: Operational and Human Performance Requirements

necessary to provide an integrated overview of how these two components comprise the training system design lifecycle. The following section provides an illustration and steps through the four stages of the design lifecycle, summarizing the interaction between the two drivers (that is, ORA and HPRA) and the impact each has on the overall design lifecycle.

Training System Design Lifecycle

The first stage in the training system design lifecycle is to understand training goals by analyzing the operational context (see Figure 9.1). In the second stage, the training goals are defined in terms of how the task is performed in the operational environment, and further into multimodal sensory information requirements, from which system design specifications can be derived. The third stage takes these data and feeds forward to the HPRA process, which decomposes the training goals from a human performance standpoint into the knowledge, skills, and attitudes (KSAs) required for successful performance of the task. From this metrics are defined that support performance measurement, which is used to assess if the system is meeting the targeted training goals. From an operational standpoint, metrics target the degree that mission performance was successful. From a human performance standpoint, the focus is on whether or not learning is occurring, with respect to the required KSAs. The final stage in the training system design lifecycle is training management. In order to ensure training goals are met, not only do trainees need to be able to practice targeted tasks and training objectives effectively (for example, in scenarios), but there needs to be an element of feedback to facilitate learning. As part of training management, scenarios and scenario manipulation variables can be developed to facilitate performance improvement on targeted tasks. Additionally, performance diagnosis based on metrics can facilitate performance summaries to support instructors in after action review and feedback to facilitate trainee performance improvements. Overviews of the two components of the training system design lifecycle, ORA and HPRA, will now be provided, along with a discussion of why each component is a necessary part of the design lifecycle.

OPERATIONAL REQUIREMENTS ANALYSIS

Operational requirements analysis involves identifying operationally driven training requirements based on the target task or mission. Operational requirements are defined by the skill set or task set that an operator is required to perform in a system and used to ensure a system operates in a manner consistent with user needs and expectations (Fairley & Thayer, 1997). Operational requirements are the primary building blocks of system requirements and thus ORA serves as the first and foremost step in ensuring that the mission's targeted spectrum of tasks is supported in the training system.

Need for ORA

Ensuring training objectives are effectively targeted through proper consideration of operational context is critical as demonstrated by the success of methods, such as event based approach to training (Fowlkes, Dwyer, Oser, & Salas, 1998), which ensure events are embedded in training to provide practice opportunities on targeted training objectives. If developers lack comprehension of mission requirements, including tasks and task requirements, there is potential for training system requirement specifications to lack vital information an operator relies on to complete a task or functionality that is critical to task performance. Further, given that training occurs in a proxy environment (not the operational environment), ORA needs to consider how that environment needs to be constructed to facilitate effective translation of inputs and output in the training environment to the real world. In other words, ORA must specify the sensory cues needed and the appropriate level of fidelity to facilitate transfer. From a practical standpoint, the generally limited resources applied to training necessitate a careful examination of the costly visual, auditory, and haptic human system interfaces that are required. When fidelity levels are too low, then trainees may not receive the multimodal information necessary to build situation awareness, make decisions, or even react to an emerging situation in a way that is similar to the operational environment. For example, a ship engineer may be training on how to react to various emergency situations. If the training control panel does not match that of his own craft, he may learn (or overlearn) how to perform the emergency procedures almost automatically—that is—without thinking about the manual steps he is going through. In that situation, if a button or knob is in a different location, or has an alternate function, he may unconsciously perform the highly trained skill incorrectly without realizing it.

In the case of providing too high of a fidelity level, it may be that expensive state-of-the-art technologies with a high wow factor are included, without real consideration of their impact on training effectiveness. In addition, it may be that training resources are spent on visuals, for example, without considering the impact of auditory cues on performance in a select domain. In these cases, the costs can be excessive compared to the actual training value gained by implementing the highest fidelity system (for example, is an expensive, fully immersive training system the most cost-effective way of training targeted skills? Is there a more cost-effective alternative to train the same skills?).

ORA seeks to ensure that fidelity requirements are based on target training goals and tasks. Otherwise, training systems may lead to ineffectively trained skills, untrained skills, negative training (that is, trained skills and procedures that will negatively affect, even impair, task performance in the real world), or unnecessary levels of fidelity in which the cost of the technology is disproportionate to the training value added.

HUMAN PERFORMANCE REQUIREMENTS ANALYSIS

Human performance requirements are those driven by human information processing and human performance, as well as human learning (knowledge and skill acquisition) needs. In terms of KSAs, knowledge refers to long-term memory stores, including declarative/semantic knowledge (facts), episodic knowledge (past events), and procedural knowledge (how to). Knowledge can be a foundation for skill performance (for example, procedural knowledge, situation awareness, and so forth) or more global/abstract, such as declarative knowledge. Skill refers to a level of proficiency on a specific task or limited group of tasks (for example, perceptual skills), which is "acquired through extended practice and training" (Ericsson & Oliver, 1995, p. 3). By decomposing tasks from the task analyses into the required knowledge and skills necessary to facilitate task performance, measures of learning can be developed to ensure effective training of the task. Additionally, by incorporating consideration of how humans learn different knowledge and skill types (for example, declarative knowledge, psychomotor skills, perceptual skills, and so forth), training management strategies can be incorporated to accelerate learning.

Need for HPRA

Consideration of human learning of key knowledge and skills in training system designs is fundamental. In fact, the definition of training is rooted in the view that training encompasses activities that are aimed at leading to skilled behavior. Thus, HPRA considers how target knowledge and skills are best learned. A prevalent finding in the training literature is that practice does not equal training. Simply building a training system that mimics the operational environment (essentially allowing practice) does not ensure learning will occur. HPRA seeks to provide the means to assess human performance and identify instructional elements, such as feedback, that can be incorporated into the training system design to accelerate learning. As such, the HPRA process involves the incorporation of training management systems, which map instructional design principles onto simulation systems, defining several components that facilitate the detection, monitoring, and diagnosis of trainee performance to drive feedback aimed at calling attention to problem performance areas and allowing remediation opportunities.

Now that overviews of ORA and HPRA have been provided, the stages of the training system design lifecycle will be discussed in detail.

Stage 1: Training Needs/Goals Identification

The first stage in the training system design lifecycle is to understand training goals by conducting a training needs analysis (TNA), through which an understanding of the operational context (that is, understanding the user/trainee and task characteristics) is attained. By addressing operational context, designers can create system requirements to target not only the appropriate tasks (for example, shoot the enemy), but the appropriate training objectives associated with each task (for example, procedural steps to arm a weapon system). TNA is a process of gathering and interpreting data from the target training community and operational environment in an effort to identify performance gaps and formulate training solutions (Cohn et al., 2007). TNA focuses on providing data concerning current versus desired performance and knowledge in an identified gap area, attitudes toward the targeted performance gap, causes of or contributing factors toward the performance gap, and potential solutions. A comprehensive TNA provides the basis from which to design training solutions that realize substantial improvements in human performance by closing the identified performance gap in a manner that is compatible with current training practices.

Training Goals

There are several different aspects of training goals that must be considered when performing a TNA, including those that describe the intended users and those that describe the intended use. One aspect is mission scope.

- For instance, is the goal to train a full mission or to simply target a specific part of a mission (part task)?

Another aspect is mission characteristics.

- For instance, given a task that involves both individual and team coordination, is the goal only to train individual skills or team skills as well?
- Is the intention for the trainee to be able to perform the task under stress or in a basic pedagogical environment?

Many training courses, particularly in the military, have clearly defined training goals, including mission scope and characteristics, described in such documents as a training and readiness manual. Training and readiness manuals often provide very detailed descriptions of the prerequisites to a course, the tasks that will be targeted in the course at both individual and team levels, and even the level of performance on the task required to successfully complete training. As described through terminal learning objectives, enabling learning objectives, and mission essential task lists, both the desired end state and the incremental steps necessary to get there are often provided, including expected performance in a live-fire environment upon completion of training. In the case where these are not available, not defined clearly, or not at a granular enough level, such requirements must be identified through observation and collaboration with the instructor to ensure the training objectives identified are in line with the organization's training goals.

Also of importance is to identify and characterize the range of target users.

- Will trainees all be novice users who will be learning the task for the first time?
- Are the trainees at a high enough expertise level that refresher training can be targeted?
- What are trainee attitudes toward the task being trained and use of training technology?

Subject matter expert (SME) interviews and questionnaires can be used to obtain information to develop user profiles and ensure the training solution is designed to be compatible with the target training community, environment, and culture. Table 9.1 provides an example of a user profile for a U.S. Marine Corps Fire Support Team (FiST) trainee.

Table 9.1. User Profile for U.S. Marine Corps Fire Support Team Trainee

Demographics

— Age 18–35, 100 percent male
— Male population indicates risk of color blindness
— English as primary language, U.S. culture
— High school graduate to college education

Knowledge, Skill Levels

— No to low level experience in FiST operations
— Declarative and procedural knowledge from classroom and practical application training, practice on target skills in live fire exercises
— Low number of deployments

Attitudes

— Perceived importance of task high due to predeployment status
— Motivated to use training technology to learn task
— Little experience with VE training systems, so few biases

Another consideration in identifying training goals is identification of training gaps (Rossett, 1987; Tessmer, McCann, & Ludvigsen, 1999). Often training systems are designed either to replace or be integrated into a curriculum with other training solutions. In such cases, training goals may be defined by training gaps that the training system can effectively bridge. These could be complete mission tasks, which existing training solutions could not target, or areas in which current training solutions are not producing sufficient training results. These gaps can provide training goals for which the training system has the opportunity to have the most impact. Through examination of the training curriculum and SME interviews, untargeted tasks can be identified. Through discussions with instructors and examination of performance records, it is possible to identify the tasks for which there are consistent patterns of suboptimal performance that are in need of augmentation or acceleration of learning. Considerations to help identify gaps by examining curriculum insertion points include the following:

- Is this system being designed to be used in a succession of training simulators that will build on each other?
- If so, is it preceded only by classroom training in which case the trainees will likely have only declarative knowledge on which to build?
- Which tasks currently are/are not targeted with each training system in the curriculum?

Through SME interviews and questionnaires, target performance gaps, current versus desired performance and knowledge in the identified gap area, attitudes toward the targeted performance gap, causes of or contributing factors toward the performance gap, and potential solutions can be identified (see the example in Table 9.2). In summary, TNA determines the who (that is, target training community), what (that is, target training tasks), when (that is, point of insertion), and where (that is, context of insertion) of the envisioned training solution.

Past field work suggests the following lessons learned:

- The identification of target trainees is critical to define the expertise of skill sets to be learned.
- The TNA should help focus ideas on content and methods of delivery.
- Dividing training gaps into must achieve, desirable, and desirable but not necessary can help drive design.
- TNA should be informed not just by existing training solutions, but also by the training and education literature.

Stage 2: User-Centered Design Specification

Once the training goals have been identified, the next stage is to develop user-centered design specifications through task analyses. The task analysis identifies how target training objectives are achieved in the operational environment. In short, task analysis involves reviewing documentation to gain familiarity with

Table 9.2. Training Needs Analysis Questions and Example Answers to Identify Performance Gaps for U.S. Marine Corps Fire Support Teams

Some questions that might be asked to determine training needs are (*cast a broad net*) as follows:

- What specific skills and abilities do your personnel need or need to improve on?
  - Team communication/coordination skills
- If you could change one thing in the manner in which your personnel currently perform their tasks, what would it be?
  - Information sharing, which involves the lead plotting of all pertinent information on a battle board and personnel checking the battle board for relevant information.
- What knowledge, skills, and attitudes would you most like for your personnel to be trained on?
  - Knowledge: coordination and deconfliction methods
  - Skills: communication skills
  - Attitudes: confidence/assertiveness

Narrow Down:

- Which tasks are currently not performed at ideal performance levels (note the most important tasks)?
  - 9 line planning
  - Suppression of enemy air defense (SEAD) planning
  - Call for fire (CFF) planning
  - Mission communication
  - Correction from mark
  - Visual acquisition and terminal control of aircraft
- What currently prevents personnel from performing these tasks at ideal performance levels?
  - Lack of taking all pertinent information into account when planning
  - Incorrect communication procedures
  - Incorrect scanning methods
  - Difficulty of detecting aircraft
- What are the ideal performance levels for these tasks?
  - Time to plan 9 line (<10 min), inclusion of all 9 lines, and coordination FiST lead
  - Time to plan SEAD (<10 min), deconfliction with aircraft path, and appropriate timeline

Discuss Potential Solutions:

- How could technology potentially support your personnel in supporting their training needs?
  - Technology solution: laptop team trainer that presents all team members with terrain view, FiST lead with virtual battle board, forward observer (FO) with virtual CFF sheet and tools and forward air controller (FAC) with virtual 9 line form and tools. Laptops can be set up next to each other to support operationally relevant team coordination.

---

practices and procedures and to identify task flows, and leveraging observation opportunities and SME interviews to characterize task knowledge requirements.

The task analysis should identify trainee information processing requirements, including precise characterization of the inputs (for example, system and environmental cues and feedback) a trainee must receive and outputs (for example, actions, responses, and communication) the trainee must convey, which provide a basis from which to derive fidelity requirements. This starts with identifying mission goals:

- What are the trainees trying to achieve in their missions or part missions being trained?
- What are the desired outcomes (for example, kill the enemy, report information, and so forth)?

The task analysis continues with the missions being decomposed into tasks and subtasks, as well as task flow, performer(s), performer responsibilities, and tools requirements.

Given the mission goal,

- What are the steps the performer(s) has to take to successfully complete the mission?
- What are the sequence and flow of information?
- Who are the key players?
- Who is primarily responsible for which tasks?
- What tools do they depend on to complete the mission?

The process of identifying the answers to these questions often starts with a review of relevant documentation. With respect to military training, this is typically the training and readiness manual, as well as other military doctrine publications, which provide doctrinal guidance and detailed information on tactics, techniques, and procedures to be employed in different mission types (for example, Joint Publication 3-09.3, Joint Tactics, Techniques, and Procedures for Close Air Support). Upfront documentation review allows a practitioner to be better prepared to collect data via observation and interviews as it facilitates development of a general framework from which to work (typically general task flow), a foundational knowledge on which to build, an understanding of the nomenclature, and an idea of gaps in knowledge to allow direct queries. After documentation review, observation of operational performance (rare opportunities in the military) or training exercises (limited opportunities) can provide extremely detailed information with respect to task decomposition into subtasks, team member roles, tools utilized, and more concrete task flow information. Instructor SME interviews are best utilized to fill gaps and drill down to very detailed levels (for example, exceptions to doctrine procedure). As an SME's time is typically extremely limited, it is critical to utilize this time wisely. If a practitioner queries an SME on basic information he or she knows is easily accessible via documentation provided, the expert may not be so generous with his or her time in the future. Additionally, if SME time is so limited that a structured interview is not possible, a second option is to develop questionnaires that the SME can complete incrementally as time provides. Table 9.3 provides a brief list of types of data resulting from a task analysis.

As a next step in developing requirements, the outcome from the task analysis can be used to feed a sensory task analysis, which is used to identify the multimodal cues and functionalities experienced during real world performance. Here is where rich contextual data are gathered and leveraged in the training system design. Sensory task analysis is conducted to determine how trainees gather information from the operational environment and how they act upon the environment in the real world. For each task and subtask, one must identify the multimodal cues (visual, auditory, haptic, and so forth) that the operator relies upon to perceive and comprehend the surrounding environment in order to successfully complete the task. From the tap on the shoulder from a teammate, to the geometry of an incoming aircraft, to the crunch of the ground beneath a tiptoeing enemy's foot, relevant multimodal information requirements must be identified for each training objective.

Table 9.3. Example Types of Data from Task Analysis

- Task descriptions, derived from observational analysis and associated documentation (for example, doctrine, field manuals, flow charts, training materials), provide a user-centered model of tasks as they are currently performed. The data to focus on include
  a. What is the general flow of task activity?
  b. What is the timing of each task step?
  c. How frequently is the task performed?
  d. How difficult or complex is the task?
  e. How important is the task to overall human-system performance?
  f. What are the consequences of task errors or omission of the task?
- Is the task performed individually or as part of a collective set of tasks, or does it require coordination with other personnel?
  g. If part of a collective set, what are the interrelationships between the set of tasks?
  h. If coordination is required, what are the roles and responsibilities of each individual in accomplishing the task?

Knowing the multimodal cues the performer depends on is not enough to capture the context of the operational environment; it is also necessary to deduce which aspects of the cues are relied upon and how the cues are used. For example, do personnel on the ground rely on merely a spot of black in the sky to detect incoming aircraft, or do they have to be able to see the wing positions in order to make fine discriminations of aircraft dynamics to assess if the aircraft is pointed at the correct target? It is important to define the task at this level of detail in order for operational requirements to be matched with KSAs to ensure cues are presented at an appropriate level of fidelity to allow successful performance of tasks and effective learning, without unnecessary technology costs.

To conduct the sensory task analysis, working from the task/subtask framework, one can use a structured interview or questionnaire to probe an SME or instructor with the following types of questions:

- For this subtask, what cues in the environment do you have to see to perform the task?
- What sounds do you have to hear?
- What physical cues do you have to feel?
- Which aspects of the cues are relied upon and how are the cues used?
- What actions do you have to perform in response?

Taken together, outcomes from the task analysis and sensory task analysis, as well as an examination of the literature (compare Milham, Hale, Stanney, & Cohn, 2005; Jones, Stanney, & Foaud, 2005; Hale & Stanney, 2006; Samman, Jones, Stanney, & Graeber, 2005), can be used to determine how the identified cues should be presented to the trainee to afford training of the task, detailing the levels of functional and physical fidelity required to allow the task to be effectively trained. *Functional fidelity* describes the degree to which a simulation imitates the information or stimulus/response options present in the real world (Swezey & Llaneras, 1997). Systems that require high levels of functional fidelity must have authentic relationships between operator inputs and outputs, but may not require spatially or physically accurate representations of system components (that is, physical fidelity). *Physical fidelity* is the degree to which a simulation imitates the multisensory (that is, visual, auditory, haptic, and olfactory) characteristics present in the real world. In systems that require high physical fidelity, it is important to capture cues beyond just the visual that are required to develop situation awareness about a mission. Further, it may be that accurate spatial and physical models are required to develop targeted skill sets. To illustrate, to target weapons handling objectives, a gun may have to be physically identical to the operational gun to allow skill sets of knowing exactly how to manually interact with the gun to facilitate automated behaviors with select weapons. If shoot/no-shoot decision making is the objective, however, then functional fidelity may suffice, where the only requirement is that there is capability to allow the trainee to input a shoot decision to an input device and receive feedback on the shot. In developing fidelity requirements, the following should be considered:

- In perceptual tasks, are multimodal cues presented in sufficiently complex environments to facilitate search and detection?
- In procedural tasks, what are the implications for multimodal cues that are not accurately spatially represented? If skills are overlearned and become automatic, is it feasible that negative transfer will occur?
- In tasks that require the development of situation awareness, are subtle real world cues represented that are early indicators of unfolding situations?
- In tasks that require the gathering and interpretation of multimodal information, how are haptic, olfactory, or other less common cues represented?
- For team training objectives, is the team represented with enough fidelity to support realistic interactions?

Based on past work (Stanney, Graeber, & Milham, 2002; Milham, Gledhill-Holmes, Jones, Hale, & Stanney, 2004; Bell et al., 2006; Milham & Jones, 2005; Jones & Bell, 2006), a list of lessons learned on collecting task data and specifying fidelity levels is provided below:

- When provided with access to current and past training curricula and to SMEs to fill in any unresolved questions, comprehensive task analysis can be accomplished with just a few one-to-three day trips to the field.
- It is beneficial to interview two to three SMEs (whether instructor or otherwise), as each informant's perception of training needs may differ slightly, and thus the core needs can be identified by cross-referencing across multiple SMEs.
- When conducting a sensory task analysis, it may be difficult for SMEs to articulate the multimodal cues they rely on as they may not realize the extent of the process they perform; therefore, it is important to have structured questions to attain this information and potential to immerse them in scenarios to elicit contextually derived responses.
- Often a task is performed under various environmental conditions (for example, day/night and high/low visibility), and it is important to extract the multimodal cues relied upon across the range of environmental conditions as they may vary.
- The goals of the task and targeted skill sets should drive the selection of fidelity.

Stage 3: Metrics Development

Metrics are a key component of the training system requirements specification process, allowing collection of relevant data and the synthesis, summary, and diagnosis of trainee strengths and weaknesses. Metrics allow identification of specific tasks and procedures with which trainees struggle to achieve sufficient performance levels. Metrics can be used to identify training gaps and redefine training goals to drive training system design. Building from the task analysis, metrics to assess performance on each subtask can be defined. Specifically, for each target task resulting from the ORA process, further decomposition into the knowledge and skills required for successful performance of the task is performed within the HPRA process. Identifying applicable knowledge and skills provides insight not only with respect to what is required for operational performance, but also how to best train each task. Specifically, there may be different levels of fidelity best suited for targeting competencies (for example, is the individual learning how to detect subtle anomalies to build situation awareness, in which case visual features are critical, or is the individual learning how to use a new display/control interface, in which case out-the-window views are not as critical as cockpit fidelity?).

Decomposing tasks into knowledge and skills can require an understanding not only of the operational tasks, but of the competency literature to create a mapping between task characteristics (for example, visual detection and auditory localization) and underlying information processing competencies/KSA constructs. To accomplish this, tasks are evaluated against a taxonomy of human performance, defining whether the skills are individual or team, perceptual, procedural, decision making, and so forth. From this, categories of tasks are classified into the type of skill set they are related to in order to further drill down into the target skill. For example, if the task analysis defines a set of perceptual tasks that is critical to task performance, the sensory task analysis can be reviewed to identify the multimodal cues that are related to performing the perceptual skill. Given the highly intensive nature of this manual process (requiring a review of literature on information processing knowledge and skills, then a mapping to domain tasks), Ahmad et al. (2007) designed a tool that begins to facilitate this mapping for tasks that rely on perceptual skills through the development of a multimodal cue, training objective, and cost matrix that utilized a sensory-perceptual objective task taxonomy (Champney, Carroll, Milham, & Hale, in press) at its core. This taxonomy links generalizable operational multimodal task characteristics with human sensory and perceptual competencies. With this type of tool, task and subtask information is gathered from the domain and broken down to a level that can be matched against the sensory and perceptual knowledge and skills targeted by the learning system to facilitate the development of metrics to monitor learning across KSAs.

Metrics often include assessment of (1) occurrence, (2) time, and (3) accuracy. For instance, did the trainee perform the required task? (for example, engage the enemy), did the trainee shoot the enemy in a timely fashion? (for example, time to shoot), and did the trainee effectively engage the enemy? (for example, accuracy of shot and kill/no kill). Metrics can be derived from documentation. However, typically, development of an effective set of performance metrics requires input from an instructor or SME. Instructors may have performance measures they employ to evaluate training. There are occasions, however, when no concrete metrics are available. For many complex military tasks, there are numerous ways that a task can be performed, and while some involve tactics more in line with doctrine than others, if the mission is successful, instructors may not drill down to technicalities. Mission outcome metrics, however, may not be granular enough to assess training system effectiveness with respect to different training objectives (that is, on which tasks or training objective is there suboptimal performance?). As a result, it may be necessary to work with instructors to identify more granular, process-level metrics. Also of importance with respect to metrics are performance thresholds. In order to gauge good versus poor performance as opposed to merely changes over time, it is necessary to identify a threshold of acceptable performance. This often proves challenging, as many times such thresholds depend on an array of environmental conditions. In these cases, attempts should be made to identify approximate performance thresholds or if/then contingency thresholds for use in evaluating trainee performance. This requires stepping through each subtask and asking the instructor to describe how he or she assesses performance on each subtask. Example questions to extract performance metric data include the following:

- How is performance on this task currently assessed by instructors (that is, what are the current performance metrics)?
- What behaviors distinguish good versus poor performance?
- What are the current versus desired performance levels on these tasks?

Once operational performance metrics are defined, these metrics can be linked to associated KSAs and, from the human performance side, metrics can be derived to evaluate learning of these KSAs. Once these linkages have been established, patterns of learning can be tracked to determine if learning on competencies is occurring over time or whether there are breakdowns at the task level (that is, the skill learning is relatively stable, with the exception of when it is conducted for a specific mission task) or at the skill level (that is, the trainee consistently is underperforming a target competency, which is affecting groups of mission tasks that utilize the skill).

The following lessons learned have been identified in past metric development efforts (Milham, Gledhill-Holmes, et al., 2004; Bell-Carroll, Jones, Milham, Delos-Santos, & Chang, 2006):

- Outcome metrics often revolve around mission success, but it is critical to define *process metrics* to determine whether or not tasks were accomplished the correct way.
- Parameters of performance (target levels) may vary based on expertise of the individual.
- Time and accuracy metrics should be clearly understood to reflect the specific goals of the mission (they tend to be inversely related).
- Given many performance metrics, it may prove beneficial to have instructors rate criticality or prioritize performance metrics.
- KSA metrics may be more meaningfully interpreted when they are mapped onto specific mission events (understanding if decision making was poor for a difficult event versus an easy event)

Stage 4: Training Management Component Development

Training management describes the process of understanding training objectives within a domain, creating training events to allow practice of targeted objectives, measuring and diagnosing performance, and providing feedback to trainees (Oser, Cannon-Bowers, Salas, & Dwyer, 1999). Done well, the training management cycle provides trainers with an opportunity to define and measure complex training objectives (for example, situation awareness), track performance gains of trainees, and help diagnose performance deficits (Pruitt, Burns, Wetteland, & Dumestre, 1997). The training management component relies on training objectives and metrics derived from the ORA, relevant KSAs and metrics derived from HPRA to facilitate development of scenario events and scenario manipulation variables, performance diagnosis methods, and performance summaries and training feedback to facilitate targeting training objectives. The primary training management components that result from the ORA process are scenarios that effectively target training objectives. Scenarios incorporate task models and procedures associated with utilization of target KSAs. In scenario based training, "the scenario itself is the curriculum" (Cannon-Bowers, Burns, Salas, & Pruitt, 1998, p. 365). To build a training scenario it is key to identify those variables that require trainees to adjust or adapt their strategies to facilitate development of expertise (Prince & Salas, 1993) on the KSAs. These variables can be used to adjust difficulty through workload, target density, or time provided to accomplish tasks. Variables can include such factors as the ambiguity of targets, distractions, or variations in other factors, such as day or night missions, or even equipment available to support the task. Each variable can be developed into an array of scenario events along a difficulty continuum, resulting in a library of scenario curriculum events from which scenarios can be developed given a targeted difficulty level and training objective. Effective scenario variables can often be identified by examining instructor-developed scenarios that range in difficulty. These variables can also be identified by a SME or instructor through a structured interview or questionnaire similar to the process for identifying metrics. For each subtask, practitioners should ask what the instructor would include in easy, medium, and hard scenarios aimed at different training objectives. Example questions to guide the development of scenarios include the following:

- Are there differences in events that would be presented at different levels of difficulty (for example, the number of enemies)?
- Are there differences in information that would be presented to trainees based on difficulty (for example, intelligence suggests there is an enemy air defense in the area versus the trainee having to detect the enemy without aid)?
- Are there stressors that would be added as difficulty increases?

These questions typically result in numerous variables and variations that can be used in scenario development. Scenarios can be developed from these variables that target different training objectives over a range of difficulty levels. Additionally, to accelerate learning, scenario selection and manipulation can be driven by learning performance. By diagnosing performance on training objectives via performance metrics, future scenario selection or online scenario manipulation can tailor to continued practice on training objectives that show performance decrements or increased/decreased scenario difficulty due to performance stabilization or performance decreases.

The primary training management components resulting from HPRA are performance diagnosis and display of this information to instructors and trainees.

Based on performance during scenarios as measured by mission performance and learning metrics, performance data can be used to support trainers by providing summaries of performance data to facilitate after action review, or the data can be used to directly provide feedback to trainees. As a trainer tool, it is important to determine the information that trainers will need to facilitate quick and to-the-point illustrations of performance breakdowns. It should decrease trainer workload and be able to illustrate key information related to training goals identified by the trainer. Trainers may require highly adaptive interfaces that can meet changing goals and foci. They may want to be able to replay performance from different perspectives or uncover patterns in performance across members of a team, individuals across an entire class, or monitor performance over time to investigate training impact on emerging needs (for example, new tactics). To get these data, it is important to identify how trainers currently conduct training, performance measurement and diagnosis, and after action review. From this, tools can be developed that support trainer needs in diagnosing trainee performance. Questions to ask trainers may include the following:

- What training goals are you assessing?
- Which metrics help you assess those goals?
- How can these data be displayed to provide a quick look to facilitate after action review?

As a tool to provide feedback directly to trainees, data must trigger interventions or provide modules that are aimed at illustrating performance, supporting the identification of errors or breakdowns, and providing learning aids to facilitate improvement on target competencies. Trainees may need self-paced tools that provide them with not only performance feedback, but adjust the training scenario itself to provide directed training aimed at decreasing gaps in performance. Questions to consider for trainees may include the following:

- What can the trainee take away from the displayed data?
- What does the trainee need to understand to improve learning on deficiencies?
- What kinds of data are provided to trainees by trainers to improve performance?

Previous work in developing training management components has resulted in many lessons learned (Jones, Bell, & Milham, 2006; Carroll, Champney, et al., 2007; Carroll, Milham, Champney, Eitelman, & Lockerd, 2007); a partial list includes the following:

- Scenario manipulation variables should be clearly mapped to training objectives to facilitate training targeting these objectives.
- It is important to understand how multiple scenario manipulations interact to affect difficulty (for example, interaction between time of day and weather conditions).
- Observing after action reviews and talking with instructors is key to determining the data to present after simulation performance; a list of all metrics is not typically useful.

- Trainees may not benefit from knowledge of results only; further remediation may be required to facilitate learning on breakdowns in performance.
- General training strategies can be used across a host of KSAs, such as providing feedback; however, the most effective learning can be achieved with diagnosis (that is, where the breakdown has occurred, what is the deviation from expert/expected performance) tied to specific feedback (for example, knowledge of correct versus observed performance).

CASE STUDY: MOT^{2}IVE SYSTEM DEVELOPMENT EFFORT

This section illustrates the training system design lifecycle by presenting its application to the design and evaluation of the Multi-Platform Operational Team Training Immersive Virtual Environment (MOT^{2}IVE) system.

Stage 1: MOT^{2}IVE Training Goals/Needs Identification

During initial MOT^{2}IVE system development, the target training audience was predeployment and deployed marines who were participating in or had completed FiST training at Tactical Training Exercise Control Group at 29 Palms, California (see Table 9.1). Having performed these skills in low fidelity training systems and/or live-fire exercises, these marines would have moderate skills and would be in the process of consolidating these skills. In order to ensure MOT^{2}IVE targeted appropriate training goals, SMEs were interviewed to assist in developing a set of training gaps known as friction points to guide training system requirements specification, so that MOT^{2}IVE could target areas in which trainees typically demonstrated performance deficiencies. The target training goals identified were those of close air support and call for fire missions to support a commander's intent as a three-man team, focusing on the friction point training gaps (for example, engagement of all targets and suppression of enemy air defense).

Stage 2: MOT^{2}IVE Task Analysis and Design Specifications

Utilizing military doctrine and training and readiness manuals, training observation, and SME interviews, the close air support and call for fire missions were decomposed into tasks and subtasks, responsible team members, and associated training gaps. Tasks and subtasks were structured in a sequential manner to indicate task flow as illustrated in the suppression of enemy air defense (SEAD) subtask example in Table 9.4.

Building from the FiST task decomposition, subtasks were further broken into the individual and team knowledge and skills required to successfully perform the tasks (Table 9.5). This was completed by first identifying a set of high level knowledge and skills thought to be most relevant to FiST operations, including the following: coordination (information exchange), coordination (mutual

Table 9.4. FiST Task Decomposition Examples

				Responsibility		
Mission Phase	**Task**	**Subtask**	**Training Gap/Goal**	**Forward Observer**	**Forward Air Controller**	**FiST Lead**
Mission Planning	**Evaluate and communicate SEAD effectiveness**	Locate where SEAD rounds hit	SEAD	1	2	
		Determine if the SEAD mission was effective	SEAD	2		2
	Communicate whether the suppression was effective to the pilot and advise whether to continue mission or not	Engage all targets		1		

performance monitoring), coordination (communication), leadership (guidance), leadership (initiative), team situation awareness (threat/friendly awareness), team situation awareness (timeline awareness), adaptive decision making (asset allocation), adaptive decision making (conflict resolution), spatial/relational knowledge/skills, perceptual knowledge, strategic knowledge and decision making, and procedural knowledge/skills. From this list, the FiST task was examined subtask by subtask through a focus group to determine which knowledge and skills were relevant to each subtask.

Through training observation at Tactical Training Exercise Control Group at 29 Palms, California, and instructor interviews, FiST subtasks were further decomposed into multimodal cue and capability requirements via a sensory task analysis to identify all information necessary to perform the close air support and call for fire tasks and interaction requirements (see Table 9.6).

Based on a FiST sensory task analysis, operational multimodal cue and capability requirements were transformed into fidelity requirements by examining human information processing requirements and mapping them to knowledge and skill training objectives for each subtask, then validating with an expert. Table 9.7 details the fidelity requirements for the tools/interfaces by which

Table 9.5. FiST Knowledge and Skills Decomposition Examples

Mission Phase	Task	Subtask	Individual Knowledge and Skills	Team Knowledge and Skills
Mission Planning	**Evaluate and communicate SEAD effectiveness**	Locate where SEAD rounds hit	Perceptual skills: visual detection of cues to indicate SEAD, auditory localization of SEAD round, detect auditory comms from indirect fire agency	
		Determine if the SEAD mission was effective	Perceptual skills: fine visual detail discrimination Spatial skills: visual distance estimation, spatial orientation; Decision making: assess effectiveness	
		Communicate whether the suppression was effective to the pilot and advise whether to continue mission or not	Decision making: Assess what actions pilot should take	Team coordination: information exchange

trainees receive and act upon the environment and then maps those to the training goals identified in the first step.

Stage 3: MOT^{2}IVE Metrics Development

In order to assess the degree to which MOT^{2}IVE facilitated the trainee meeting mission goals, operational performance metrics had to be identified. Through instructor interview, metrics that could be used to discriminate good and poor performance levels were identified, including thresholds separating these two

Table 9.6. FiST Multimodal Cue and Capability Requirements Example

Task	Subtask	Required Multimodal Cues	Required Capability
Evaluate and communicate SEAD effectiveness	FAC/FO locates where SEAD rounds hit	Visual: – SEAD round on deck – Magnified view through binoculars – Map Auditory: – Sound of SEAD round flying over head and hitting deck Communications from indirect fire agency	Ability to scan terrain within
	FAC/FOs determine if the SEAD mission was effective	Visual: – Terrain – SEAD round on deck – Damage to EAD	
	FAC communicates whether the suppression was effective to the pilot and advises whether to continue mission or not	Auditory: – Communications from pilot	– Ability to communicate with pilot

performance levels. Table 9.8 provides example metrics for the SEAD evaluation task.

Next, operational performance metrics were mapped to associated knowledge and skills (K&S) to assess the degree that MOT^{2}IVE facilitated learning. If tracked over multiple examples or over time, patterns in learning on these K&S dimensions can be determined (Table 9.8). With MOT^{2}IVE, these metrics were used to develop a performance measurement and diagnostic tool, described in the next section, which tracked learning trends over time, KSA training objectives, and mission phases (see Figure 9.3). When examined with subtask performance, the outputs of this tool can be used to determine if decrements in performance are due to a mission level task or lack of knowledge of a skill set.

Stage 4: MOT^{2}IVE Training Management Component Development

To monitor trainees' performance, and to track progress toward training goals, training management strategies and components were developed to inject key events into scenarios, calculate metrics, diagnose breakdowns in performance, and to illustrate these breakdowns to instructors. Scenarios were developed by creating scenario manipulation variables associated with each friction point

Table 9.7. FiST Fidelity Requirements Example

Task	Subtask	Knowledge & Skills—Training Objective	Human Information Processing Requirements	Fidelity Requirements
Evaluate and communicate SEAD effectiveness	FAC/FO locates where SEAD rounds hit	Perceptual skills	Visual: – Detect SEAD round on deck potentially outside field of view Audio: – Localize: sound of SEAD round flying over head and hitting deck – Detect: communications from IDF agency Response: – Ability to scan terrain	Visual display – Low desolution – Wide field of view – Audio: – Spatialized
	FAC/FO determines if the SEAD mission was effective	Perceptual skills Spatial skills	Visual: – Discriminate distance and depth perception – Discriminate: visually localize – Discrimination of fine detail	Visual display – High resolution – Normal field of view –

training gap across an array of difficulty levels (that is, crawl, walk, and run). These variables and difficulty variations were used to develop scenarios. An example is provided in Table 9.9.

The performance assessment and diagnostic tool (PAST; Carroll, Champney, et al., 2007) training management component was developed to measure and diagnose performance by identifying the root errors associated with mission failures and how they propagated through to mission outcomes. This tool illustrates the chain of events leading up to each error to identify earlier mistakes or contributing factors that may have led to performance errors. Detailed logic was created to chronologically link specific metrics to root causes, and metrics were categorized by the training goal they were related to (see Figure 9.4).

This case study has illustrated the successful application of the training system design lifecycle to the development of requirements for a U.S. Marine Corps FiST trainer (MOT^{2}IVE), which has been embraced by the operational

Table 9.8. FiST Performance Metric Examples

Task	Subtask	Performance Metrics	Performance Thresholds	Knowledge and Skills
Evaluate and communicate SEAD effectiveness	FAC/FO locates where SEAD rounds hit	– Did FAC/FO visually locate SEAD round? – Time to locate SEAD rounds	– Yes –<2 s	– Spatial knowledge – Perceptual knowledge
	FAC/FO determines if the SEAD mission was effective	Did FAC/FO correctly evaluate effectiveness of SEAD? Time to evaluate SEAD effectiveness	– Yes – <2 s	– Spatial knowledge/ skill – Perceptual knowledge/ skill – Decision making
	FAC communicates whether the suppression was effective to the pilot and advises whether to continue mission or not	– Did FAC communicate effectiveness of SEAD to pilot?	– Yes	– Team coordination (information exchange)

community and incorporated into multiple U.S. Marine Corps training curricula due to its operational relevance.

CONCLUSION

This chapter discusses in detail two components of the training system design lifecycle: operational requirements analysis and human performance requirements analysis. Through the ORA process discussed herein, system requirements can be informed by training needs/gaps the system must address, as well as the multisensory information and interaction requirements necessary to target these training gaps. Additionally, the HPRA process seeks to ensure systems are designed to effectively support training through development of metrics to assess whether training goals are being met and, with the provision of training management strategies, to assist in attainment of these goals. Both processes are concerned with understanding the goals of the system from the trainee's

Friction Point Summary | Timeline | Performance Summary

Training Objectives Summary

	Training Objective	FO	FAC	FiST Lead	Team
Individual	Spatial Knowledge	77% (24/31)	78% (29/37)	94% (30/32)	95% (19/20)
	Perceptual Knowledge	100% (4/4)	67% (8/12)	100% (7/7)	100% (4/4)
	Strategic Knwldg & Decision Making	100% (3/3)	100% (4/4)	80% (4/5)	100% (3/3)
	Procedural Knowledge / Skills	86% (6/7)	75% (3/4)	71% (5/7)	50% (1/2)
	Team Coordination	(0/0)	(0/0)	100% (1/1)	(0/0)
	Team Performance / Communications	####	####	####	####
	Affective and Attitudinal	####	####	####	####
Team	Team SA	89% (17/19)	85% (11/13)	81% (13/16)	78% (7/9)
	Adaptive Decision Making	85% (22/26)	84% (27/32)	90% (26/29)	88% (21/24)
	Coordination	60% (9/15)	60% (9/15)	83% (10/12)	(0/0)
	Leadership	0% (0/1)	0% (0/1)	83% (5/6)	0% (0/1)

Figure 9.3. FiST Knowledge and Skill Training Objective Learning Trends

perspective, developing the environment to support and monitor knowledge and skill development, and using measures of performance to provide trainer tools or trainee tools to reduce decrements. By grounding the development of training systems in ORA and HPRA, training effectiveness can be built into the system

Table 9.9. FiST Scenario Variation Example

Training Gap/ Goal	Scenario Manipulation Variable	Crawl Variation	Walk Variation	Run Variation
SEAD	Type of enemy air defense, location of enemy air defense, and whether enemy air defense was preidentified or unidentified	One enemy air defense that is identified to FiST in scenario (for example, commanders intent briefing, fire support control center info, and so forth)	One enemy air defense not identified to FiST, so FiST has to detect it	Multiple enemy air defenses, one identified to FiST in scenario and one not identified, so FiST has to detect it

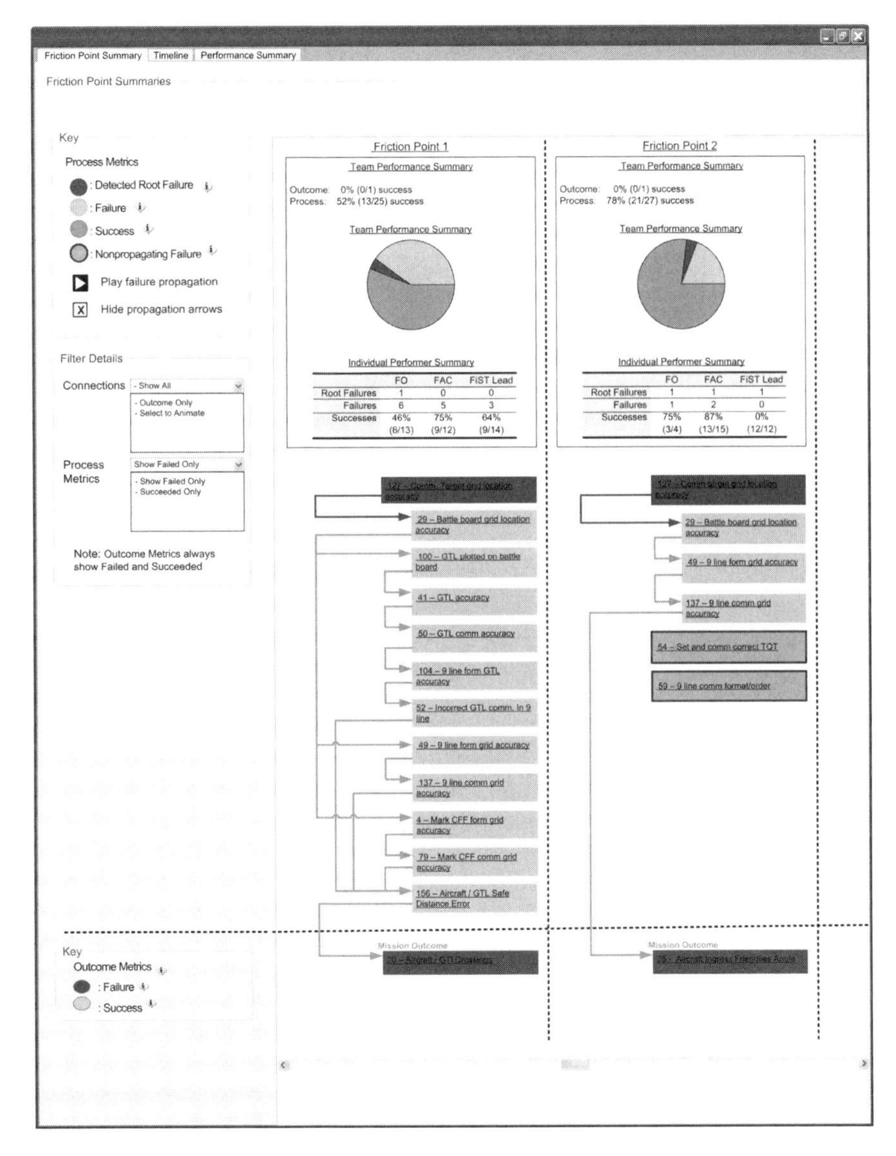

Figure 9.4. Performance Assessment and Diagnostic Tool Root Error Diagnostic Tree

from the ground up. Doing this is critical to ensure that overall training goals are met by providing trainees with opportunities to conduct operationally relevant missions and supporting the learning of the knowledge and skills that are needed to successfully achieve mission outcomes.

REFERENCES

Ahmad, A., Carroll, M., Champney, R., Milham, L., Parrish, T., & Chang, D. (2007). *Virtual reality training system development guidance tool for multimodal information fidelity level selection: Tool for the Optimization of Multimodal Cues for the Advancement of Training System Design—TOMCAT* (Phase 1 Final Rep., Contract No. N00014-07-M-0231). Arlington, VA: Office of Naval Research.

Bell, M., Jones, D., Chang, D., Milham, L., Becker, W., Sadagic, A., & Vice, J. (2006). *Fire Support Team (FiST) task analysis surrounding eight friction points* (VIRTE Program Rep., Contract No. N00014-04-C-0024). Arlington, VA: Office of Naval Research.

Bell-Carroll, M., Jones, D., Milham, L., Delos-Santos, K., & Chang, D. (2006). *Multi-level timeline logic, metric definitions and DIVAARS tags* (VIRTE Program Interim Rep., Contract No. N00014-04-C-0024). Arlington, VA: Office of Naval Research.

Cannon-Bowers, J. A., Burns, J. J., Salas, E., & Pruitt, J. S. (1998). Advanced technology in scenario based training. In J. A. Cannon-Bowers & E. Salas (Eds.), *Making decisions under stress: Implications for individual and team training* (pp. 365–374). Washington, DC: American Psychological Association.

Carroll, M. B., Champney, R., Jones, D., Milham, L., Delos-Santos, K., & Chang, D. (2007). *Metric toolkit deliverable: Performance Assessment and Diagnostic Tool (PAST) design, development and evaluation* (VIRTE Program Technical Rep., Contract No. N00014-04-C-0024). Arlington, VA: Office of Naval Research.

Carroll, M. B., Milham, L., Champney, R., Eitelman, S., & Lockerd, A. (2007). *ObSERVE initial training strategies report* (Program Interim Rep., Contract No. W911QY-07-C-0084). Arlington, VA: Office of Naval Research.

Champney, R. K., Carroll, M. B., Milham, L. M., & Hale, K. S. (in press). Sensory-perceptual objective task (SPOT) taxonomy: A task analysis tool. *Proceedings of the Human Factors and Ergonomics Society 52nd Annual Meeting.* Santa Monica, CA: Human Factors and Ergonomics Society.

Cohn, J. V., Stanney, K. M., Milham, L. M., Jones, D. L., Hale, K. S., Darken, R. P., & Sullivan, J. A. (2007). Training evaluation of virtual environments. In E. L. Baker, J. Dickieson, W. Wulfeck, & H. O'Neil (Eds.), *Assessment of problem solving using simulations* (pp. 81–105). Mahwah, NJ: Lawrence Erlbaum.

Ericsson, K. A., & Oliver, W. L. (1995). Cognitive skills. In N. J. Mackintosh & A. M. Colman (Eds.), *Learning and skills* (pp. 19–36). London: Longman Group Limited.

Fairley, R. E., & Thayer, R. H. (1997). The concept of operations: The bridge from operational requirements to technical specifications. In M. Dorfman & R. H. Thayer (Eds.), *Software engineering* (pp. 44–54). Los Alamitos, CA: IEEE Computer Society Press.

Fowlkes, J., Dwyer, D. J., Oser, R. L., & Salas, E. (1998). Event-based approach to training. *The International Journal of Aviation Psychology, 8*(3), 209–221.

Goode, H. H., & Machol, R. E. (1957). *Systems engineering: An introduction to the design of large-scale systems.* New York: McGraw-Hill.

Hale, K. S., & Stanney, K. M. (2006). Enhancing spatial awareness with tactile cues in a virtual environment. *Proceedings of the Human Factors and Ergonomics Society 50th Annual Meeting* (pp. 2673–2677). Santa Monica, CA: Human Factors and Ergonomics Society.

Jones, D., & Bell, M. (2006). *MOT2IVE V2 system requirements* (VIRTE Program Interim Rep., Contract No. N00014-04-C-0024). Arlington, VA: Office of Naval Research.

Jones, D., Bell, M., & Milham, L. (2006). *VIRTE demo 3 scenario manipulation variable matrix* (VIRTE Program Interim Rep., Contract No. N00014-04-C-0024). Arlington, VA: Office of Naval Research.

Jones, D. L., Stanney, K., & Foaud, H. (2005, August). *Optimized spatial auditory cues for virtual reality training systems.* Paper presented at the 2005 APA Convention, Washington, DC.

Milham, L., Gledhill-Holmes, R., Jones, D., Hale, K., & Stanney, K. (2004). *Metric toolkit for MOUT* (VIRTE Program Rep., Contract No. N00014-04-C-0024). Arlington, VA: Office of Naval Research.

Milham, L., Hale, K., Stanney, K., & Cohn, J. (2005, July). *Using multimodal cues to support the development of situation awareness in a virtual environment.* Paper presented at the 1st International Conference on Virtual Reality, Las Vegas, NV.

Milham, L., & Jones, D. (2005). *MOUT room clearing cross reference of sensory cues (operational and metaphoric) to environmental information required for tasks: Interim technical report* (VIRTE Program Interim Rep., Contract No. N00014-04-C-0024). Arlington, VA: Office of Naval Research.

Milham, L. M., Cuevas, H. M., Stanney, K. M., Clark, B., & Compton, D. (2004). *Human performance measurement thresholds* (Phase I Final Rep. No. N00178-04-C-3019). Dahlgren, VA: Naval Surface Warfare Center.

Oser, R. L., Cannon-Bowers, J. A., Salas, E., & Dwyer, D. J. (1999). Enhancing human performance in technology-rich environments: Guidelines for scenario-based training. In E. Salas (Ed.), *Human/technology interaction in complex systems* (Vol. 9, pp. 175–202). Stamford, CT: JAI Press.

Prince, C., & Salas, E. (1993). Training and research for teamwork in the military aircrew. In E. L. Wiener, B. G. Kanki, & R. L. Helmreich (Eds.), *Cockpit resource management* (pp. 337–366). Orlando, FL: Academic Press.

Pruitt, J. S., Burns, J. J., Wetteland, C. R., & Dumestre, T. L. (1997). ShipMATE: Shipboard Mobile Aid for Training and Evaluation. *Proceedings of the Human Factors and Ergonomics Society 41st Annual Meeting* (pp. 1113–1117). Santa Monica, CA: Human Factors and Ergonomics Society.

Rossett, A. (1987). *Training needs assessment.* Englewood Cliffs, NJ: Educational Technology Publications.

Samman, S. N., Jones, D., Stanney, K. M., & Graeber, D. A. (2005, July). *Speech, Earcons, Auditory Spatial Signals (SEAS): An auditory multi-modal approach.* Paper presented at the HCI International Conference, Las Vegas, NV.

Schmorrow, D., Solhan, G., Templeman, J., Worcester, L., & Patrey, J. (2003, June). *Virtual combat training simulators for urban conflicts and performance testing.* Paper presented at the International Applied Military Psychology Symposium, Brussels, Belgium. Retrieved April 8, 2003, from http://www.iamps.org/praha/Paper-Schmorrow2.doc

Stanney, K. M., Graeber, D., & Milham, L. (2002). *Virtual environment landing craft air cushion (VELCAC) knowledge acquisition/engineering* (VIRTE Program Rep., Contract No. N0001402C0138). Arlington, VA: Office of Naval Research.

Stanney, K. M., & Zyda, M. (2002). Virtual environments in the 21st Century. In K. M. Stanney (Ed.), *Handbook of virtual environments: Design, implementation, and applications* (pp. 1–14). Mahwah, NJ: Lawrence Erlbaum.

Swezey, R. W., & Llaneras, R. E. (1997). Models in training and instruction. In G. Salvendy (Ed.), *Handbook of human factors* (pp 514–577). New York: Wiley.

Tessmer, M., McCann, D., & Ludvigsen, M. (1999). Reassessing training programs: A model for identifying training excesses and deficiencies. *Educational Technology Research and Development, 47*(2), 86–99.

Chapter 10

BUILDING VIRTUAL ENVIRONMENT TRAINING SYSTEMS FOR SUCCESS

Joseph Cohn

There are many steps between the first ideation of a system and the last hammer blow to the final to-be-delivered system. These include the following:

- Concept generation,
- Requirements definition,
- Metrics development,
- System validation,
- Return on investment determination.

These steps are interdependent and collectively represent the different elements that comprise the training system development cycle. While it is possible to enter this cycle at any step—for example, a development team may be asked to calculate a return on investment for a commercial off-the-shelf product to determine if it satisfies an identified need—more often than not this cycle is entered at either the concept generation step or the requirements definition step.

Concept generation teams are often tasked with a very general mandate to identify leap-ahead technologies or to anticipate new challenges before they emerge. So, when a design and development team enters this cycle at the concept generation step, they often do so with a vague notion of what it is they are trying to solve. This is often a costly and time consuming approach, but because of the intellectual freedom it provides, it may also be the one most likely to produce a leap-ahead solution because they may have wide latitude in coming up with viable solutions. Starting at this step represents a high risk, and (potentially) high reward decision. Contrastingly, when a team enters this cycle at the requirements definition step, the team members are typically provided with a performance challenge to solve. Here, they must use the specified parameters as the starting point for developing their solutions. This approach is therefore more constrained, but, because of its bounded nature, it is more likely to return a viable solution for a given investment. Starting at this step represents a moderate risk, moderate reward decision.

The focus of this chapter is on system development cycles that begin with the requirements definition step. We first explore different ways of developing requirements; next, we look at different approaches for generating concepts that will satisfy these requirements and for building metrics to help developers evaluate progress and success; we follow this with a short discussion on system validation and return on investment calculations and conclude by using a real world example of how these different steps may be applied to developing a notional virtual environment (VE) based infantry training system.

REQUIREMENTS

A requirement is simply a description of what a system/product/tool must do (Ross & Schoman, 1977). There are many different ways of addressing requirements, but, typically, four basic questions must be answered:

1. What must the system do?
2. How will the system be built?
3. Why build this particular system?
4. When should success be declared?

Some of the more common methods for addressing these questions include

- Knowledge elicitation,
- Knowledge analysis, and
- Knowledge validation.

Knowledge Elicitation

There are two main objectives that any knowledge elicitation effort must achieve: (1) identifying the critical gap that the to-be-developed system must address and (2) obtaining the necessary information to characterize the target users and their associated environment. Both objectives may be met with the following:

- Obtaining and reviewing all relevant user community documents (training manuals, system documentation, newsletters, and so forth),
- Conducting subject matter expert (SME) interviews in order to
    a. Understand performance gap(s) and
    b. Determine where to insert a given solution,
- Using questionnaires to
    c. Determine consensus points relating to proposed performance gap(s) and
    d. Characterize attitudes toward proposed performance gap(s) and proposed system solutions,
- Convening focus groups to drive consensus at points of dissension,

- Using both questionnaires and SME interviews to develop user community profiles that characterize
  - e. The targeted community in terms of demographics, knowledge, skills, and abilities and
  - f. The targeted environment in terms of the culture and operational context into which the envisioned solution will be integrated.

Knowledge Analysis

Once the necessary information is gathered, it must be processed and formatted to support data-driven design decisions. The most common, general tool for formally representing this information is the task analysis (Kirwan & Ainsworth, 1992; Chipman, Schraagen, & Shalin, 2000), which comes in many different forms and is used for a wide range of purposes, including

- Identifying task flows,
- Characterizing user knowledge requirements,
- Determining information processing requirements (that is, inputs to the system and outputs from the system), and
- Developing system fidelity requirements.

Knowledge Validation

Gathering and representing the information is only part of the challenge. Typically, the team that interacts with the user community is not the team that develops the system that will be delivered to the user community. Consequently, the user information must be represented in such a manner that it supports the development team as it designs and constructs the actual solution. This requires developing models and formalizations that

- Characterize task interdependencies in terms of workflow and timelines,
- Capture performance specifications, and
- Develop use case scenarios that link these models and formalizations so that developers will be better able to anticipate the impact of design trade-offs on user performance.

Once these models and formalizations are constructed, they may then be validated. This includes returning to the same user groups that provided the initial information and reviewing with them the captured knowledge; it may also involve identifying other user groups, from the same community, who have not yet provided their inputs, providing an additional source for comment and review.

CONCEPT GENERATION

Developing concepts to create solutions that will satisfy a given set of requirements is often the most difficult step of the system development cycle. The critical challenge is taking the various pieces of information—from user needs to system specification—that form the bulk of the requirements development step and envisioning a new approach that will satisfy these needs. The main obstacle is that any design and development team is limited by its own understanding of the problem and solution spaces. The concept generation step focuses on enabling teams to move outside of their collective decision loops and consider both spaces from new perspectives.

One approach for arriving at new solutions is to seed new concepts by understanding current solutions and identifying their inefficiencies (Scanlan, 2007). An existing solution will be made up of different components—be they individual technologies, such as a laptop's hard drive, battery, and so forth, or individual methodologies, such as during action or after action feedback. Each of these components has a set of operating parameters, which combine to create that system's capabilities. Changes to a component will impact system efficiency. Therefore each of these components can be analyzed in terms of the specified requirements and assessed in terms of their pros and cons.

Some general approaches for moving through the concept generation phase include the following (after MacLean, 2006; Durfee, 1999):

- *Brainstorming:* developing a variety of concepts that provide the seed material for follow-on discussions,
- *Scenario Generation and Role-Playing:* exploring different concepts in terms of how they might be applied by the user community to solve the identified gap and stepping through the design and development of actual tools derived from these concepts, and
- *Prototyping:* developing simple, low cost "models" (for example, storyboards) to quickly demonstrate how the concepts being considered will impact, and be impacted by, the user.

When done properly, concept generation will not only lead to new solutions that satisfy identified requirements, but it will also provide *testable hypotheses* that can be used later, during the validation step, in order to demonstrate the benefits of using the newly developed system.

METRICS

The requirements step provides the information that will be used by the design and development teams as they construct their solutions; using this information, the concept generation step provides a way for developers to envision possible solutions. As system development proceeds, it is important to ensure that progress may be continually assessed and that the answer to "when should you declare success?" may be clearly obtained through the cycle, not just at time of system delivery. In order to do this, we must not only develop metrics that quantify end-user performance on the system, but that also support the system designers and developers who are creating the system, and decision makers who represent the broader interests of the organization (or community) for which the system is being developed. These metrics will be developed based to a large extent on the data captured during the requirements step and bounded by the concept generation step.

Metrics provide the means for discussing system merits. Done well, a comprehensive set of metrics will provide system designers and developers with a continuous snapshot of system performance to weigh progress against requirements; they will also help system users determine system effectiveness, enabling them to provide constructive feedback to the design and development team, and they will help decision makers (such as those who must purchase the final product) establish long-term returns on investment. Kirkpatrick (1998) defined four levels of metrics that focus on evaluating training systems across these different levels and that are captured over different time horizons:

1. Reaction: immediate responses to the system, in terms of usability, acceptance, and satisfaction (Nielsen, 1993; Stanney, Mollaghasemi, Reeves, Breaux, & Graeber, 2003); best used for obtaining quick snapshots of system progress and alignment with user specified goals.
2. Learning: near-term responses of individual users following exposure to the system; best used as part of a near-term effectiveness evaluation paradigm to ensure that the identified performance gap is bridged at the level of the individual user in response to his or her interaction with the system (Lathan, Tracey, Sebrechts, Clawson, & Higgins, 2002).
3. Behavior: midterm responses of the community in response to large-scale interaction of individual members with the system; best used as a midterm assessment of the system impact across the user community.
4. Results: long-term impact to the larger organization as a result of sustained, community-wide exposure to the system; best used as part of a return on investment assessment.

Interleaved with these levels are different time requirements for capturing these metrics. Some of these levels may be satisfied over relatively short time spans, while others may require longer time investments in order to be fully populated. As we will see later, the notion of time plays an important role in determining return on investment—how long one is willing (or able) to wait to populate a set of metrics will directly impact the usefulness of a given set of metrics.

Building Metrics: Different Types

Regardless of the role that a particular metric may be used for (quick assessment of system usability or longer term determination of system impact), all metrics must be built by first determining the kinds of information they must represent (Department of Energy, 2005):

- Process or outcome: Process metrics capture behaviors that evolve during an event, while outcome metrics capture behaviors following the event (Salas, Burke, Fowlkes, & Priest, 2004);

- Measures of performance (MOPs) or measures of effectiveness (MOEs; Sproles, 2000): MOPs measure system performance, providing the engineer's view of the system, while MOEs measure user performance, providing the user's view of the system.

Figure 10.1 illustrates the interrelationship between these types of metrics. Both the system and the user views may be captured using either process or outcome measures. Time to populate each type of metric becomes critical when considering the cost-benefit trade-offs with using a given metric to satisfy a particular level of metric. The longer it takes to capture the metric, the greater the cost and the less likely it will be used in final assessment.

Building Metrics: The Process

Developing metrics is as much an art as a science. A general process for creating any metric will build on the earlier development cycle steps and incorporate new ones, including the following:

1. Metric Qualification: Using information gathered during both the concept generation and requirements definition steps to identify intended effects. This is oftentimes best done using a prose/descriptive format rather than a mathematical or algorithmic one (for example, how accurate will a user be in performing a given task?).
2. Metric Quantification: Using the prose/descriptive format to develop pseudo-representations (for example, the user is able to fire 20 rounds in 5 s with 70 percent accuracy).
3. Metric Development: Translating (1) and (2) into an actual metric. This may be done by crafting pen/paper assessments, tapping into the actual system's data outputs, or

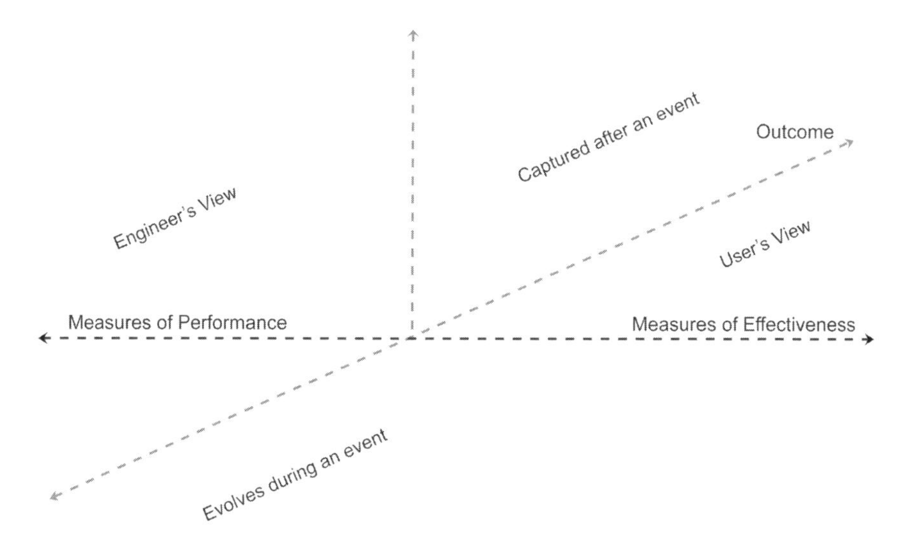

Figure 10.1. Interrelationship between Different Types of Metrics

by capturing the user's output using an external system (for example, a motion capture system to observe user behavior).

When building metrics, it is important to ensure that any final measure has the following attributes (adapted from Rosenberg & Hyatt, 1997):

- Specific: Measures must provide clear and focused indications of an identified behavior.
- Assessable: Measures must be statistically analyzable.
- Attainable: Measures must characterize properties that are neither too hard nor too easy to capture and that are expected to vary with different contexts and conditions.
- Realistic: Measures must clearly link back to identified desired effects.
- Timely: Measures must be recordable within a time frame that allows the resultant information to inform the desired level of user.

RETURN ON INVESTMENT

Return on investment (ROI) calculations are meant to answer a single question: Is the system development effort worthwhile? Recalling the four different levels of metrics, the answer to this question may be calculated in different ways using different metrics and providing different answers:

- At the reaction and learning levels, ROI may tell developers that a particular system configuration is not providing the desired level of response, suggesting that an alternative set of components may be better.
- At the behavior level, ROI may provide community leaders with deeper insight into the relationship between a given system and the desired impact on the end-user community's long-term performance.
- At the results level, ROI may enable decision and policy makers to determine if the investment in the system is worth the overall cost

It is important to understand who the end users will be and what information they will need from the calculation. Reaction level metrics provide a wealth of information pertaining to specific aspects of the system as it is being developed (for example, usability assessments, demographic information, and so forth)—a lot of information pertaining to a little bit of the larger system and how it will be used. This information will be used primarily by system developers to make who must have ready access to such minutiae in order to fine-tune the system. On the other hand, results level metrics typically provide a summative assessment of the system as a whole, long after the system has been developed—a little bit of information that will be applied to decisions concerning the entire system. This information will be used primarily by key decision makers as they make determinations on whether to continue using the system, to request large-scale upgrades or modifications, or to change how the system is used within the organization.

This division of metrics leads to a trade-off in how ROI assessments are used, which can be qualified using a notional cost of waiting function. ROIs based on reaction and learning level metrics can be performed in a relatively short amount of time; consequently, the cost of waiting to capture these metrics is small. On the other hand, ROIs based on behavior or results level metrics, which often take months if not years to capture, have much larger associated cost of waiting—when these metrics are finally quantified, it is likely too hard to make any significant system design changes. Consequently, a major challenge in performing ROIs is finding the right balance between the level of metric and the intended use of the ROI (see Figure 10.2). Depending on one's perspective, the cost of waiting to obtain a given level of metric will vary with time. Here, the cost to wait is evaluated from the perspective of an organizational decision maker, for whom the results level is the most critical metric. Populating this metric requires a significant investment in time, which may introduce other trade-offs.

VALIDATION

How a system is validated depends to a large extent on what purpose it was built to serve. Because validation may oftentimes take the form of an experiment, in which the utility of a given system is compared to the utility of other systems, across a set of metrics, it is crucial to develop a testable hypothesis—or

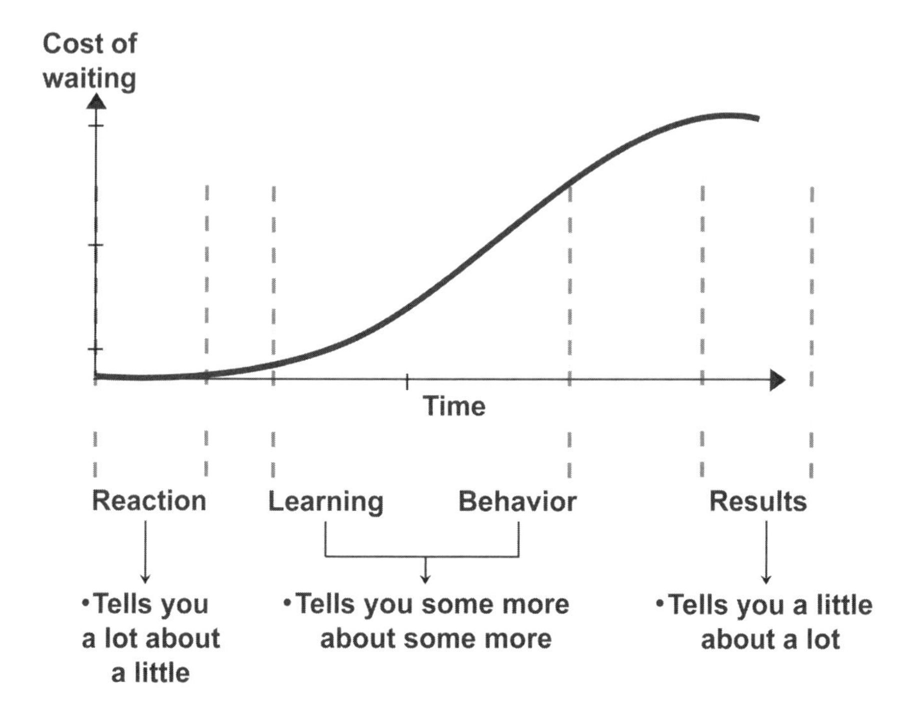

Figure 10.2. Cost of Waiting Function

hypotheses—relating to the anticipated impact of the system. For example, if one were to propose building a training system that was meant to improve situation awareness, one hypothesis for validating this system would be to compare performance using this system against performance using the traditional approach(es). As mentioned earlier, these hypotheses are meant to be formulated during the concept generation step.

In terms of building VE simulations, a generic hypothesis would be that training on a given system provides a level of improvement that translates to enhanced performance of the real world task being simulated above that that would be obtained using other types of training interventions (Lathan et al., 2002). One approach to validating this hypothesis is to show that training in a VE produces enhanced performance on real world tasks, using the transfer of training (ToT) method (Lathan et al., 2002). This method estimates simulator effectiveness as a comparison of two groups of trainees, an experimental group that receives simulator training and a control group that receives all of its training in the real world. Performance between the two groups is then compared along a set of metrics (Boldovici, 1987; Cohn, Helmick, Meyers, & Burns, 2000). The specific nature of a given ToT depends on the results of the earlier system development cycle steps. Requirements help define who may serve as the subject pool, and how; metrics provide insight into the expected differences in behaviors between individuals who receive VE training and those who do not receive that training when performing operational tasks in the real world environment, and even return investment provides a pathway for making use of the final results.

EXAMPLE

To see how these different steps may be linked, let us now turn to a real world example, developing a VE based tool that will train small teams of infantry in the art of room clearing. In keeping with the flow of this chapter, let us assume that we enter the system development cycle at the requirements step given the following information:

1. What must the system do?
   - Allow users to immediately immerse themselves without prior training
   - Provide an environment for learning new room clearing methods, practicing already learned ones
   - Permit use by infantrymen with a wide range of knowledge, skills, and abilities
2. How will the system be built?
   - Multimodal sensory stimulation (spatialized audio, haptic feedback, and wide field tracking)
   - Portable and low energy consumption
3. Why build this particular system?
   - Live training is becoming increasingly costly and dangerous
   - Training time and space are becoming more difficult to find

4. When should success be declared?
    - Engineer's View: Achieve desired system specifications
    - User's View: Long-term retention and recall of skills; reduced time to train; increased throughput in training pipeline.

Our challenge then is to develop new concepts for achieving these requirements, build the right metrics to help assess progress, and conduct a return on investment assessment and a validation study.

Concepts

Our challenge is to build a more effective training tool using VE technologies. This tool should be easy to use, low cost, and performance enhancing. These criteria allow us to develop a set of hypotheses, two of which are highlighted below:

1. Hypothesis 1: Our VE will reduce the total cost of training.
2. Hypothesis 2: System fidelity has a significant impact on how well the VE will deliver the desired training.

Using our brainstorming tools, we start by decomposing existing VE systems into their components and capabilities. These include tracking systems, display technologies, and human computer interaction technologies. Considering how these systems may be improved upon provides one an inroad to achieving our requirements.

Figure 10.3 illustrates one way of conceptualizing the system components. In line with our hypotheses, we consider both high end and low end components, which may either be pulled directly from the commercial marketplace (commercial off-the-shelf products) or require significant development efforts (total new product). After role-playing and paper prototyping different combinations of solutions—and referencing the knowledge models derived from interviews with the end user—we propose a middle-of-the-road solution that trades high end display and tracking systems for low end ones in order to develop an entirely novel type of natural human computer interface.

Metrics

For each of the four levels of metrics—reaction, learning, behavior, and results—a range of specific measures could be developed. Table 10.1 presents some of these in terms of both MOPs and MOEs and process and outcome. At the most basic level, reaction, consider metrics that characterize the display system. Process metrics at this level include how quickly the scene is updated (times per second; MOP because it refers to the system properties) and how much information the user is able to continually glean from this display (field of view; MOE because it refers to the user's experience). Outcome measures include how well the system holds up under sustained use (ruggedness; MOP) and how well the

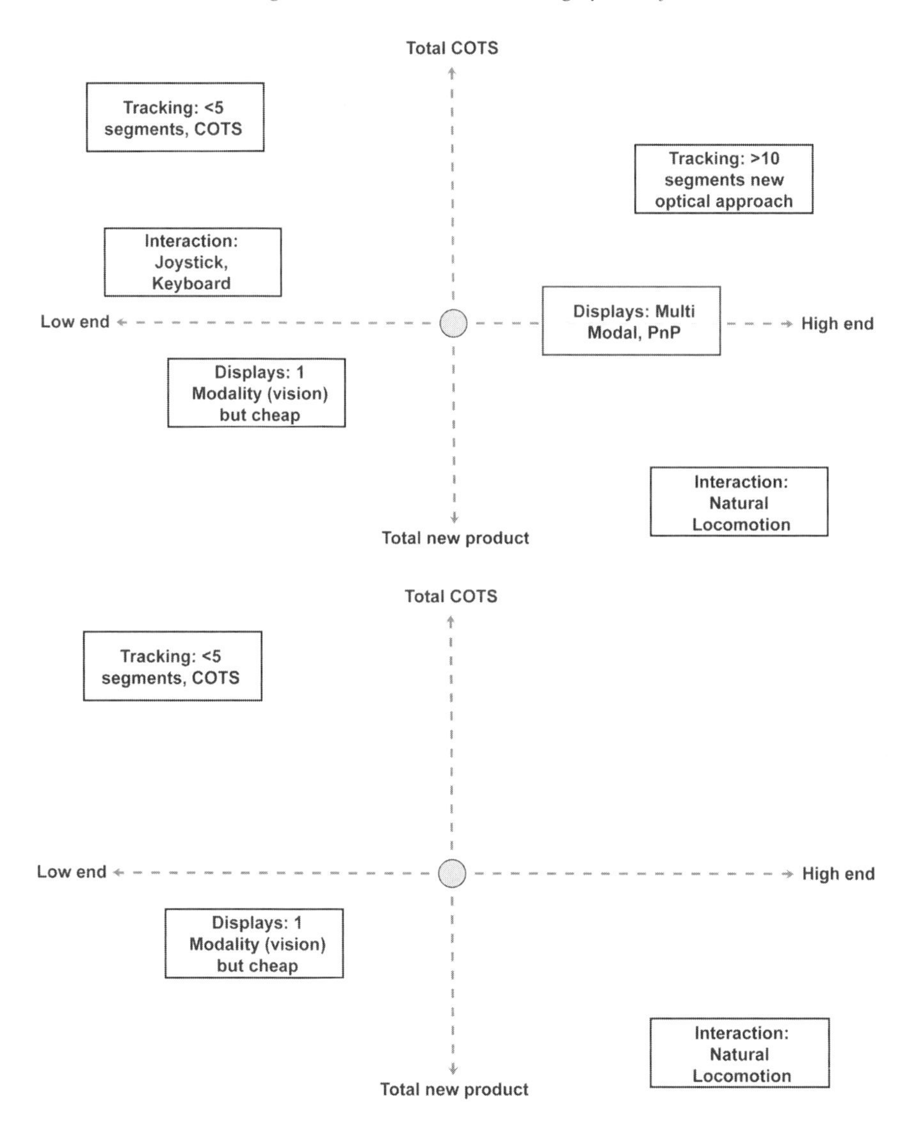

Figure 10.3. Results of the concept generation step. (Top) Possible solutions for achieving the desired requirements. (Bottom) Proposed solutions for achieving the desired requirements based in part on guiding hypotheses.

user holds up under continued exposure to an artificial visual stimulus (side effects; MOE). Scaling up to the next level, learning, consider metrics that capture how the scene unfolds as the system provides training. Process metrics include how well the system moves the user from scene to scene without jarring and abrupt transitions (scene transitions; MOP) and how effectively the user exploits the training being provided (maintaining situational awareness; MOE).

Table 10.1. Examples of MOPs and MOEs for a Virtual Environment Training System, Assessed at Each of the Four Levels of Training Effectiveness Evaluation, in Terms of Processes and Outcomes

		Factor	Process	Outcome
Reaction	MOP	Display	Update rates	Rugged
	MOE		Field of view	Side effects
Learning	MOP	Scenario	Scene transitions	Event tagging
	MOE		Maintain situational awareness	Combined scores
Behavior	MOP	Curriculum	Technology integration	Streamlined training
	MOE		Smoother team interactions	Faster training completion times
Results	MOP		Reduced cost of overall training	
	MOE		Increased survivability, faster and larger throughput	

Outcome metrics include how well the system was able to capture any long key events occurring within training scenarios (event tagging; MOP) and how much better the user was at performing a set of tasks following completion of the training (combined scores; MOE). Continuing to the next level of metric, behavior, consider how the system impacts multiple users over the long term. Process metrics at this level include evaluating the degree to which these new training tools are integrated into the organizations larger training plan (technology integration; MOP) and the degree to which making this training more readily available impacts how individuals work together (smoother team integration; MOE). Outcome metrics may focus on the ease with which this training is delivered (streamlined training; MOP) and how delivering training using VE technology reduces the amount of time the organization devotes to delivering training (faster training completion times; MOE). Finally, at the highest level of metric, results, the distinction between process and outcome becomes blurred, with corresponding metrics emphasizing such long-term benefits as reduced cost of overall training (MOP) and increased survivability (of users when in actual combat; MOE).

Return on Investment

Our initial hypotheses can be combined into a single statement: using higher end VEs reduces training cost compared to other systems or to no training at all. Specifically, we suggest that using a high end VE system will train users faster to an acceptable level of performance over other approaches. Consider a marksmanship task, in which a passing score is 80 percent of total rounds fired landing within a particular region of a target. Assume it is determined that without any training students require eight hours of practice on the actual task to achieve this performance level. Further assume that when students are given access to a high end VE for four hours, it is determined that they subsequently require three hours of training on the real task to achieve the desired performance level, a savings of training time of five hours. In contrast, when students use the low end VE for four hours, it is determined that they subsequently require four hours of training on the real task to achieve the desired performance level, a savings of training time of four hours. To assess the effectiveness of using VEs in this application, we use the training effectiveness ratio (TER; Wickens & Hollands, 2000). The TER is defined as

$$\text{TER} = \frac{\text{amount of training time savings}}{\text{time in training program}}$$

A TER less than 1 indicates the VE is less effective than the current approach, a TER greater than 1 indicates the VE is more effective than the current approach, and a TER equal to 1 indicates no difference in effectiveness. In this example,

$$\text{TER(low-end)} = \frac{4}{4} = 1.0$$

while

$$\text{TER(high-end)} = \frac{5}{4} = 1.25$$

Here the TER for the low end VE is 1, indicating essentially no benefit in terms of time (with the VE students still devoting a total of eight hours of time to training, divided equally between practicing in the VE and on the actual task), while the TER for the high end is 1.25, indicating some benefit in terms of time saved (with the VE students devoting only seven hours of time to training).

This is only a partial assessment of the effectiveness of either VE system. When metrics from other levels, such as behavior and results, are considered, the benefits (or costs) of using the VE system become evident. Such factors as the cost of range time (that is, the number of personnel required to ensure safety, the cost of rounds, wear and tear on the actual weapons, and so forth), as well as the availability of time on the range (sure to be highly restrictive) compared to the availability of time on the VE (depending on how rugged the system, potentially 24/7), compared to the cost of VE development, production, and tech support, may provide additional insights into which VE solution, if any, is ideal.

Validation

The data used to populate the ROI calculations must come from experimentation. Here, in order to capture the depth of data needed to create the trade-off curves, the ideal experiment design would proceed in several phases. Phase 1 would involve iterative usability and human factors assessments. These would be embedded in the system development plan to occur around the point at which key system functionalities were planned for delivery. The outcome of these assessments would then be rank-ordered recommendations for redesign that the developers could then incorporate into their next system build. Phase 2 would include the ToT types of studies. As Roscoe (1980) suggests, in order to develop the actual curves, the design would require exposure of multiple groups of trainees to incrementally longer training regimens in the VE, and the subsequent assessment of performance in a suitable transfer environment (here, training regimen may be defined either in terms of number of trials or trial length; the distinction depends on how the actual training is sequenced). Phase 3 would involve longer-term tracking of the impact of the training system on the organization.

SUMMARY

There is no simple recipe for ensuring successful development of a VE training system, and there are no simple equations for calculating success. Training systems that produce enhanced performance in laboratory settings may fail to yield similar results in the field, or they may fail to be adopted by the community for other reasons having nothing to do with their training effectiveness, but everything to do with their failure to become successfully integrated into the community's concept of operations. Nevertheless, certain steps may be identified which, when followed, will likely increase the likelihood of a system being effective in achieving its performance enhancement goals and successful in becoming part of a user community's training repertoire. The steps discussed in this chapter provide one way of realizing this goal.

REFERENCES

Boldovici, J. A. (1987). Measuring transfer in military settings. In S. M. Cormier (Ed.), *Transfer of learning: Contemporary research and learning* (pp. 239–260). San Diego, CA: Academic Press, Inc.

Chipman, S. F., Schraagen, J. M., & Shalin, V. L. (2000). Introduction to cognitive task analysis. In J. Maarten Schraagen, S. F. Chipman, & V. L. Shalin (Eds.), *Cognitive task analysis* (pp. 3–23). Mahwah, NJ: Lawrence Erlbaum.

Cohn, J. V., Helmick, J., Meyers, C., & Burns, J. (2000, November). *Training-transfer guidelines for virtual environments (VE).* Paper presented at the 22nd Annual Interservice/Industry Training, Simulation and Education Conference, Orlando, FL.

Department of Energy. (2005, October). *How to measure performance: A handbook of techniques and tools.* Retrieved June 6, 2007, from http://www.orau.gov/pbm/handbook/

Durfee, W. (1999). *Concept generation.* Retrieved June 3, 2007, from http://www.me.umn.edu/courses/me4054/lecnotes/generate.html

Kirkpatrick, D. L. (1998). *Evaluating training programs.* San Francisco: Berrett-Koehler Publishers.

Kirwan, B., & Ainsworth, L. K. (1992). *A guide to task analysis.* London: Taylor & Francis.

Lathan, C. E., Tracey, M. R., Sebrechts, M. M., Clawson, D. M., & Higgins, G. A. (2002). Using virtual environments as training simulators: Measuring transfer. In K. M. Stanney (Ed.), *Handbook of virtual environments: Design, implementation, and applications* (pp. 403–414). Mahwah, NJ: Lawrence Erlbaum.

MacLean, K. (2006). *Concept generation and prototyping.* Retrieved June 3, 2007, from http://www.ugrad.cs.ubc.ca/~cs344/resources/supp-concGen&Prototype.html

Nielsen, J. (1993). *Usability engineering.* Boston: Academic Press.

Roscoe, S. N. (1980). *Aviation psychology.* Ames: Iowa State University Press.

Rosenberg, L., & Hyatt, L. (1997). Developing a successful metrics program. Retrieved June 7, 2007, from http://satc.gsfc.nasa.gov/support/ICSE_NOV97/iasted.htm

Ross, D. T., & Schoman, K. E. (1977). Structured analysis for requirements definition. *IEEE Transactions On Software Engineering, SE-3*(1), 6–15.

Salas, E., Burke, C. S., Fowlkes, J. E., & Priest, H. A. (2004). On measuring teamwork skill. In J. C. Thomas (Ed.), *Handbook of psychological assessment: Industrial and organizational assessment* (pp. 428–442). Hoboken, NJ: John Wiley & Sons.

Scanlan, J. (2007). *Concept generation.* Retrieved June 1, 2007, from http://www.soton.ac.uk/~jps7/Lecture%20notes/Lecture%202%20Concept%20generation.pdf

Sproles, N. (2000). Coming to grips with measures of effectiveness. *System Engineering: The Journal of the International Council on Systems Engineering, 3*(1), 50–58.

Stanney, K. M., Mollaghasemi, M., Reeves, L., Breaux, R., & Graeber, D. A. (2003). Usability engineering of virtual environments (VEs): Identifying multiple criteria that drive effective VE system design. *International Journal of Human-Computer Studies, 58*(4), 447–481.

Wickens, C. D., & Hollands, J. G. (2000). *Engineering psychology and human performance* (3rd ed.). Upper Saddle River, NJ: Prentice Hall.

Chapter 11

LEARNING TO BECOME A CREATIVE SYSTEMS ANALYST

Lemai Nguyen and Jacob Cybulski

The important role of creativity has increasingly been recognized in requirements engineering (RE), an early stage in the lifecycle of systems development. Although creativity plays an important role in the discovery, exploration, and structuring of the conceptual space of the requirements problem, creativity has not yet been accepted as an essential ingredient of teaching and learning in RE. This chapter describes a novel approach to learning in RE that synthesizes different dimensions of constructivist learning and creativity education theory to support creative problem exploration and solving in RE. This learning approach will be illustrated through a training environment consisting of face-to-face classroom and online activities, as well as, computer based simulation.

LEARNING CREATIVE REQUIREMENTS ANALYSIS

The development and introduction of a new information system to a business or military organization is an opportunity for innovating or reinventing that organization's practice, processes, or products in order to leverage their benefits and create value. Requirements engineering is an early process in the systems development lifecycle where innovation plays an especially important role. In general, RE involves the creation of a vision for the future system through the discovery, analysis, modeling, and validation of user requirements. Specifically in the context of this book, RE involves the elicitation, modeling and analysis, specification, and verification and validation (see Volume 1, Section 2 Perspective) of training system requirements. During this process, the systems analyst (requirements engineer) works with various systems development teams and stakeholders often including the management, business people and users (for example, educators and learners of the training system), technology vendors, and possibly the organization's business partners and/or customers.

The description of the requirements engineering topic is covered extensively in terms of process models, requirements elicitation and modeling techniques, support tools, approaches to validating and managing requirements, and documentation and templates. Interested readers are directed to see various textbooks (for example, Robertson & Robertson, 2005; Dennis, Wixom, & Tegarden, 2004; Kotonya & Sommerville, 1998; Sommerville & Sawyer, 1997) or research reviews (for example, Nuseibeh & Easterbrook, 2000; Gervasi, Kamsties, Regnell, & Achour-Salinesi, 2004; Opdahl, Dubois, & Pohl, 2004). Tremendous effort has focused on describing and supporting the systems analyst in the construction of a requirements specification that reflects the real world problem situation through understanding and solving the problem as perceived by the user. This chapter, however, focuses on an alternative view of requirements engineering—creativity—and proposes an approach to training creative systems analysts.

Recently, it has been argued that to be effective the systems analyst should also be an inventor (Robertson, 2005), and it is essential that the RE process itself is creative as well (Nguyen & Swatman, 2006). These two emerging arguments open a challenge to the RE community: how best to train and learn to be a creative systems analyst. This chapter addresses this challenge by describing the creativity aspects of RE, discusses advantages and limitations of current education approaches in RE, and proposes a new approach to learning to become a creative systems analyst.

TEACHING AND LEARNING IN REQUIREMENTS ENGINEERING

Overview of Current Teaching and Learning Approaches

Overall, there are three major approaches to learning RE: (1) taking an industry-intensive course (often ranging from half a day to several days), (2) taking requirements engineering as one of the subjects in a tertiary course (graduate diploma, graduate, or postgraduate degrees), or (3) workplace learning (often working alongside expert systems analysts). Each of these learning approaches has advantages and disadvantages. Analysis of these advantages and disadvantages supports the need to incorporate and promote creative thinking into these learning approaches.

Learning RE through an Industry-Intensive Course

Industry-intensive courses (or workshops) are often provided by various professional associations and consulting or training companies. These are often instructor led and sometimes can be delivered via a computerized learning system. These courses aim at providing formal knowledge (about processes, techniques, notations, and tools) over a short period of time with small illustrative exercises to allow learners to apply and acquire some practical skills. Limitations of such courses are the unrealistic setting of exercises and a condensed delivery of rich materials (RE knowledge). Due to these limitations, the learner often faces a gap between the knowledge acquired from the course and its application in practice, or a mismatch between "approved" practice and "actual" practice (Nguyen, Armarego, & Swatman, 2005).

Learning RE through a Subject(s) Included in a Tertiary Course

Students who are enrolled in such degree programs as Information Systems (IS), Software Engineering, or Computer Science often learn RE in a course such as Systems Analysis and Design or in a specific Requirements Engineering course. Such subjects are commonly offered in a single semester with a wide range of classroom, as well as self-paced, learning activities. Typically, lectures and tutorials (or laboratory work) allow the teacher to transfer formal knowledge (processes, techniques, notations, and sometimes tools) and allow students to apply the knowledge received though illustrative exercises or discussion questions. Assignments are often used to enable students to self-learn by drawing and applying relevant knowledge to a given problem. Common advantages of the tertiary learning approach include the acquisition of rich knowledge and opportunities to work on practical exercises repeatedly during a semester. Many problem based assignments are conducted in groups; therefore, they allow learners to interact with each other to discuss and share their learning. Common limitations of this approach include the controlled setting of class activities (especially time), unrealistic practical exercises, and assignments with predefined, teacher-designed problem space (see, for example, Minor & Armarego, 2004).

To overcome the lack of realistic practical exercises and assignments, many universities provide learners with a project based or an industry placement course, often scheduled near the completion of their qualifications. In such courses, learners are engaged in small, self-managed projects with an assigned client or work with a team of professionals at their workplaces. While such project based or industry placement courses support experiential learning, they are also a major source of problems, including inadequate provision of teaching and technical resources, elevated teaching costs, lack of available industry partners/projects, and uncertainties from the workplace environment, which may interfere with the curriculum program and course syllabus set by the teacher and/or the university.

Learning RE at the Workplace (On the Job Training)

Many practitioners learn on the job by working alongside more capable experts. Advantages of this approach to learning include the learner's participation in realistic cases, adoption of real roles and responsibilities, and acquisition of experience in dealing with real clients in real organizational settings; all of these provide a rich experiential learning environment that enables an authentic vocational knowledge acquisition process. However, limitations of this approach include a lack of access to formal (and appropriate) knowledge, a lack of a pedagogical process taking the learner from simple to complex tasks, and a reluctance of industry participants to share their knowledge (Billett, 1995). Due to many business impediments and management's reluctance to accept the high risks associated with innovative ideas (Cybulski, Nguyen, Thanasankit, & Lichtenstein, 2003), the workplace cannot be treated as a safe learning "playground" in which a flexible and constraint-free environment would allow the learner to try out

Table 11.1. Characteristics of Learning Approaches

Learning Approaches	Characteristics
Industry intensive courses	• Instructor led, classroom activities, formal structured knowledge, unrealistic setting, and condensed materials within a short time
Formal tertiary courses	• Instructor led, a range of activities from classroom to project based, formal structured knowledge, and rich knowledge on a semester basis • Project based learning attempts to address the unrealistic setting at the expense of providers' resources
Workplace learning	• Self-learning, real setting, and rich experiences gained • Lack of formal structured knowledge, lack of a pedagogical approach, high risk to "experiment" ideas

different (and potentially dangerous) strategies when learning different concepts and techniques. See Table 11.1.

Discussion

A range of approaches to learning RE have been developed and adopted in professional training and higher education. Each comes with its own benefits and limitations. Formal education (course based learning) aims primarily at the acquisition of RE processes (analysis and modeling), techniques, notations, requirements management, and other general abilities (such as communication and team skills); however, it lacks exposure to realistic and collaborative industry projects (Minor & Armarego, 2004). A literature survey by Dallman (2004) noted a lack of learning support for creative thinking, cognitive flexibility, and metacognitive learning strategies in current formal education. Workplace learning, while providing realistic projects, lacks access to formal knowledge and pedagogical processes (Billett, 1995). At the same time, practitioners who are well positioned to effectively transfer their professional experience to the RE learners are not well informed of creativity techniques that may apply to their relevant RE practice (Maiden & Robertson, 2005).

The Creativity Problem-Based Learning framework (Armarego, 2004) was developed to integrate cognitive flexibility, metacognitive learning strategies, and constructivist learning elements; to allow the learner to learn in a situated experiential environment; and to provide cognitive apprenticeship by working with an expert coacher. Armarego's approach provides a rich and flexible learning environment to enable authentic knowledge acquisition and encourage creative thinking. While benefits have been reported (Armarego, 2004), it is yet unclear how the framework supports the inclusion of creativity theory. Creativity processes and techniques to generate ideas and solutions, to extend the conceptual space, and to evaluate the creative outcome can be included within such a framework in an informed and structured way.

CREATIVITY IN RE

There seem to be two distinct views of the RE process within the RE community. The first view, held by many authors, considers problem solving in RE as a systematic, structured, and evolutionary process, during which the problem is gradually explored, refined, and structured into the requirements model. Various methods have been proposed to guide the systems analyst to decompose the user's problem and compose the requirements model using different decomposition approaches, modeling techniques, and notations (for example, see Jackson, 2005; Kotonya & Sommerville, 1998; Dennis, Wixom, & Tegarden, 2004). The second—and new—view of the requirements process emerged from action research and case studies (Nguyen & Swatman, 2003; Nguyen, Carroll, & Swatman, 2000; Nguyen, Swatman, & Shanks, 1999), which reveal episodes of insight-driven reconceptualization and restructuring of the requirements model during the generally incremental development of the model. These restructuring episodes can be characterized as "Aha!" moments during which the systems analyst unexpectedly sees a new perspective of the problem and, as a result, restructures the requirements model significantly. These studies confirm the Gestalt psychology theory of insight and restructuring in ill-structured problem understanding and solving (Mayer, 1992; Ohlsson, 1984). Furthermore, this new view of the requirements process emphasizes that the problem in RE is not given (not there waiting to be elicited), but instead emerges as the systems analyst enters the situation, learns, explores, and discovers different problem areas when interacting with the situation and various stakeholders. Hence, the RE process itself can be seen as a constructivist process. Nguyen and Shanks (2006b) noted two analogical views of the design process held within the design studies community—where the design process is seen either as a rational problem solving process (Simon, 1992) or a constructivist process (Schön, 1996). These two views represent two forces of problem solving: the enforcement of a structured process to avoid chaos and errors, as opposed to relaxation of constraints in dealing with the emergent problem space by taking advantage of opportunistic cognitive behaviors and heuristics of participating professionals (Nguyen & Shanks, 2006b). Nguyen and Shanks further suggested that these two views are complementary and need to be integrated to support a collaborative process consisting of cycles of structured building of and opportunistic restructuring of the requirements model.

Robertson (2005) set a challenge to RE practice to recognize the importance of discovery and invention of new ideas in the requirements acquisition process rather than simply relying on passive elicitation and analysis of what users say they need. This challenge spurred a review of the role of the systems analyst during the elicitation process, which now has been described as the requirements discovery process. A series of creativity workshops in RE were conducted by Maiden and his colleagues at City University, London, United Kingdom (Maiden & Robertson, 2005; Maiden, Manning, Robertson, & Greenwood, 2004; Maiden & Gizikis, 2001), in which they demonstrated how various creativity techniques, such as brainstorming, domain mapping, analogy reasoning, and constraint removal, to name a few, can be incorporated within structured RE processes to discover and explore ideas and requirements. Other creativity techniques were also suggested to be incorporated within the requirements elicitation by other researchers (Mich, Anesi, & Berry, 2004; Schmid, 2006). Nguyen and Shanks (2006a) reviewed different characteristics of the creative processes in the creativity literature and design studies, related them to the RE process, and called for an integrated process and tool environment to support the systems analyst in adopting creative techniques and tools capable of exploring and structuring the problem space in RE. From a combination of collaborative and cognitive perspectives, a group of researchers at the University of South Australia currently investigate and develop an ICT-enabled[1] environment to support creative team problem solving using the distributed cognition theoretical foundation (Blackburn, Swatman, & Vernik, 2006).

Another challenge in supporting creativity in RE in an organizational setting was discussed at length by RE practitioners and business and IT managers participating in a focus group (Cybulski et al., 2003). The management practice and organizational culture strongly influence not only the development, but also the appraisal and adoption of creative IT-enabled solutions to business problems. According to Nguyen and Shanks's (2006a) creativity framework for RE, novelty, value, and surprisingness can be used as three characteristics to recognize and evaluate the creative outcome in RE. Novelty refers to the extent that the new system is different from existing systems. Value refers to the usefulness, correctness, and fit (appropriateness) of the system in the context of use. Surprisingness refers to the unexpected features of the system. Research is currently under way to define ways to assess these characteristics. To support creativity in RE, it is also important to appreciate changes to norms that have traditionally been accepted within, practiced by, and grounded in the organizational culture (Regev, Gause, & Wegmann, 2006). In a similar vein, the Creativity in Requirements Engineering framework classifies and describes various individual and organizational factors that influence creativity by the systems analysts (Cybulski et al., 2003; Dallman, Nguyen, Lamp, & Cybulski, 2005).

Overall, creativity has recently received increasing interest within the RE research community. Creativity techniques and tools can be integrated within various requirements engineering approaches to inform and support systems analysts in their collaborative effort to invent and develop requirements for new information systems, including virtual environments for training and education in the military. While such integrated approaches promise potential benefits, the primary focus of the chapter is the creative systems analyst as an expected outcome of a training environment. Creativity plays an important role for those systems analysts who want to add novelty and value while exploring and constructing the problem space and subsequently when solving the problem. However, fostering creativity in RE practice and teaching creative methods in RE education have so far received very little attention (Dallman, 2004; Armarego,

[1]ICT: information and communication technology.

2004; Nguyen et al., 2005). As a result, experienced practitioners and RE learners are not well informed of how to practice RE creatively.

LEARNING TO BECOME A CREATIVE SYSTEMS ANALYST

Learning Process

Throughout the RE literature, there has been a common agreement that the RE process can be characterized as application domain specific, technical, and contextual, that is, embedded within a specific organization and social setting (Sutcliff & Maiden, 1998; Jackson, 2005; Coughlan & Macredie, 2002; Checkland & Scholes, 1999; Goguen, 1997). Further, RE can be seen as a process of solving "wicked" problems that involve technical issues, social complexity, and dynamics (Conklin, 2006). The problem solving activity in RE requires both problem understanding as well as problem solving (Visser, 1992). The problem solver continually interprets the problem situation, constructs a knowledge representation of the problem, and forms and evaluates possible solutions. This intertwining process of problem understanding and solving is reflected in the incremental structuring and occasional restructuring of the requirements model. There are important implications for the systems analyst to be viewed as a learner.

- The emergence of the problem situation suggests that RE itself is a learning process, more specifically, a constructivist learning process during which the learner constructs his or her knowledge by structuring and reflecting upon the emergent problem (Gero, 1996; Schön, 1996; Armarego, 2004; Robillard, 2005).
- Creativity plays an important role for the exploration, construction, and expansion of the problem space. Indeed, creativity is defined as an internal process of exploration and transformation of the conceptual space in an individual mind (Boden, 1991, 1998).
- The problem in RE is of a technical as well as social nature (Conklin, 2006). Therefore the systems analyst's learning process takes place in a domain specific, social, and collaborative context.

The above implications led us to believe that the fundamental objectives of RE education must also be reevaluated, which led us to grounding the RE learning approach in a synthesis of the constructivist learning and creativity education theories.

Based on Piaget's (1950) theory, constructivist learning refers to the authentic and personal building up of knowledge. This knowledge building process occurs in the individual learner's mind through two mechanisms: assimilation and accommodation. Assimilation occurs when the learner interprets and incorporates new learning into an existing conceptual framework representing his or her knowledge of a topic area. Accommodation occurs when the learner could not fit the new learning into his or her existing framework; as a result, he or she reframes (restructures) the existing conceptual framework. These two mechanisms are consistent with the structuring and restructuring activities in RE (Nguyen & Swatman, 2006). Vygotsky (1978) stresses the important role of a combination of collaboration among learners (through which the learner receives feedback and coaching) and practical exercises (through which the learner constructs knowledge and gains skills). These underpinning theories of constructivist learning have been synthesized into the three dimensions of endogenous, exogenous, and dialectic constructivism (Moshman, 1982):

- *Endogenous* dimension: The learner learns through an *individual* construction of knowledge. Accommodation and assimilation are two mechanisms that enable the endogenous construction of knowledge. The teacher can play a facilitator role, but the learner takes a more active role and assumes ownership of his or her learning and knowledge building.
- *Exogenous* dimension: The learner learns from a combination of *formal instructions* and *realistic and relevant exercises* through which he or she refines knowledge through instructions and feedback received from the teacher when undertaking practical exercises.
- *Dialectic* dimension. The learner learns through *collaboration and interaction* with teachers (experts) and peers through realistic experiences. The *scaffolding* provided by the more capable collaborators is especially important.

Dalgarno (2005) developed a three-dimensional learning environment that incorporated elements from these three different dimensions of constructivist learning. His successful application of this learning environment in teaching chemistry encouraged us to pursue a rich RE learning environment in which the learner will be supported with elements from the above constructivism dimensions. The learning should take place through a range of learning activities: knowledge acquisition from formal instructions, practical exercises and project based realistic experiences, as well as collaborative and individual construction of knowledge.

While extrapolating this view of constructivist learning, we have examined the issue of creativity education, where there has been an argument about whether creativity is a domain-specific or domain-general ability. There has been a strong view that creativity is inherently associated with a certain type of intelligence and that domain expertise is required to identify to what extent a creative product extends a domain knowledge boundary (Solomon, Powell, & Gardner, 1999; Gardner, 1993). Therefore, creativity should be seen as domain specific and creativity education should be adapted to a specific domain. However, Root-Bernstein and Root-Bernstein (2004) argued that creativity should rather be seen as domain general because it is inherently associated with commonly intuitive and metacognitive capabilities; therefore, creativity education should target intuitive and metacognitive learning. Baer and Kaufman (2005) argued that creativity includes both domain-general as well as domain-specific capabilities. They developed the Amusement Park Theoretical (APT) model for creativity education. Their APT integrates both domain-general and domain-specific creativity elements. Based on this theory, domain-general creativity elements include intelligence, motivation, and environment, that is, creativity-supported culture; whereas domain specific creativity elements are categorized from a general thematic area to a domain and microdomain subarea. APT has been suggested as having potentials in creativity education in RE (Nguyen & Shanks, 2006a). We adapt APT specifically to the RE domain:

- At the level of general thematic creativity: intelligence can be determined as problem understanding and solving and social and communication skills; individual motivation should be recognized and linked to learning objectives, and learning environment elements need to be identified and linked to the constructivism dimensions.
- At the level of RE specific creativity: business knowledge (for example, training programs and processes in military), technology knowledge, analysis and modeling techniques and tools, and creativity techniques and tools should be integrated to generate creative ideas and to recognize and evaluate creative products.

Having synthesized and adapted the above constructivism dimensions and APT theory to RE, we propose a learning environment including the elements shown in Figure 11.1 to support a constructivist learning approach that incorporates creativity learning for systems analysts.

A creativity-supported culture is identified as an element at the general creativity level in APT (Baer & Kaufman, 2005). This element is adapted in our approach as a simulated learning environment to support different constructivism dimensions (through promoting flexibility in framing and reframing knowledge and collaborative creativity). Different levels of (domain-general and domain-specific) creativity elements are integrated within this learning environment. This adaptation assists the learner in recognizing and understating constructivism

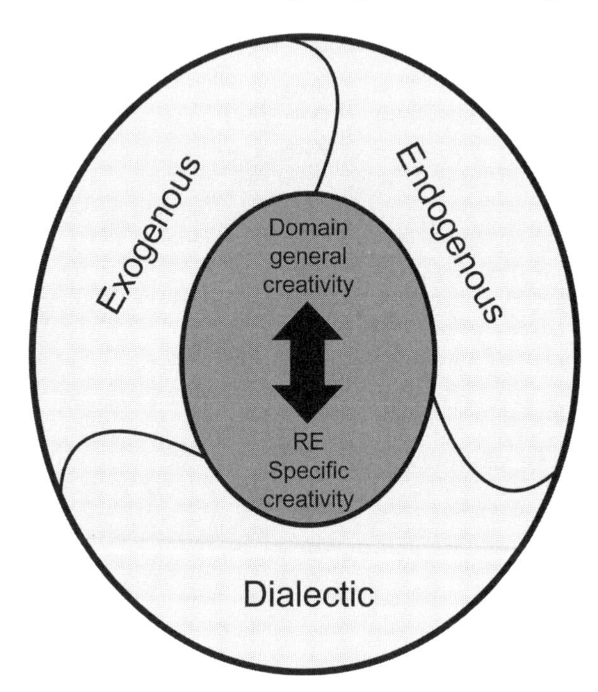

Figure 11.1. Incorporating Creativity Learning within Constructivist Learning

dimensions supported by a particular learning program and taking advantage of how the program could support his or her learning process. For example, a course based program (formal education) would potentially support the exogenous dimension of learning, in which case, the learner should apply formal instructions from the learning program while working on RE exercises. Based on the feedback from the instructor, the learner should refine and clarify the knowledge received. However, with RE workplace learning, the learner should apply the accommodation and assimilation mechanisms proactively in more realistic experiences and seek collaboration and formal approval of knowledge constructed from time to time. As different constructivism dimensions are integrated within our proposed environment, the learner needs to recognize them and take advantage of their integration (see the next section).

At the level of general problem solving creativity, our proposed approach includes the following:

- Elements for intelligence building, such as problem understanding and solving capabilities (for example, problem recognition, strategy planning, idea generation and brainstorming, solution formulation and evaluation, and so forth).
- Elements for identifying and communicating motivations. Individual learner's motivations and learning objectives need to be identified and communicated with the teacher (or coach) to align reward mechanisms to suit individual learning objectives and motivations.
- Support for social interactions and collaboration in problem based projects, within and cross-team communication, and with facilitator(s). A combination of social software and face-to-face interactions can be used to facilitate electronic communication and collaboration to allow the learner to acquire social and communication skills (team building, negotiation, exchange of information, group collaborative support, and so forth).

At the RE specific level, our proposed approach integrates the following:

- Support for the learner to learn through relevant experiences (small exercises, case studies, and projects) through providing appropriate knowledge and instructions—processes (such as Waterfall, Rapid Application Development, Agile development, and so forth), elicitation techniques (such as scenario based interviews and observation), modeling techniques (such as use case, object oriented, data flow diagram, entity relationship, and so forth), and requirements management tools (see Volume 1, Section 2 Perspective).
- Support for individual as well as collective creativity. Creativity techniques, such as brainstorming, imagination, search for ideas, idea association, analogical thinking and play, as well as the use of creativity tools, will be integrated within the RE lifecycle.
- Support for flexible cognitive processes and support for monitoring the evolutionary structuring and insight-driven restructuring of the requirements model.

The next section will illustrate the proposed conceptual framework in a case in which creativity was incorporated within a constructivist learning approach demonstrated by students undertaking an RE subject in various master degree programs, including business, commerce, and information systems.

A Case of Learning Creative RE Using a Simulated Learning Environment

A project in RE (Cybulski, Parker, & Segrave, 2006a, 2006b) was designed to enhance and enrich the learners' abilities to discover and elicit information systems requirements (from both business and technology viewpoints)—an essential skill of the information systems professional. While teaching requirements elicitation is common in information systems and software engineering schools, such teaching is usually limited to conducting simple interviews and formalizing the collected information into a requirements specification. The more challenging requirements elicitation skills, which unfortunately are very often neglected, include detection of conflicting and redundant information, handling omission of essential facts, and dealing with the absence of management approval and customer feedback to fully validate the collected and analyzed requirements. Through our project, the learner was engaged in various activities to overcome the above-mentioned problems, to independently and collaboratively seek solutions to these problems, and to apply some creative approaches to dealing with the shortcomings of the specified requirements. In this way, the project was designed to support endogenous and dialectic constructivism and creativity learning at a general thematic level of (business) problem solving creativity.

Our RE project (codenamed FAB ATM) required the learners to work in teams to produce specification of a banking product (a new generation of automated teller machines [ATMs]). In the initial stages of the project, the teams used a computer simulation (henceforth called FAB ATM simulation) of a virtual meeting room, where the learners had an opportunity to meet with the simulated staff of a hypothetical banking organization (FAB—the First Australian Bank). During the meetings, the teams conducted a series of interviews with a view to collecting requirements for their project. The simulated interviews allowed project teams to first design interview questionnaires and then engage simulated interview participants in a lengthy conversation (see Figure 11.2). The requirements elicited in the process of such interviews represented distinct viewpoints of the bank staff, for example, a technical officer or a branch manager. After the interviews, the learners had to analyze the collected requirements and identify redundancies, conflicts, and omissions; all of which had to be reconciled, removed, or filled in with information obtained in the process of self-directed research. Results of these activities were eventually presented to the bank manager (role-played in the real world) for validation and the final approval and subsequently sealed in the form of a consistent specification document.

Immersive educational simulations, games, and role-playing were central to the conduct of our RE project. To support different constructivism dimensions, we used a blended simulation learning environment, where some activities were conducted in classes (lectures and tutorials), some in project teams (face-to-face meetings, online discussion boards, and chat rooms), and yet others with the use of a virtual (meeting) environment—named Deakin LiveSim (Cybulski et al., 2006a, 2006b). Through formal classes (lectures and tutorials), the blended learning environment supported exogenous constructivism. Through a range of inter- and cross-team communication activities and project-related consultation

Figure 11.2. Simulated RE Interviews with Multiple Participants in a Corporate Environment

(provided by teachers), the blended environment supported dialectic constructivism. Through simulated interviews, documentation collection, and interpretation, the learners acquired general problem understanding and problem solving skills. In the provided qualitative feedback, the learners generally praised the experiences gained with the simulation; for example, they said,

> The interview CD is really a very good idea, which offered us a virtual interview environment via multiple media technology.
>
> The interview stimulation program did offer us a chance to be a part of interview, to touch it, to feel it and to experience it.
>
> The actual interview simulation session was very informative and convenient allowing some flexibility in the actual interview technique.

To incorporate RE creativity learning, a typical RE lifecycle was adopted in the FAB ATM simulation project (see Figure 11.3). The process took the learners through the learning "funnel," which leads them from the fuzziest (completely open to imagination and creativity—for example, extending ATM with share trading, Web based, or human-touch user interfaces) to the most formal and constrained knowledge (which requires breaking of technology and business dogmas to arrive at some workable solutions). Individual learners started their project work

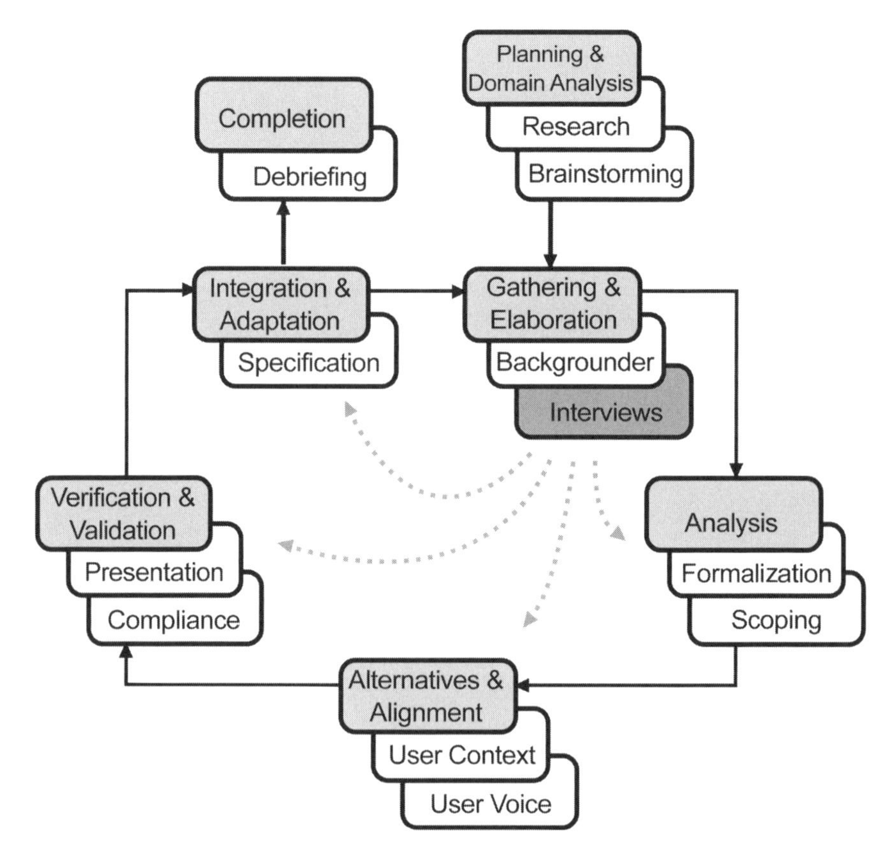

Figure 11.3. RE Activities Supported by the Simulated Learning Environment

by investigating the problem domain (research) and then in groups exploring and planning their projects (that is, both individual and collective brainstorming and scenarios solution exploration). These were followed by gathering requirements (via a face-to-face backgrounder and interviews with simulated people), analyzing the discovered and elicited requirements (to include their formalization and scoping), and later investigating information systems requirements, alternative solutions, and possible business alignment issues (using user context and user voice analysis teaching). Their next stage involved the verification and validation of requirements (using a formal presentation and feedback collection), integration of new requirements with adapted legacy requirements (in a specification document), and finally the project completion. All learners were also asked to reflect upon, elaborate, and document knowledge and experience gained.

In the FAB ATM project, the simulation was used to confront the learner's (often unstoppable) creativity, imagination, preconceptions, and ideas with "reality." The information was gathered by asking questions and listening to the answers provided by the simulated people, observing their body language and passing judgment on the degree of trust that could be vested in them, taking notes, working with vastly incomplete data and working under pressure of time, and engaging in independent investigation and collaboration with team members and the simulated people.

The tasks that specifically demanded learners to invoke their creative problem solving can be found across the entire project and the RE lifecycle, but it could be specifically located in a number of problem domains, that is, in aiming at business/IT alignment, coping with the richness of the stakeholder base, overcoming deficiencies of the legacy system, setting requirements for technology reuse, dealing with technology selection and innovation, and facing the challenges of the imminent business change.

While this blended simulated learning environment (such as that used in the FAB ATM project) cannot completely replace student placement in a real organization, it provides the learners with a safe environment in which they can experiment with different possible outcomes (Cybulski et al., 2006a, 2006b). The FAB ATM project adopts a partial view of reality, which can be referred to as "circumscribed" reality. Such circumscribed reality simulations attach only key aspects of authenticity to their objects and environment. While they sacrifice some degree of reality, at the same time, they never cross the threshold of acceptability to the learner. The FAB ATM simulation provides learners with rich interactivity, which relies on a state machine implemented in Macromedia Flash and which in real time combines video fragments of live people to deliver conversational characters with meaningful behavior. While media form and interaction are simple for the learner, the complexity is created in the learner's mind rather than in the technology used to support the environment. Finally, any educational computer simulation ought to be part of a larger educational framework with many aspects of learner experience. Hence, our FAB ATM computer simulation supported endogenous constructivism.

In addition, our FAB ATM simulation provided the teacher with an opportunity to be in control of educational outcomes (by defining objectives to be reached) and processes (by setting tasks to be undertaken, stages to be completed, and methods to be used by learners) and to achieve the comparability of gained experience, which is hard to attain in the real-life projects and student placement situations. The FAB ATM project provided us with many opportunities to apply innovative and effective learning styles (see Figure 11.4). In addition to the traditional ways of learning by "being told" in lectures, "by discovery" in tutorials and "by doing" in projects, the FAB ATM project also provided avenues for learners to learn by experiencing work and by taking on professional roles of business consultants and systems analysts. All these learning styles are actively pursued in lectures (via demonstrations), tutorials (via discussions), and projects (via a virtual environment). This is achieved by learners being immersed in an authentic and believable simulation environment (such as that used in FAB ATM simulation) and by conducting realistic tasks that allow students to learn "by observing" people's behavior in a complex corporate setting, "by playing" the professional roles, and "by communicating" and "by collaborating" with their team members

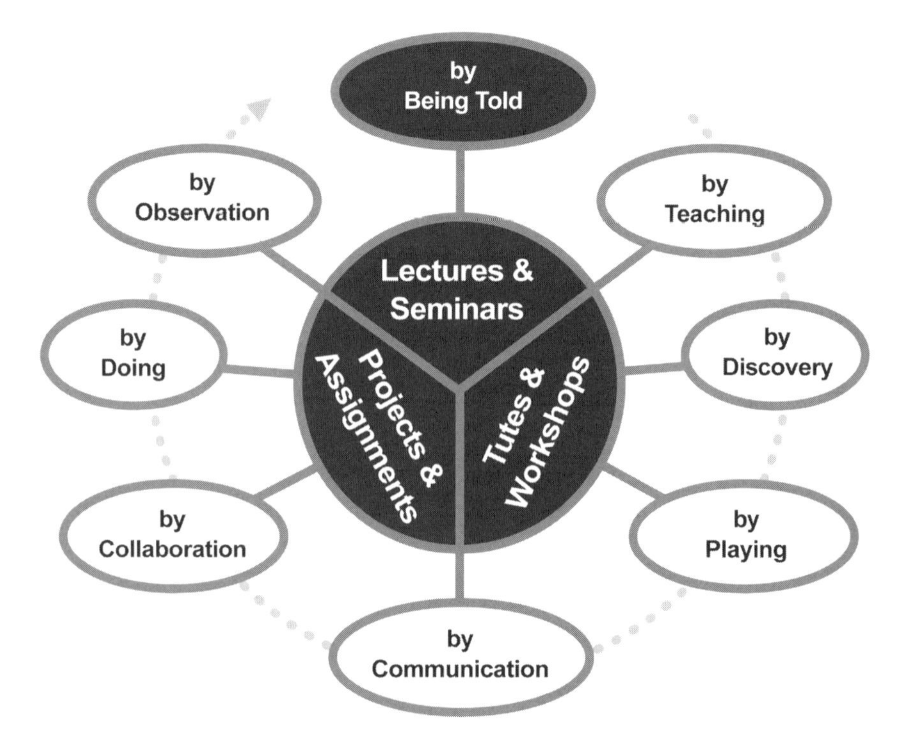

Figure 11.4. Learning Styles Supported by the Simulated Learning Environment

and with the simulated characters (in both virtual and real contexts). Finally, learners also took on the responsibility of teaching each other in face-to-face meetings and online discussion. The richness of the available learning styles offered RE teachers alternative paths to students' minds, to the seamless creation of new knowledge and skills, and most importantly, to the effective development of professional experience. Through all these, our learning environment supported elements of endogenous, dialectic, and exogenous constructivism learning and teaching. While in our FAB ATM project, the simulation system simulated a technology innovation project in the banking domain, it was based on the typical RE lifecycle. Therefore, the system has the potential to simulate a requirements project in other domains, for example, to develop requirements for a VE training system.

The blended approach to educating creative systems analysts provided us with an opportunity to arrive at a compromise between educational outcomes (acquiring knowledge, developing skills, embracing creativity, and gaining experience) and environmental constraints (time, costs, labor, and quality). We relaxed the confines of the problem settings to foster students' creativity and then confronted them with the reality of which rigidity could be overcome only by breaking technical and business dogmas in the creative fashion. By circumscribing the learner's reality, we used a combination of simulated reality and virtuality to immerse individuals in the authentic and believable problem situation, yet we were able to control educational outcomes and provide the safety of the protected educational context. We used a variety of media and learning approaches to support the learning process, not only to facilitate students to gain skills, knowledge, and creativity, but also to achieve these objectives creatively.

CONCLUSION

This chapter weaves a story of requirements engineering education. As many other good stories, the chapter provides a lesson to learn for the reader, be you a practitioner, an educator, or a student. As the nature of information systems changes, so does the role of systems analysts, who are now required to act not only as human repositories for users' wishes, demands, and requirements, but also to become inventors, innovators, and learners and be the facilitators of such innovativeness and scholarship among their clients and users. Thus, the shift from requirements elicitation to requirements discovery poses new challenges for RE practice, and a major problem is the apparent lack of the creative knack in the systems analysts' skill portfolio. This is also a challenge for RE educators, who need to expose their students to the authentic and believable situations in which learners can be immersed in realistic problems and in which they can truly experience the processes of domain learning and problem solving and the wickedness of the social and organizational complexities, which constantly redefine the problem and rescope its many solutions. Games, role-playing, and simulations could become part of the answer to the newly posed challenges. However, yet again we may find that it is of fundamental importance for learners and educators to be inventive and open to self-improvement and learning. And so, we, too, need to take the creative path of risk and innovation and to employ new and exciting approaches to using learning and teaching technologies.

REFERENCES

Armarego, J. (2004, December). *Learning requirements engineering within an engineering ethos.* Paper presented at the 9th Australian Workshop on Requirements Engineering (AWRE'04), Adelaide, Australia.

Baer, J., & Kaufman, J. C. (2005). Bridging generality and specificity: The Amusement Park Theoretical (APT) model of creativity. *Roeper Review, 26,* 158–163.

Billett, S. (1995). Workplace learning: Its potential and limitations. *Education + Training, 37*(5), 20–27.

Blackburn, T., Swatman, P., & Vernik, R. (2006, May). *Cognitive dust: Linking CSCW theories to create design processes.* Paper presented at the 10th Computer Supported Cooperative Work in Design (CSCWD'06), Nanjing, China.

Boden, M. A. (1991). *The creative mind: Myths and mechanisms.* New York: Basic Books, Inc.

Boden, M. A. (1998). Creativity and artificial intelligence. *Artificial Intelligence, 103,* 347–356.

Checkland, P., & Scholes, J. (1999). *Soft systems methodology in action: A 30-year retrospective.* New York: Wiley.

Conklin, J. (2006). *Dialogue mapping: Building shared understanding of wicked problems.* Hoboken, NJ: John Wiley and Sons Ltd.

Coughlan, J., & Macredie, R. D. (2002). Effective communication in requirements elicitation: A comparison of methodologies. *Journal of Requirements Engineering, 7*(2), 47–60.

Cybulski, J., Nguyen, L., Thanasankit, T., & Lichtenstein, S. (2003, July). *Understanding problem solving in requirements engineering: Debating creativity with IS practitioners.* Paper presented at the Pacific Asia Conference on Information Systems (PACIS 2003), Adelaide, Australia.

Cybulski, J., Parker, C., & Segrave, S. (2006a, December). *Touch it, feel it and experience it: Developing professional IS skills using interview-style experiential simulations.* Paper presented at the 17th Australasian Conference on Information Systems (ACIS 2006), University of South Australia, Adelaide.

Cybulski, J., Parker, C., & Segrave, S. (2006b, December). *Using constructivist experiential simulations in RE education.* Paper presented at the 11th Australian Workshop on Requirements Engineering (AWRE'06), University of South Australia, Adelaide.

Dalgarno, B. (2005). The potential of 3D virtual learning environments: A constructivist analysis. *e-Journal of Instructional Science and Technology (e-JIST), 5*(2).

Dallman, S. (2004). *What creativity do students demonstrate when undertaking requirements engineering.* Melbourne, Australia: Deakin University.

Dallman, S., Nguyen, L., Lamp, J., & Cybulski, J. (2005, May). *Contextual factors which influence creativity in requirements engineering.* Paper presented at the 13th European Conference on Information Systems (ECIS 2005), Regensburg, Germany.

Dennis, A., Wixom, B. H., & Tegarden, D. (2004). *Systems analysis and design with UML Version 2.0: An object-oriented approach* (2nd ed.). Hoboken, NJ: John Wiley & Sons.

Gardner, H. (1993). *Frames of mind: The theory of multiple intelligences* (10th anniversary ed.). New York: Basic Books.

Gero, J. S. (1996). Creativity, emergence and evolution in design: Concepts and framework. *Knowledge-Based Systems, 9*(7), 435–448.

Gervasi, V., Kamsties, E., Regnell, B. O., & Achour-Salinesi, C. B. (2004, June). *Ten years of REFSQ: A quantitative analysis.* Paper presented at the International Workshop on Requirements Engineering: Foundation for Software Quality (REFSQ'05), Riga, Latvia.

Goguen, J. A. E. (1997). Towards a social, ethical theory of information: In social science research. In G. Bowker, L. Gasser, L. Star, & W. Turner (Eds.), *Technical systems and cooperative work: Beyond the Great Divide* (pp. 27–56). Mahwah, NJ: Lawrence Erlbaum.

Jackson, M. (2005). Problem frames and software engineering. *Information & Software Technology, 47*(14), 903–912.

Kotonya, G., & Sommerville, I. (1998). *Requirements engineering: Processes and techniques.* West Sussex, England: John Wiley & Sons.

Maiden, N., & Gizikis, A. (2001). Where do requirements come from? *IEEE Software, 18*(5), 10–12.

Maiden, N., Manning, S., Robertson, S., & Greenwood, J. (2004, August). *Integrating creativity workshops into structured requirements processes.* Paper presented at the 2004 Conference on Designing Interactive Systems, Cambridge, MA.

Maiden, N., & Robertson, S. (2005, September). *Integrating creativity into requirements engineering process: Experiences with an air traffic management system.* Paper presented at the 13th IEEE International Conference on Requirements Engineering (RE'05), Paris, France.

Mayer, R. E. (1992). *Thinking, problem solving, cognition* (2nd ed.). New York: W. H. Freeman and Company.

Mich, L., Anesi, C., & Berry, D. M. (2004, June). *Requirements engineering and creativity: An innovative approach based on a model of the pragmatics of communication.* Paper presented at the Requirements Engineering: Foundation of Software Quality (REFSQ'04), Riga, Latvia.

Minor, O., & Armarego, J. (2004). Requirements engineering: A close look at industry needs and model curricula. *Proceedings of the 9th Australian Workshop on Requirements Engineering* (AWRE'04; pp. 9.1–9.10). Available from http://awre2004.cis.unisa.edu.au/

Moshman, D. (1982). Exogenous, endogenous, and dialectical constructivism. *Developmental Review, 2,* 371–384.

Nguyen, L., Armarego, J., & Swatman, P. (2005, December). *Understanding the requirements engineering process: A challenge for practice and education.* Paper presented at the 7th International Business Information Management Association, Cairo, Egypt.

Nguyen, L., Carroll, J., & Swatman, P. A. (2000, January). *Supporting and monitoring the creativity of IS personnel during the requirements engineering process.* Paper presented at the 33rd Hawaii International Conference on System Sciences (HICSS-33), Maui, HI.

Nguyen, L., & Shanks, G. (2006a, December). *A conceptual approach to exploring different creativity facets in requirements engineering.* Paper presented at the 17th Australasian Conference on Information Systems (ACIS 2006), Adelaide, Australia.

Nguyen, L., & Shanks, G. (2006b, September). *Using protocol analysis to explore the creative requirements engineering process.* Paper presented at the Information Systems Foundations: Theory, Representation and Reality, 3rd Biennial ANU Workshop on Information Systems Foundations, Canberra, Australia.

Nguyen, L., & Swatman, P. A. (2003). Managing the requirements engineering process. *Requirements Engineering, 8*(1), 55–68.

Nguyen, L., & Swatman, P. A. (2006). Promoting and supporting requirements engineering creativity. In A. H. Dutoit, R. McCall, I. Mistrik, & B. Paech (Eds.), *Rationale management in software engineering* (pp. 209–229). Berlin, Germany: Springer-Verlag.

Nguyen, L., Swatman, P. A., & Shanks, G. (1999). Using design explanation within formal object-oriented method. *Requirements Engineering, 4*(3), 152–164.

Nuseibeh, B. A., & Easterbrook, S. M. (2000). Requirements engineering: A roadmap. *Proceedings of the Conference on the Future of Software Engineering* (pp. 35–46). New York: ACM Press.

Ohlsson, S. (1984). I. Restructuring revisited: Summary and critique of the Gestalt theory of problem solving. *Scandinavian Journal of Psychology, 25,* 65–78.

Opdahl, A. L., Dubois, E., & Pohl, K. (2004, June 7–8). *Ten years of REFSQ: Outcomes and outlooks.* Paper presented at the International Workshop on Requirements Engineering: Foundation for Software Quality (REFSQ'05), Riga, Latvia.

Piaget, J. (1950). *The psychology of intelligence.* New York: Routledge.

Regev, G., Gause, D. C., & Wegmann, A. (2006, September). *Creativity and the age-old resistance to change problem in RE.* Paper presented at the 14th IEEE International Requirements Engineering Conference (RE'06), Minneapolis, MN.

Robertson, J. (2005, January/February). Requirements analysts must also be inventors. *IEEE Software, 22*(1), 48, 50.

Robertson, S., & Robertson, J. (2005). *Requirements-led project management: Discovering David's slingshot.* Boston, MA: Addison-Wesley.

Robillard, P. N. (2005, November/December). Opportunistic problem solving in software. *IEEE Software, 22*(6), 60–67.

Root-Bernstein, R., & Root-Bernstein, M. (2004). Artistic scientists and scientific artists: The link between polymathy and creativity. In R. J. Sternberg, E. G. Grigorenko, & J. L. Singer (Eds.), *Creativity: From potential to realization* (pp. 127–151). Washington, DC: American Psychological Association.

Schmid, K. (2006). A study on creativity in requirements engineering. *Softwaretechnik-Trends, 26*(1). Retrieved August 2006, from http://pi.informatik.uni-siegen.de/stt/26_1/

Schön, D. A. (1996). Reflective conversation with materials. In T. Winograd (Ed.), *Bringing design to software* (pp. 171–184). New York: ACM Press.

Simon, H. A. (1992). *Sciences of the artificial.* Cambridge: MIT Press.

Solomon, B., Powell, K., & Gardner, H. (1999). Multiple intelligences. In M. A. Runco & S. R. Pritzker (Eds.), *Encyclopedia of creativity* (Vol. 2, pp. 273–283). San Diego, CA: Academic Press.

Sommerville, I., & Sawyer, P. (1997). *Requirements engineering: A good practice guide.* Chichester, England: John Wiley & Sons Ltd.

Sutcliff, A., & Maiden, N. (1998). The domain theory for requirements engineering. *IEEE Transactions on Software Engineering, 24*(3), 174–196.

Visser, W. (1992). Designers' activities examined at three levels: Organisation, strategies and problem-solving processes. *Knowledge-Based Systems, 5*(1), 92–104.

Vygotsky, L. S. (1978). *Mind and society: The development of higher mental processes.* Cambridge, MA: Harvard University Press.

SECTION 3

PERFORMANCE ASSESSMENT

SECTION PERSPECTIVE

Eduardo Salas and Michael A. Rosen

Performance measurement and assessment are fundamental components for effective training. They drive the provision of feedback and decisions about remediation. Specifically, in order to maximize learning by providing corrective feedback and to decide what future training is needed, the trainee's current competency must be assessed and diagnosed. This is as true for training that occurs in virtual environments (VEs) and simulations as it is for training that occurs in a classroom. However, while performance assessment is rarely a simple task in any context, there are unique and challenging issues when assessing performance for training in VEs. Therefore, the chapters in this section are dedicated to exploring a wide variety of performance assessment issues in VEs for training. The goal of this Section 3 Perspective is to place these contributions into context by providing an overview of the fundamentals of performance assessment and measurement in VEs and training. Specifically, we address three main goals. First, we provide an overview of the need for performance assessment and measurement in VEs for training, as well as some of the major challenges to doing this effectively. Second, we describe the concept and process of performance diagnosis, the goal of assessment and measurement in training. Third, we provide a review of some guiding principles to developing performance assessment and measurement systems that are diagnostic of trainee competencies.

PERFORMANCE IN VEs FOR TRAINING: WHY DO IT, AND WHAT IS SO HARD ABOUT IT ANYWAY?

Simulations have long been used to prepare people for tasks and conditions of performance, particularly those that were important, infrequent, or dangerous. For example, archaeological evidence suggests that leather birthing models (much like modern plastic birthing simulators) were used to teach maneuvers to assist in childbirth in prehistoric times (Macedonia, Gherman, & Satin,

2003). Additionally, the board game chaturanga, the ancestor of modern chess developed in seventh century India, provided a simulation of battle and was used to develop strategic and tactical thinking in military commanders. The historical and archaeological record is filled with examples such as these all the way up to and including modern simulation and training beginning in the early part of the last century (for example, flight simulation for the training of military pilots).

Using simulations for developing skills, therefore, is not new or novel by any account. However, the modern use of simulations and VEs for training is distinct from this longer historical tradition in two important ways: increased sophistication of the learning environment and increased use for *systematic* training. Both of these differences have important implications for performance assessment and measurement.

First, the technological sophistication of the simulations and VEs used for modern training vastly exceeds preceding learning tools and environments. Spurred primarily by the geometrically increasing power of computers and their ability to represent more and more of the real world with higher levels of physical fidelity, the gap between simulated and simulation is ever narrowing. This increased sophistication of the technology used for training in simulations and VEs offers an increasingly robust palette for the design of learning environments. It has been clearly demonstrated that more physical fidelity (that is, a closer approximation of the physical detail of a targeted real world task) is not always necessary for reaching desired learning outcomes (Hays & Singer, 1989). In fact, some intentional deviations from exact replication of the real world task can produce better training results for some tasks (for example, above real time training for pilots; Lane, 1994) and some trainees (for example, increasing fidelity as the learner progresses from novice to expert; Dreyfus & Dreyfus, 1986). Ultimately though, the greater power and control afforded the developer of learning environments, the greater potential for increased training effectiveness. However, this poses a significant challenge. The more complex the task or learning environment, the more complex the performance of the trainees will become, complicating the matter of measuring performance and generating feedback during practice. Essentially, the complexity of the learning environment can now equal or surpass that of the real world task environment. Consequently, many of the measurement problems associated with measuring performance in the real world (Arvey & Murphy, 1998) are present in the VEs.

Second, VEs are used for *systematic* training and not just experiential learning. That is, VEs are used as part of a training delivery method for the acquisition of specified knowledge, skill, and attitude (KSA) competencies that underlie effective performance. This is an explicit process of identifying the type and level of performance the trainees should acquire during training, designing scenarios that afford opportunities to practice and acquire these competencies, and providing trainees with performance feedback. This is in contrast to *unguided* practice where the learner may not be practicing correct forms of performance or receiving feedback. Practice and experience are the primary means by which expertise develops, but this practice must be structured, guided, and accompanied by feedback (Ericsson, Krampe, & Tesch-Romer, 1993).

Taken together, these two points highlight (1) the great promise of training with VEs and (2) a great challenge to achieving this promised effectiveness. That is, training with VEs can potentially accelerate the development of expertise, resulting in a workforce with more individuals functioning at higher levels of effectiveness; however, in order to do this, practice opportunities in VEs must be systematically engineered and trainees must receive corrective feedback. In the following sections, we discuss performance diagnosis, the process of measuring performance to determine the causes of effective and ineffective performance. This information is necessary to make good decisions during the training process.

PERFORMANCE DIAGNOSIS: HOW DO YOU ASSESS PERFORMANCE IN VEs?

Performance is the actions involved in completing a task (Fitts & Posner, 1967); it is not the results or outcomes of these actions, but the actions themselves (Campbell, 1990). This is an important distinction to make, especially in the context of training. It is the processes of performance, the actions taken while completing a task (that is, how the task is done), that are trained, not the outcomes (that is, the results of all of the actions). Consequently, the processes of performance are what must be assessed during training.

Most fundamentally, performance assessment involves (1) capturing performance in some manner (quantitative or qualitative data), (2) comparing that representation of performance to some standard, and (3) making decisions based upon the comparison between observed performance and the standard or target performance. In the context of training in VEs, an idealized version of this process proceeds as follows:

1. Trainee performance is measured during practice activities within the VE.
2. These metrics are compared to prespecified learning objectives and lists of competencies targeted for training to determine the degree to which the trainee possesses these targeted competencies and has met the learning objectives (that is, has the trainee learned the appropriate performance?).
3. The results of this comparison are used to make decisions about what feedback to give and what future training is required to ensure that the trainee has reached the specified learning objectives.

The above three-step process appears simple enough, but can be quite complex to carry out effectively. In many cases, difficulty results from such issues as the multidimensional and dynamic nature of performance being trained and measured, and such practical constraints as the availability of observers and trainers. Additionally, many potential purposes drive the development and implementation of performance assessment or measurement systems for use in VEs. These include selection of personnel, tests for certification or qualification, feedback and remediation during training, assessment of interventions, validation of virtual environments, and others. Developing any measurement system involves trade-offs, and the purpose of the measurement system determines how these trade-offs are made.

For training in VE, the ultimate purpose of performance measurement is diagnosis; that is, performance measurement should allow for correct inferences to be made about the causes of effective and ineffective performance (Salas, Rosen, Burke, Nicholson, & Howse, 2007; Cannon-Bowers & Salas, 1997). Knowing why an individual or team performed a certain way provides the necessary information for generating and providing the feedback needed for learning. In order to do this, a performance measurement system must capture a broad array of performance measures, a performance profile. This performance profile is illustrated in Figure SP3.1 and is briefly described here. For a more detailed discussion of these issues, see Salas and colleagues (2007).

Different measures have different "informational yields" (Swing, 2002); that is, some measures are more informative than others, but no one measure tells the entire story. Therefore, for a measurement system to be diagnostic of the reasons behind an observed performance, it should incorporate a broad range of measures from different sources. This performance profile can then be used to make more informed decisions during training. To that end, a performance profile should have measures that (1) capture performance at multiple levels of analysis (for example, team, individual, and multiteam systems), (2) are grounded in the context of the task being trained (for example, they are not abstract performance dimensions), (3) are linked to the competencies (that is, the KSAs targeted for training, (4) are descriptive of the processes of performance, (5) are captured

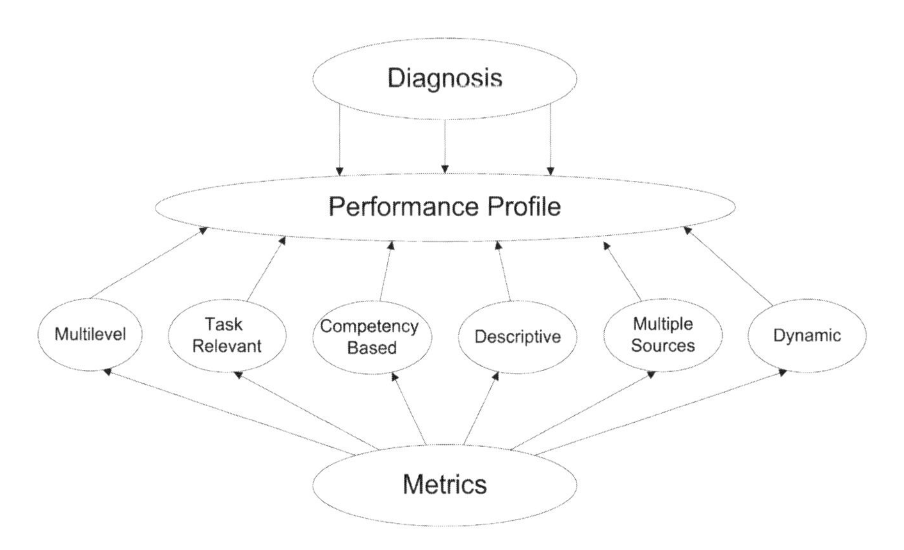

Figure SP3.1. Characteristics of Measures Involved in Creating a Performance Profile (From Salas et al., 2007)

from multiple sources, and (6) capture performance over time. These measures are captured during practice activities (step one of the process outlined above). The resulting data are used to determine the degree to which trainees possess the competencies targeted for training (step 2). This assessment is used to make decisions about feedback and remediation targeted at a specific trainee's needs (that is, the deficiencies in competencies that may exist). In the following section, we provide a description of some guiding principles for carrying out performance diagnosis.

GUIDING PRINCIPLES FOR PERFORMANCE DIAGNOSIS IN VEs

In the preceding sections, we have identified why performance assessment is critical for training in VEs and some general challenges to effective performance measurement and assessment in VEs. In this section we summarize existing best practices and guidelines for developing diagnostic measurement systems in scenario based training (for example, Salas, Rosen, Held, & Weissmuller, in press; Rosen, Salas, Wilson, et al., 2008; Oser, Cannon-Bowers, Salas, & Dwyer, 1999).

Measures Should Be Based in Theory

Theory provides guidance in the form of descriptions of performance. These descriptions are useful for deciding what to measure, one of the most critical and difficult tasks in designing a performance measurement system. For example, if a VE curriculum is designed to build decision-making skills during urban combat, relevant models of decision making (for example, recognition-primed decision-making model; Klein, 1998) provide a basis for understanding the critical processes involved (for example, situation assessment, pattern recognition, and mental simulation). Similarly, if the VE curriculum is designed to develop teamwork skills, theoretical models are available to guide the development of performance measures for capturing the essential components of performance (for example, communication, mutual support, leadership, situation monitoring, and team orientation; Salas, Sims, & Burke, 2005).

Measures Should Capture Competencies and Be Designed to Meet Specific Learning Outcomes

Performance measures should be driven by the competencies being trained and the targeted learning outcomes. Theory can be used to develop a general understanding of what must be measured, but subsequently the KSAs targeted for acquisition in a given scenario must be explicitly defined. Most types of performance (for example, teamwork) are far too complex to be fully mastered in one scenario; therefore, the specific competencies targeted must be measured as specifically as possible. For example, if a scenario provides opportunities to practice leadership (but not the other aspects of teamwork), then it makes little sense to measure and provide feedback on the other aspects of teamwork that are not required for performance during a scenario.

Measures Should Capture Multiple Levels of Performance when Appropriate

Performance is often interdependent. Individuals rarely act alone in modern organizations. Consequently, when interdependent work is being trained in VEs, the performance measurement system must be able to make differentiations between individual taskwork and coordinated teamwork (Cannon-Bowers & Salas, 1997). For example, if a VE is used to train teamwork for medical teams responding to trauma, the performance measurement system must distinguish between the individual competencies necessary for taskwork (for example, can the medic start an IV [intravenous therapy]) and the team level aspects of performance (for example, communication). If an IV is late being started, was it because the medic was inefficient or because communication was ineffective? To answer this and related questions, it is necessary to measure multiple levels of performance.

Measures Should Be Linked to Events in the VE

As noted earlier, the complexity of the task environments represented by VEs rivals that of the real on-the-job environment. Assessing performance on the job is notoriously difficult due to this complexity (Arvey & Murphy, 1998). Consequently, many of the same challenges arise in VEs. However, a significant advantage in VEs is that the training developer has a significant amount of control over the environment. This control can be used to create opportunities for performance diagnosis by inserting critical events into the scenario. Trainee responses to these events are indicative of the presence or absence of targeted competencies (see Rosen, Salas, Silvestri, Wu, & Lazzara, 2008; Fowlkes, Dwyer, Oser, & Salas, 1998).

Measures Should Focus on Observable Behaviors

One of the primary advantages of using VEs for training is that they afford the opportunity for dynamic practice, that is, for the trainee to exhibit the actual processes and behaviors of performance. Focusing on the observable behaviors during these practice opportunities has several advantages for the present purposes. First, it greatly increases the reliability of ratings made by observers (Bakeman & Gottman, 1997). Second, it increases the likelihood that performance measurement processes can be captured automatically. Additionally, sensors and input devices of various types frequently are used for individuals to interface with and function in the VE; this input to the system is in the form of observable behaviors exhibited by the trainees. This is input that is already captured automatically and can be used as performance measurement with careful planning.

Use Multiple Measures from Multiple Sources

Triangulation is a valuable strategy for dealing with the complexity of performance in VEs. That is, viewing performance from multiple "angles" (that is, different measurement tools) will provide a more robust understanding of what is happening and allow for more certain inferences to be made about the causes of performance. For example, Campbell, McCloy, Oppler, and Sager (1993) propose that performance is determined by three factors: declarative knowledge, procedural knowledge and skill, and motivation. Therefore, if a trainee does not exhibit the targeted performance, it could be because of low levels in one or more of these factors. Different situations (for example, high declarative and procedural knowledge, but low motivation; high declarative knowledge and motivation, but low procedural) require different feedback and future training to correct the deficiencies. Determining which underlying case is causing the observed performance deficiency requires measuring all three factors.

Focus Measures on the Processes of Performance

Consider the case of training naval aviators to land on an aircraft carrier. The ultimate criterion of interest is, "Did the pilot land the plane safely on the carrier?" However, measuring just this variable, the outcome of the pilot's performance processes, does not provide enough information to provide detailed corrective feedback. If the pilot does not land safely (that is, is waived off or crashes into the carrier or water), it definitely indicates that there is a problem, but provides no information on how to fix it. Therefore, the processes of performance should be measured during practice activities and not just the outcomes.

Train Observers and Structure Observation Protocols

When observations are necessary, and they frequently are in training with VEs, steps must be taken to ensure that the data generated by observers are reliable and valid. Two fundamental approaches to doing this include training raters and providing structured tools to guide observation. First, rater training helps to ensure that the judgments made by one rater are the same (or highly similar) to those made by a second observer rating the same performance. Second, structured observation protocols help to guide the observers' attention to the critical aspects of performance, reducing their overall level of workload and increasing the accuracy of ratings. These approaches seek to eliminate bias from the observer, reduce the overall level of error in the data, and subsequently increase the quality of data upon which decisions are made.

Facilitate Post-Training Debriefs and Training Remediation

The overarching goal for most performance measurement systems in VEs for training is to maximize learning, that is, to increase the rate at which individuals acquire the targeted competencies. Therefore, the data captured during practice activities should enable rapid decision making about feedback and remediation. It should provide trainers with the fuel for specific, timely, and targeted feedback (for example, visual aids for after action reviews).

CONCLUDING REMARKS

We hope the chapters contained within this section provide the community with the motivation, information, and needed tools to ensure VEs promote learning when used for training purposes. We also hope the principles provided in this Section 3 Perspective are validated, expanded, modified, and refined as research is conducted and as practitioners try to apply them. Wc all can benefit from a collaboration between the researchers and the practitioners involved in designing and delivering VE systems for training.

ACKNOWLEDGMENTS

This work was partially supported by the Office of Naval Research Collaboration and Knowledge Interoperability (CKI) Program and ONR MURI Grant No. N000140610446 (Dr. Michael Letsky, Program Manager).

REFERENCES

Arvey, R. D., & Murphy, K. R. (1998). Performance evaluation in work settings. *Annual Review of Psychology, 49,* 141–168.

Bakeman, R., & Gottman, J. M. (1997). *Observing interaction: An introduction to sequential analysis* (2nd ed.). Cambridge, United Kingdom: Cambridge University Press.

Campbell, J. P. (1990). Modeling the performance prediction problem in Industrial and Organizational Psychology. In M. D. Dunette & L. M. Hough (Eds.), *Handbook of Industrial and Organizational Psychology*. Palo Alto, CA: Consulting Psychologists Press.

Campbell, J. P., McCloy, R. A., Oppler, S. H., & Sager, E. (1993). A theory of performance. In N. Schmitt & W. Borman (Eds.), *Personnel selection in organizations* (pp. 35–70). San Francisco, CA: Jossey-Bass.

Cannon-Bowers, J. A., & Salas, E. (1997). A framework for developing team performance measures in training. In M. T. Brannick, E. Salas, & C. Prince (Eds.), *Team performance and measurement: Theory, methods, and applications* (pp. 45–62). Mahwah, NJ: Erlbaum.

Dreyfus, H. L., & Dreyfus, S. E. (1986). *Mind over machine: The power of human intuition and expertise in the era of the computer*. New York, NJ: The Free Press.

Ericsson, K. A., Krampe, R. Th., & Tesch-Romer, C. (1993). The role of deliberate practice in the acquisition of expert performance. *Psychological Review, 100*(3), 363–406.

Fitts, P. M., & Posner, M. I. (1967). *Human Performance*. Belmont, CA: Brooks/Cole.

Fowlkes, J. E., Dwyer, D. J., Oser, R. L., & Salas, E. (1998). Event-based approach to training (EBAT). *The International Journal of Aviation Psychology, 8*(3), 209–221.

Hays, R. T., & Singer, M. J. (1989). *Simulation Fidelity in Training System Design*. New York: Springer-Verlag.

Klein, G. (1998). *Sources of power: How people make decisions*. Cambridge, MA: MIT Press.

Lane, N. E. (1994). *Above Real-Time Training (ARTT): Rationale, effects, and research recommendations* (No. NEL-TR-94-01). Orlando, FL: Naval Air Warfare Center Training Systems Division.

Macedonia, C. R., Gherman, R. B., & Satin, A. J. (2003). Simulation laboratories for training in obstetrics and gynecology. *Obstetrics & Gynecology, 102*(2), 388–392.

Oser, R. L., Cannon-Bowers, J. A., Salas, E., & Dwyer, D. J. (1999). Enhancing human performance in technology-rich environments: Guidelines for scenario-based training. In E. Salas (Ed.), *Human/Technology Interaction in complex systems* (Vol. 9, pp. 175–202). Stamford, CT: JAI Press.

Rosen, M. A., Salas, E., Silvestri, S., Wu, T., & Lazzara, E. H. (2008). A measurement tool for simulation-based training in emergency medicine: The simulation module for assessment of resident targeted event responses (SMARTER) approach. *Simulation in Healthcare, 3*(3), 170–179.

Rosen, M. A., Salas, E., Wilson, K. A., King, H. B., Salisbury, M., Augenstein, J. S., Robinson, D. W., & Birnbach, D. J. (2008). Measuring team performance for simulation-based training: Adopting best practices for healthcare. *Simulation in Healthcare, 3*(1), 33–41.

Salas, E., Rosen, M. A., Burke, C. S., Nicholson, D., & Howse, W. R. (2007). Markers for enhancing team cognition in complex environments: The power of team performance diagnosis. *Aviation, Space, and Environmental Medicine Special Supplement on Operational Applications of Cognitive Performance Enhancement Technologies, 78* (5), B77–85.

Salas, E., Rosen, M. A., Held, J. D., & Weissmuller, J. J. (in press). Performance measurement in simulation-based training: A review and best practices. *Simulation & Gaming: An Interdisciplinary Journal.*

Salas, E., Sims, D. E., & Burke, C. S. (2005). Is there a big five in teamwork? *Small Group Research, 36*(5), 555–599.

Swing, S. R. (2002). Assessing the ACGME general competencies: General considerations and assessment methods. *Acad Emerg Med, 9*(11), 1278–1288.

Part VII: Purpose of Measurement

Chapter 12

MEASUREMENT AND ASSESSMENT FOR TRAINING IN VIRTUAL ENVIRONMENTS

Jared Freeman, Webb Stacy, and Orlando Olivares

This chapter of the handbook presents the fundamental knowledge of measurement and assessment and strategies for using them to improve the design of training in virtual environments (VEs). We define the role of assessment in training venues (VEs included), the components of assessment, the functions measurement and assessment can serve in training, what is assessed, how assessments are made, and how the quality of measures themselves can be evaluated. We close by describing promising future directions for measurement and assessment in VEs.

THE IMPORTANCE OF MEASUREMENT AND ASSESSMENT

Virtual environments are often rich worlds in which to practice skills. This does not necessarily mean that VEs train effectively. To train, VEs must provide *frequent practice focused* on essential skills, plus *feedback.* These conditions grow expertise (Ericsson, Krampe, & Tesch-Romer, 1993), and they are necessary for learning in any venue: classroom training, multimedia training, intelligent tutoring systems, Web based training, and live exercises. Measurement and assessment are essential to all three of these conditions to determine how much practice is required, to select experiences that train deficient skills, to populate feedback, and for other functions. Assessment helps ensure that VE systems train effectively.

VEs rarely have sufficient measurement and assessment capabilities to support training, however. In 1997, the Office of the Inspector General of the U.S. Department of Defense issued an audit stating that the military had not demonstrated the training value of its simulators, despite huge investments in the technology. More than a decade later, few VEs assess the knowledge and skills of trainees, though many document the effects that trainees achieve (for example, the number of enemies killed or the health of the synthetic patient after virtual surgery).

WHAT IS AN ASSESSMENT?

We define assessment as a value judgment concerning the adequacy of the learner's performance. An assessment concerning mastery of a *training objective* (for example, to triage virtual patients accurately) is generated by applying a *performance standard* (for example, scores above 95 percent denote expertise) to a *measurement* (for example, 98 percent) computed from *data* concerning the behavior of an *agent* in *context.* Each of these terms is essential to the concept of assessment and to its practice. VEs that fail to address all of these aspects in their assessments either require the instructor to do so manually or they leave the trainee to infer, often erroneously (Dunning, Johnson, Ehrlinger, & Kruger, 2003), the state of his or her own skill and the lessons to be learned.

An *agent* is the person or organization associated with an assessment. We identify agents with assessments in order to know whom to credit (or blame) or to whom to provide feedback that improves performance. Just who is an agent varies with the function of assessment. If the objective is to improve trainee performance, an assessment might be associated with the trainee, such as a human pilot (or team of pilots) in training. If the objective is to assess the expertise of observers or the agreement between them, an assessment would be associated with the observer(s) monitoring trainee performance. If the objective is to improve the quality of software agents (or nonplayer character, in gaming terms), an assessment would be associated with the agent or its author. Such distinctions are important in complex VEs in which, for example, a pilot's failure to fire on a ground target might be attributed to (1) poor performance by the pilot, (2) poor observation by the trainer, or (3) failure of a synthetic agent to clear the pilot to fire on the target.

The *performance context* is the setting in which the agent(s) take the measured action (or inaction). The context may be defined by the state of the simulated ambient environment (for example, day/night or raining/not raining). It may be defined by the entities in that environment (for example, four incoming enemy aircraft, one friendly wingman, and two enemy surface-to-air missile (SAM) installations). It may be defined by the actions and interactions of those entities (for example, the formation in which enemy aircraft approach or the destruction of the wingman by enemy missiles). It may be defined by all of these and by their relationships in time and space (for example, the proximity of the enemy SAM, the order in which one encounters the enemy SAM and aircraft, or the latency between those encounters).

Data are observations of action. Data may be provided directly by the trainee (for example, in a survey), an observer (such as a trainer), a measurement instrument (such as an eye-tracking device), or the simulation system. Data relevant to behaviors or skills—which VEs are designed to elicit—may concern the state of an object, its actions, or its interactions. For example, data concerning the location of a missile strike may be generated by the simulator as latitude, longitude, and altitude above sea level. Data concerning an avatar of a human operator may concern its location, utterances, posture or gesture, and actions.

A *measure* is a formula that transforms these data so that they lie on a defined measurement scale. Its product is a *measurement.* For example, a formula for calculating the accuracy of a missile strike may compute the distance between the location of a target and the impact location of a missile. Scales are of several types. A nominal scale consists of discrete categories (for example, a missile "hit" or "missed" its target). An ordinal scale consists of ordered, discrete categories (for example, the miss was within 50 m (meters) of the target, 50–200 m, or more than 200 m). An interval scale consists of ordered, equidistant points (for example, the bombs struck 10 m apart). A ratio scale consists of ordered, equidistant points with a meaningful definition of zero, to support ratio calculations (for example, the missile struck 40 m from the target, twice the allowed distance of 20 m). The units of measures for scales (length, mass, time, temperature, information, area, and so forth) are more or less exhaustively enumerated in the Suggested Upper Merged Ontology of Measures (Teknowledge, 2007).

A *performance standard* partitions the measurement scale into meaningfully labeled sections. It imbues the measurement with value and utility relative to some *training objective* or performance objective: 80 percent of missiles fired struck the target = journeyman performance. Often, the standard is conditioned on the context. For example, 60 percent accuracy may qualify as journeyman performance in difficult terrain, adverse weather conditions, or when achieved under fire.

An assessment is rarely specified in all of its complexity (above) in simulation based training. This is unfortunate. Assessment necessarily includes the components described above. If they are not specified explicitly, they are implicit; trainers and trainees instantiate them by assumption. It is difficult to understand such assessments, and control their quality, because assumptions may vary between scorers. Thus, one assessor may consider an environmental factor important that another ignores; one may assess trainee actions differently from another; one may draw a different line between novices and journeymen, journeymen and experts. Such assessment conditions require evaluators and the designers of evaluation systems to do the best they can; often this is good enough to ensure learning, but these conditions can produce unreliable assessments.

Two strategies help to ensure that assessments are reliable. The first is to define measures systematically to ensure that they bear on the objectives of training and refine them experimentally to pare away those that cannot be taken reliably (MacMillan, Entin, & Morley, in press). The second is to specify each of these aspects of measures in a format or formal language (Stacy, Ayers, Freeman, & Haimson, 2006), so that they can be critiqued by experts for completeness and correctness, implemented unambiguously (even automatically), and refined over time. To some extent, however, the decision of how tightly to define an assessment—no less than the decision of what to measure and how—depends on how it will be used. We turn next to the functions of assessments in VE for training.

WHAT FUNCTIONS DOES ASSESSMENT SUPPORT IN VE TRAINING SYSTEMS?

Measurement and assessment serve two functions in training: instruction and management of instruction. We discuss these below, beginning with assessment for instruction, which is the primary focus of this chapter.

Assessment in Training

Assessments are content for training *feedback*. It is common to supplement assessments in feedback with descriptions or replays of the assessed behaviors, their context, and their effects. Feedback may also include inferences from assessed behavior concerning the knowledge and skill of the trainee.

Objective assessments in feedback inform the trainee about his or her skill and are catalysts for subjective assessment. For example, one feedback system for authoring and delivering debriefs to distributed military teams (Wiese, Freeman, Salter, Stelzer, & Jackson, in press) presents objective (automatically generated) assessments using stoplight icons (red and green) beside each training goal. Selecting an assessment replays the events in the air mission scenario at the moment of measurement. Instructors and trainees can then subjectively assess their performances in context.

Measurements and assessments can support *diagnosis* of performance failures. Measurement systems (human or machine) that consider the context of performance can discriminate whether an error is due to an actor (trainee, observer, or synthetic entity) who failed to respond when the context required it or to poor scenario design or control, which failed to present the conditions that would test trainee knowledge and skill or which possessed extraneous or inappropriate design elements that inadvertently obscured the relevant conditions (for example, distracted the trainee). Measurement systems that represent causal relations between events can help diagnose the root causes of error. A very detailed and rich specification of assessment context can subsume this causal model. For example, it might specify that a pilot may release weapons only after receiving the communication "free and clear" from a ground controller. If the controller fails to issue that permission and the pilot fails to release weapons, then the root cause of a failure to release weapons lies with the controller, and this is evident in the assessment of both the controller (who receives a failing assessment on issuing the permission) and the pilot (who receives no credit or demerit for failing to release weapons because the context lacked the required permission). The complexity of diagnosis rises with the complexity of the training environment, the number of participants, and the distribution of their actions over time. Consider the case of a VE in which four pilots fly in formation (that is, coordinating with each other) on a bombing run, controlled by a ground surveillance team, which in turn coordinates with a team that supplies intelligence imagery and other information. This is an example of the direction in which some VE training is headed: training distributed, multifunctional teams of teams. Semi-automated measurement and assessment may be required to assess the interactions between trainees in these settings because the coordination between distributed trainees simply may not be observable by a human trainer in any one location. Assessment techniques and technology to support such training are largely in the research stage. Meanwhile, assessment technology may serve trainers best by cueing them to make assessments or diagnoses at moments when coordination is occurring across distributed trainees.

Assessments can support the *prognosis* of performance success and failure. Such predictions can help human and automated controllers of VEs to adapt the difficulty of the training experience so that it provides an appropriate challenge within what Vygotsky (1978) called the zone of proximal development, the range of challenge that the trainee can master with appropriate support.

Assessments can drive instructional *prescriptions.* Prescriptions are recommendations, such as the suggestion that a trainee read a specific manual before resuming training, engage in certain part-task skills training, or execute a particular scenario next. Prescriptions select and schedule training. Thus, they address the requirement, given at the beginning of this chapter, that training provide frequent practice of essential or targeted skills. The benchmarked experiential system for training provides one example of this capability. The authors (Levchuk, Shebilske, & Freeman, 2007; Gildea, Levchuk, Freeman, Narakesari, & Shebilske, 2007; Shebilske, Gildea, Levchuk, & Freeman, 2007) developed optimizations and other benchmarks of performance in a complex team, air warfare task, assessed performance against those benchmarks, entered those assessments into a decision aid that modeled the team knowledge state and the impact of specific training events on that state (both knowledge state and training effects were represented probabilistically using a partially observable Markov decision process), and used the output of that aid to select one of tens of scenarios available for training. This integration of assessments with a mathematical model of instruction strategy produced reliably greater learning, in a laboratory experiment, than did a scenario selection strategy that implemented a hierarchical part-task approach.

Assessments can drive *training events and context.* An assessment that the trainee is performing at or above the highest measurable level in a scenario (a "ceiling effect") may drive VE controllers (human or automated) to introduce events or a context that increases the level of challenge to the trainee or challenges a different and presumably unmastered skill. These manipulations can take many forms. A VE training system that employs a human trainee as pilot, a synthetic wingman, a synthetic coach, and a synthetic simulation controller (Bell, Ryder, & Pratt, in press) might increase the challenge for a trainee assessed to be an expert by handicapping the performance of the synthetic wingman, decreasing the guidance provided by the coach, or, as a synthetic simulation controller, introducing a thunderstorm to diminish the quality of the environment, decreasing the capabilities of the aircraft, increasing the number of SAM sites that threaten the pilots, and so forth. Such manipulation is common in intelligent tutoring systems, but it is rare for VEs to be imbued with the intelligence to make instructional decisions. One explanation may lie in the complex chain of events this invokes. Changes to context influence which measures are taken and the standards by which performance is assessed, which in turn shape feedback during and after the action.

Assessment for Training Management

Measurement and assessment can also serve managerial functions. They can help designers to evaluate and incrementally improve software agents, such as synthetic pilots, coaches, and simulation controllers (above). A training system can archive assessments of the state of these agents, the actions of agents, and the actions of trainees; analyze these; and refine the logic or the mathematics of agents so that they make finer distinctions between trainee actions, more accurate diagnoses of failure, better predictions of training outcomes, and improved selection of feedback content and presentation.

Assessments can help training managers to select, assess, and train VE controllers, who must manage those conditions during scenario trials; observers, who must take reliable measures and assessments of trainee performance; and trainers, who must convey essential knowledge and skills to trainees or cue them to exercise them in VE training.

Assessments can help instructional designers to evaluate VE scenarios, which must present appropriate conditions for learning. Instructional designers should recognize that the effectiveness of a training scenario is partly a function of the trainee; individual differences often moderate the training-outcome relationship. Measures of individual differences can be used to assess trainee characteristics (for example, reaction time, reading speed, and visual acuity) before training, and these characteristics can be used to select good training content given that trainee's characteristics.

Finally, assessments can help software engineers to verify that a simulator is performing as designed. We have witnessed at least one training exercise in which simulation controllers launched an enemy fighter aircraft against human trainees, observed trainees react to it as if it was a commercial aircraft, and later discovered that the software engineers had mistakenly applied the label for that fighter aircraft to a model of a commercial airliner. A very complex measurement scheme is needed to capture these rare errors, one that correlates data concerning the supposed characteristics of an object (for example, the flight characteristics of an enemy fighter aircraft) with the characteristics as they are perceived by the trainee (for example, through tactical displays).

WHAT IS ASSESSED IN VIRTUAL ENVIRONMENTS?

The objects of assessment in VE training are, at the first order, the end state of a task or mission, the process by which it was executed, the capabilities or attributes of the performing agents, and the state of entities in and out of the VE. At the second order are assessments of trends, such as rate of improvement or learning.

A measure of the end state of a task or mission is referred to as a measure of effectiveness (MOE). MOEs address "what" was done and typically include the occurrence, accuracy, or quality of outcomes and/or the timeliness with which they were achieved. A measure of the steps executed in response to an event or in pursuit of an outcome is referred to as a measure of performance (MOP). MOPs address "how" the outcome was achieved. The distinction between an MOE and an MOP is, thus, largely a function of what one considers an outcome, that is, of the training objectives. One trainee's MOP may be another's MOE.

Assessments of the capabilities or attributes of the trainee are often inferred from MOPs or fine-grained MOEs. They typically concern the state of declarative knowledge (knowledge of what) and procedural knowledge (knowledge of how). Declarative knowledge is commonly and probably best assessed using conventional instruments (for example, pen and paper multiple-choice tests), provided that they recreate realistic cues to elicit recognition or recall of the associated knowledge of interest, where the trainee must interact with an object to demonstrate his or her declarative knowledge. Medical trainees do this when they practice diagnostics using a real mannequin or virtual patient. Procedural knowledge is appropriately evaluated in VEs because VEs provide rich cues to skill selection, controls for skill execution, and a realistic tempo for sequences of actions. Such assessments may describe the individual or the team or organization that collectively demonstrates the coordination of the target skills.

The state of trainees themselves is sometimes measured using instruments that assess personality, intelligence, or other relatively stable individual differences. Trainee physiological state is, increasingly, measured using techniques such as eye tracking and electroencephalography to assess otherwise unobservable cognitive states, such as arousal, anxiety, concentration, and insight (Bowden, Jung-Beeman, Fleck, & Kounios, 2005). These factors may moderate training effects, and so assessment on these factors can be used to customize training to the individual.

Second order measures and assessments combine data from the primary measures and assessments above. Assessments of efficiency consider the number of effects achieved or processes executed per resource expended or per unit time. Assessments of reliability address accuracy over trials. Measures of learning represent the rate of change in effectiveness, process, or capabilities over time or training trials.

Other forms of training assessment are possible, of course. The discussion above addresses assessments of skill or learning in training scenarios. This corresponds to the second level—called learning—in Kirkpatrick's (1994) hierarchy. VE training can be assessed from trainee reactions (level one) and often is assessed only in this way. It can be assessed from records of performance on the job (level three) and from the results (for example, the return on investment or rate of mission success) that it has on the organization as a whole (level four).

In basic research concerning VEs, the characteristics of the environment itself are often measured. Central to this mission is understanding whether the VE delivers the distinctive experience it is intended to deliver: a sense of presence (Mikropoulos & Strouboulis, 2004), which is enabled by immersion and interactivity (Schloerb, 1995; Steuer, 1992). Presence enhances training of interpersonal communication skills in a medical VE (Johnson, Dickerson, Raij, Harrison, & Lok, 2006). Presence in VE training is even more clearly critical when the "real world" environment is largely virtual, as is the case in telemedicine or endoscopic surgery (Zhang, Zhao, & Xu, 2003). Presence is a metric of the ecological validity of VEs, but it is not necessarily correlated with training effectiveness. For example, many experienced aviators believe that high fidelity motion systems are essential for flight training simulators, but the research evidence indicates that they are not (Lintern, 1987). There is even some evidence that the use of a motion system in a flight-training simulator can interfere with transfer to real flight (Lintern & McMillan, 1993). Recent research attempts to model this non-monotonic relationship between fidelity and training effectiveness (Estock, Alexander, Gildea, Nash, & Blueggel, 2006) to support the design and acquisition of VEs.

Measurement of the environment also has value in training applications. Measurement can identify the cues to which trainees are exposed and (fail to) respond. This is particularly important in VEs in which trainees create their own experiences, environments in which it is not possible to plan measurement opportunities in advance. For example, Forterra Systems Inc.'s Online Interactive Virtual Environment presents an urban environment in which trainees command avatars to execute military operations, and trainers or their confederates control civilian and insurgent avatars who populate the virtual city. There is no formally scripted sequence of events (that is, software code). An assessment engine was developed (Haimson & Lovell, 2006) to identify emergent situations in which trainees could exercise doctrinal maneuvers. These assessments serve as bookmarks used by trainers during debriefing to leap to situations in which knowledge of doctrinal maneuvers can be discussed and evaluated.

HOW ARE ASSESSMENTS MADE IN VE TRAINING?

Three methods are used to make assessments in VE training: trainee responses to questions and other probes, observer data collection, and collection by automated instruments. These methods have been categorized as subjective measurement (trainee response) and objective measurement (observational and automated measurement; Wilson & Nichols, 2002). What constitutes a measure as objective or subjective has been a point of discussion and contention for many years (see Annett, 2002a, 2002b; Wilson & Nichols, 2002), with subjectivity carrying with it the connotation, at least for ergonomists, of being second-rate, soft science (Annett, 2002b). Nonetheless, subjective measurement is many times the only way in which the experiences and opinions of participants/trainees can be captured. Examples are measuring the side effects and aftereffects of VE participation in VRISE (virtual reality induced symptoms and effects; Cobb, Nichols, Ramsey, & Wilson, 1999) or the sense of presence (Steuer, 1992). In contrast, there are situations in which trainee/participants are not aware or good judges of effects that they are objectively measured to have experienced (for example, postural instability) (Wilson & Nichols, 2002), and here subjective measures are deprecated.

Measurement and assessment is typically conducted by human observers, even in the largest and most expensive simulation systems—those that most faithfully reproduce the physical controls (such as a cockpit), environment (for example, landscapes and weather), and object attributes (for example, missile flight characteristics). Assessment in these settings is done largely by observers using paper or electronic forms. Observers may validate that a specific training experience has been delivered, rate or evaluate trainee actions, and record their observations and judgments concerning second order characteristics (such as efficiency) and inferences about unobservable states (such as situational awareness). The reliance on observers is in part an historical artifact of assessment practices in real (not virtual) exercise environments. In part it is a necessity: real and artificial environments are rarely well instrumented for automated measurement and assessment, and it is often difficult and expensive to automatically assess the complex tasks executed in these environments. It is common in military simulation exercises, for example, to have a large cadre of domain experts score a larger assembly of trainees. Naturally, an added benefit of using observers to assess performance in a VE is that those same measures (if developed appropriately) can also be used in live training exercises (Wiese, Nungesser, Marceau, Puglisi, & Frost, 2007). This ensures that trainees are assessed similarly across environments, and it can facilitate analysis of the effectiveness of live and virtual training environments.

Subjective measures and assessments are sometimes gathered from trainees using paper or electronic surveys to assess individual differences, reactions to training, and evaluations of their own performances. The latter are not necessarily reliable (Dunning et al., 2003), but often have the valuable side effect of encouraging trainees to recall key training events and think critically about them. Trainees may also be asked to assess the state or performance of other teammates, the team, or the scenario events. In this case, they are acting as observers, ones who have information and perspectives that may differ significantly from those of trained observers.

Automated measures and assessments are taken by instrumentation within the VE or external to it. Such physiological measurement systems as eye tracking and functional near-infrared imaging typically are not often tightly integrated with each other or with the VE. Instrumentation within the VE typically taps the system's databases or data bus. Many military systems use the high level architecture (HLA) that enables measurement instruments to read simulation data (for example, the state of aircraft and their interaction with missiles) and write measures or assessments to the bus as a federate with a status similar to that of any trainee operating a station on the system. For example, a measurement technology developed for assessing human performance in military simulations (Stacy et al., 2006) subscribes to specific data, captures those data when they are delivered across the HLA bus, computes measures of performance from those data using rules typically derived from experts or authoritative sources, and makes measurements and assessments available to debriefing systems using Web services after the training simulation system has been shut down.

HOW CAN ONE ASSESS ASSESSMENT?

The quality of measures and assessments is a function of their feasibility, their utility for the function at hand, and their statistical properties.

The *feasibility* of a measure or assessment determines whether it can be made at all, and made efficiently. Assessments in some experimental training settings—often those involving virtual environments—are subject to federal regulations that protect the privacy and health of human subjects. To be taken at all, assessments in these settings must pass muster with an institutional review board concerned with disclosure to subjects, safe conditions of administration, and anonymous or secure storage of data about individuals. The feasibility of assessment is also a function of the cost of developing measures and assessments (below), validating them, administering them, and using them. For example, the cost of developing automated measures may include modifying a simulator data model to distribute data concerning the state of a display or control with which the user interacts to demonstrate a skill and developing software that computes measurements and assessments from those data. This can take months given the complexities of simulation hardware and software, the social and political environment of a simulation engineering team, and the business processes designed to prevent spurious changes to the simulation system. Feasibility is also a function of the cost of administering measures. This cost is low when assessment is automated. The price of observer assessment is driven by the cost of hiring and training a sufficient number of domain experts to make measurements and assessments reliably.

The *utility* of an assessment is determined by whether it fulfills its function. Measures and assessments developed primarily for instructional functions, for example, must be meaningful to trainees and actionable, thus instructionally effective. To achieve this, developers of measures must often analyze performance in the domain at hand, using observation, interviews, surveys, and document analysis. For example, the Mission Essential Competency technique developed and used by the U.S. Air Force employs group interview techniques and confirmatory surveys to define critical competencies, knowledge, and skills in a domain. Experiments can refine this understanding of key competencies. Methods of developing measures and assessments from these findings (MacMillan et al., in press; Carolan et al., 2003) specify the observable behaviors that denote these competencies in a mission simulation. At a minimum, these methods of defining assessments ensure that the trainers and trainees understand them. Instructional science and art can be applied to ensure that these meaningful assessments help trainees to learn, whether the assessments are delivered directly as feedback or influence the content or schedule of training (as described above). Assessments that fulfill their function are said to have consequential validity (Messick, 1995).

Measures should also have certain statistical properties: reliability and validity, with reliability being a precondition for validity.

A measure is reliable if repeated administration to a subject produces similar scores (for example, of accuracy of communication in foreign language training) or if different evaluators apply the measure to the same data and produce similar scores. The former definition of reliability is confounded with the concept of learning, in which scores on measures rise with training. Thus, the reliability of a measure must be tested in conditions that preclude learning (for example, by controlling expertise or time on task).

A measure can be valid in several respects (Anastasi, 1988). Criterion validity exists when a measure accurately assesses current state or accurately predicts a future state. A piloting test has criterion validity if it predicts job performance ratings, rank, or some other indicator of proficiency. Content validity exists when a measure addresses all of a given domain, but only that domain. For example, an assessment of aircraft piloting skill has content validity if it addresses flight preparation, takeoff, cruise, landing, and taxi, and it omits passenger relationship management. Construct validity exists when a measure assesses the unobservable, theoretical construct that it claims to measure (for example, intelligence). Measures with construct validity should correlate with measures that are known to be related to the construct, should not correlate with measures known to address distinct constructs, and they should respond as predicted to the state of other constructs. (These tests are known as convergent, divergent, and nomological validation, respectively).

Consequential validity (Messick, 1995) exists when an assessment fulfills its intended function for the individual or the organization, as when a test of customer service skills, when used to select new employees, improves customer service. There is an axiological component to consequential validity; the assessments that an organization performs and the functions that assessment serves are dynamic. They change as the organization evolves and as its missions change. Thus, it is a significant challenge for organizations to continuously align their assessment efforts with their states and missions.

Unfortunately, it is unusual for VE training developers to test the reliability and validity of assessments and measures used in applied (not experimental) training. Often, the apparent relevancy of a measure—its face validity—suffices as an endorsement of the measure or assessment. This is disappointing in part because measures developed for their face validity alone often have deep statistical flaws and limited utility. For example, observers may give all teams passing assessments regardless of their performance because the performance standards are vaguely defined, the measures are based on observer inferences and not observable events, or for other reasons.

WHAT IS THE FUTURE OF ASSESSMENT FOR VE TRAINING?

The growth of VE training systems presents opportunities to improve how, how frequently, and how well we assess human performance. We envision a future for VE training in which (1) assessment-driven simulation ensures that virtual environments provide training, not just practice and (2) simulation-driven assessment ensures that simulations provide sound conditions for evaluating human performance.

Assessment-driven simulation will address the problem—common in large-scale military simulation acquisitions and in PC (personal computer) gaming environments repurposed for training—that the virtual environment is designed to provide practice, but without regard for its effectiveness as a training device. Large-scale systems (full motion aircraft cockpit simulators, for example) are designed to maximize the fidelity of controls, displays, and the physics of entities represented in the environment on the assumption that trainers will put these to good use. PC gaming environments (such as *Neverwinter Nights* and *Second Life*) are designed to maximize engagement using dramatic visuals, storylines, and rapid reward schedules. Neither class of system generally supports the core functions of instructional design: (1) creating conditions for learning and practice of essential skills and (2) assessing competency on those skills. We look forward to a future in which training requirements drive the design of simulation controls, displays, and physics models. Scenario authoring tools will help instructors put these features to use by constraining the design of scenarios to relevant conditions, ensuring, for example, that scenarios for training medics to respond to mass casualties present a mix of cases that challenges, for example, skills in diagnosis and personnel tasking and does so at a pace that presents a challenge requiring learning. Recent research explores this vision (Stacy, Walwanis, & Colanna-Romano, 2007) by formally representing aspects of instructional design (for example, representing training objectives, conditions, and measures in XML [Extensible Markup Language] schemas) and applying constraint logic programming techniques to schedule instructional experiences before and during simulation runs.

Simulation-driven assessment will resolve the difficulty of accessing and using simulator data to measure and assess human performance. Current simulations typically provide few assessments, and few measures, mainly MOEs, not MOPs. Large training institutions supplement these assessments with observer assessments and debriefs structured (in the best cases) to elicit subjective assessments. The data required to assess processes often are represented only in replay "video," but they are not available directly to measurement systems. For example, data may be available that a missile is fired at an aircraft and that the pilot maneuvers to avoid it, but rarely are data available concerning the instruments with which the pilot perceives that event (the auditory signal from a radar warning receiver or the visual signal on a radar display) or controls with which the pilot makes a maneuver. Thus, it is possible to automatically assess that a pilot makes avoidance maneuvers, but the details required to diagnose failures are lost. We look forward to a future in which a rich data stream is available with which to automate some of the measurement and assessment work that human observers now take on. This will free observers to apply their expert judgment to evaluate other trainee competencies (for example, verbal communications) that cannot reliably be assessed automatically or that require subtle interpretation only humans can make.

CONCLUSION

This chapter described foundations of measurement and assessment for training in virtual environments. We defined an assessment as a complex object (a value judgment on a measurement computed from data generated by an agent acting in context to attain a training objective), an object that can be formally specified to clarify or even automate assessment. We described ways in which researchers and trainers can apply measures and assessments in training (as feedback, to diagnose and forecast performance, and to drive instruction) and training management (to evaluate software agents, human training staff, scenarios, and the simulator itself). We specified that the proper objects of measurement and assessment are the trainee (his or her declarative knowledge, procedural knowledge, physiological state, and individual psychological differences) and the training environment itself. We described the fundamental approaches (subjective and objective) to measurement and assessment, and the standards for assessing assessment (on feasibility, utility, reliability, and validity). Finally, we defined future directions for measurement and assessment, in which assessment drives simulation and in which simulation drives assessment. We hope that this framework helps training researchers and designers to instrument VEs so that they measure human performance in context and apply these measures in ways that enhance the power of VEs to train.

ACKNOWLEDGMENTS

We thank several colleagues who contributed insights and editorial comments: Cullen Jackson, Ph.D., Emily Wiese, Ph.D., Eileen Entin, Ph.D., and Diane Miller.

REFERENCES

Anastasi, A. (1988). *Psychological testing.* New York: Macmillan Publishing Co.

Annett, J. (2002a). Subjective rating scales: Science or art? *Ergonomics, 45,* 966–987.

Annett, J. (2002b). Subjective rating scales in ergonomics: A reply. *Ergonomics, 45,* 1042–1046.

Bell, B., Ryder, J., & Pratt, S. (in press). Communications and coordination training with speech-interactive synthetic teammates: A design and evaluation case study. In D. Vincenzi, J. Wise, P. Hancock, & M. Mouloua (Eds.), *Human factors in simulation and training.* Mahwah, NJ: Lawrence Erlbaum.

Bowden, E. M., Jung-Beeman, M., Fleck, J., & Kounios, J. (2005). New approaches to demystifying insight. *Trends in Cognitive Sciences, 9,* 322–328.

Carolan, T., MacMillan, J., Entin, E., Morley, R. M., Schreiber, B. T., Portrey, A., Denning, T., & Bennett, W., Jr. (2003). Integrated performance measurement and assessment in distributed mission operations environments: Relating measures to competencies. *Proceedings of the Interservice/Industry Training, Simulation & Education Conference* [CD ROM]. Arlington, VA: National Training and Simulation Association.

Cobb, S. V. B., Nichols, S. C., Ramsey, A. D., & Wilson, J. R. (1999). Virtual reality-induced symptoms and effects (VRISE). *Presence: Teleoperators and Virtual Environments, 8,* 169–186.

Dunning, D., Johnson, K., Ehrlinger, J., & Kruger, J. (2003). Why people fail to recognize their own incompetence. *Current Directions in Psychological Science, 12*(3), 83–87.

Ericsson, K. A., Krampe, R. T. H., & Tesch-Romer, C. (1993). The role of deliberate practice in the acquisition of expert performance. *Psychological Review, 700,* 379, 384.

Estock, J. L., Alexander, A. L., Gildea, K. M., Nash, M., & Blueggel, B. (2006). A model-based approach to simulator fidelity and training effectiveness. *Proceedings of the Interservice/Industry Training, Simulation, and Education Conference* [CD ROM]. Arlington, VA: National Training and Simulation Association.

Gildea, K., Levchuk, G., Freeman, J., Narakesari, S., & Shebilske, W. (2007). BEST: A benchmarked experiential system for training. *Proceedings of the 12th Annual International Command & Control Research & Technology Symposium*. Retrieved August 29, 2007, from http://www.dodccrp.org/events/12th_ICCRTS/CD/iccrts_main.html

Haimson, C., & Lovell, S. (2006). Pattern recognition for cognitive performance modeling. In K. Murray & I. Harrison (Eds.), *Capturing and using patterns for evidence detection: Papers from the 2006 Fall symposium* (Tech. Rep. No. FS-05-02; pp. 120–126). Menlo Park, CA: Association for the Advancement of Artificial Intelligence.

Johnson, K., Dickerson, R., Raij, A., Harrison, C., & Lok, B. (2006). Evolving an immersive medical communication skills trainer. *Presence, 15,* 33–46.

Kirkpatrick, D. L. (1994). *Evaluating training programs: The four levels.* San Francisco: Berrett-Koehler.

Levchuk, G., Shebilske, W., & Freeman, J. (2007). *A model-driven instructional strategy: The benchmarked experiential system for training (BEST).* Manuscript submitted for publication.

Lintern, G. (1987). Flight simulation motion systems revisited. *Human Factors Society Bulletin, 30*(12), 1–3.

Lintern, G., & McMillan, G. (1993). Transfer for flight simulation. In R. Telfer (Ed.), *Aviation instruction and training* (pp. 130–162). Aldershot, United Kingdom: Ashgate.

MacMillan, J., Entin, E. B., & Morley, R. (in press). Measuring team performance in complex and dynamic military environments: The SPOTLITE Method. *Military Psychology.*

Messick, S. (1995). Validity of psychological assessment: Validation of inferences from persons' responses and performances as scientific inquiry into score meaning. *American Psychologist, 50*(9), 741–749.

Mikropoulos, T. A., & Strouboulis, V. (2004). Factors that influence presence in educational virtual environments. *Cyberpsychology & Behavior, 7,* 582–591.

Schloerb, D. W. (1995). A quantitative measure of telepresence. *Presence, 4,* 64–80.

Shebilske, W., Gildea, K., Levchuk, G., & Freeman, J. (2007). Training experienced teams for new experiences. *Proceedings of the Human Factors and Ergonomics Society 51st Annual Meeting* [CD-ROM]. Santa Monica, CA: Human Factors and Ergonomics Society.

Stacy, W., Ayers, J., Freeman, J., & Haimson, C. (2006). Representing human performance with Human Performance Measurement Language (HPML). *Proceedings of the Interservice/Industry Training, Simulation and Education Conference* [CD ROM]. Arlington, VA: National Training and Simulation Association.

Stacy, W., Walwanis, J. M., & Colanna-Romano, J. (2007). Using pedagogical information to provide more effective scenarios. *Proceedings of the Interservice/Industry Training, Simulation & Education Conference* [CD ROM]. Arlington, VA: National Training and Simulation Association.

Steuer, J. (1992). Defining virtual reality: Dimensions determining telepresence. *Journal of Communications, 42,* 73–93.

Teknowledge. (2007). *Overview of the SUMO.* Retrieved August 29, 2007, from http://ontology.teknowledge.com/arch.html#Measure

Vygotsky, L. S. (1978). *Mind and society: The development of higher mental processes.* Cambridge, MA: Harvard University Press.

Wiese, E. E., Freeman, J., Salter, W. J., Stelzer, E. M., & Jackson, C. (in press). Distributed after action review for simulation-based training. In D. A. Vincenzi, J. A. Wise, M. Mustapha, & P. A. Hancock, (Eds.), *Human factors in simulation and training.* Mahwah, NJ: Lawrence Erlbaum.

Wiese, E. E., Nungesser, R., Marceau, R., Puglisi, M., & Frost, B. (2007, November). Assessing trainee performance in field and simulation-based training: Development and pilot study results. *Proceedings of the Interservice/Industry Training, Simulation and Education Conference* [CD ROM]. Arlington, VA: National Training and Simulation Association.

Wilson, J. R., & Nichols, S. C. (2002). Measurement in virtual environments: another dimension to the objectivity/subjectivity debate. *Ergonomics, 45,* 1031–1036.

Zhang, G., Zhao, S., & Xu, Y. (2003). A virtual reality based arthroscopic surgery simulator. *Proceedings of the IEEE International Conference on Robotics, Intelligent Systems and Signal Processing* (Vol. 1, pp. 272–277). Los Alamitos, CA: IEEE.

Chapter 13

TRAINING ADVANCED SKILLS IN SIMULATION BASED TRAINING

Jennifer Fowlkes, Kelly Neville, Razia Nayeem, and Susan Eitelman Dean

Achieving the coordination of numerous weapons systems and personnel units is a principal goal of the U.S. Marine Corps (USMC). The USMC is small compared with the other branches of the U.S. Armed Services, owns fewer resources, and thus relies heavily on an ability to bring those resources to bear in a strategically coordinated manner, that is, to achieve combined arms. Accordingly, significant USMC training resources are directed toward preparing personnel to execute combined arms operations. Distributed, simulation based training systems, in which elements of combined arms teams are represented and linked, are among the critical enabling technologies for training combined arms and other distributed operations.

Despite the importance of simulation, it is difficult to achieve pedagogically sound training using simulation based training for reasons pertaining to complexity of the training environment and the difficulty in training such higher order skills as coordination and adaptability. The objective of the present chapter is to describe the challenges to conducting simulation based training for large teams and then to present instructional strategies that can be used to help address the challenges. The strategies are based on the results of a cognitive task analysis to identify training needs for Fire Support Teams combined with principles for advanced learning. The Fire Support Team is one of the USMC's resources for providing combined arms. The cognitive task analysis was used to elicit detailed knowledge from seven experienced instructors with respect to a series of challenging combined arms events. The strategies identified can be used to augment existing simulation systems for combined arms training and potentially for training other large tactical teams. In addition, it is hoped that the strategies can be used to guide the design of instructional features in emerging systems.

COMPLEX TRAINING ENVIRONMENTS

The complexity of the combined arms environment is one of the challenges to providing effective simulation based training. Training combined arms teams may involve tens, or even hundreds, of system operators. As Lane and Alluisi (1992) noted,

> The players in this simulated battlefield environment are not only the weapon system operators, but also the commanders, staffs, logisticians, support units, intelligence personnel, and decision makers at all levels—in short, all the combat, combat support, and combat-service support elements assigned to the battle force and its support.
>
> (p. 23)

Indeed, simulation based training systems used for combined arms represent complex environments that, because of their emergent properties, can undermine effective training that is predicated on controlling training experiences to meet training objectives. That is, training event flow is largely determined by the give-and-take, real time interactions of players and simulated entities in the exercise. While such environments are probably representative of the actual battlefield, they are difficult to engineer as effective training exercises because task content is largely left to chance (Fowlkes, Lane, Dwyer, Willis, & Oser, 1995).

There are other consequences of complexity for training. Table 13.1 lists elements of complex environments taken from Feltovich, Hoffman, Woods, and Roesler (2004) and identifies some of the implications for training. Some of the implications follow: effective training and measurement systems must take into account the goal-directed nature of team performance and the integrated manner in which individuals and subteams support overall team goals; training, assessment, and feedback must occur at the individual, subteam, and large team levels; there are multiple ways in which a system can break down, and there are multiple ways in which effective performance can occur; and, finally, to be truly diagnostic, training and measurement systems must take into account the context in which performance occurs in order to be valid. These factors would tax any training system.

The scenario based training (SBT) method (Cannon-Bowers, Burns, Salas, & Pruitt, 1998) is one approach that can be used to exert some control over the training environment and to address some of the implications in Table 13.1. Scenario based training is founded on the notion of arranging or constraining scenarios, in a realistic fashion, to optimally support learning. The critical underlying assumptions of SBT are that (a) each of the training components must be linked to achieve an effective learning environment and (b) training opportunities are not left to chance. Thus, in SBT, events are systematically identified or introduced to provide known opportunities to support training. The SBT framework does not ameliorate all of the issues associated with the complexity shown in Table 13.1, but makes many of them more tractable, including controlling task context, being able to anticipate important training and measurement

Table 13.1. Characteristics of Complex Systems and Implications for Training

Element of System Complexity[a]	Implication for Training
Processes occur continuously	Relevant performance strengths and weaknesses • are detected by assessing quantitative and qualitative changes over time versus at specific points. The latter is easier. • may manifest at any time, challenging human and machine measurement systems to stay on guard.
Multiple processes occur at any one time	Instructor personnel must monitor, control, and assess multiple personnel or subteams to be trained, as well as the entities with which they are interacting, imposing a heavy workload and methods for optimally focusing attention.
There are interdependencies among the systems represented	Training and assessment systems must unravel the multiple contributions to performance outcomes. In addition, knowledge from multiple domains (for example, from ground and air systems) is needed to critique performance.
Diverse explanatory principles are needed to account for performance	The complex contributions to performance include trainees (at team, subteam, and individual levels), trainee perspective, status of simulation systems (fully functional, partially functional, and nonfunctional), trainee familiarization with the capabilities and limitations of the simulation system, and a host of scenario context factors such as information ambiguity.
Explanatory principles rely on a total system view	Judgments of performance strengths and weaknesses of an individual or team must consider the context in which the performance took place and the interactions with the other system components. It is not as easy to isolate the causes of performance effects within complex systems.
Cases in the domain display variability	Optimal or fixed solution paths are difficult to identify because of the myriad ways that performance can break down as well as recover. This challenges both human and machine performance assessment systems.

[a]From Feltovich et al. (2004).

opportunities, and reducing instructor workload. Below, we describe the basic SBT method. We then describe how it can be used as a framework for addressing the training of higher order skills within simulation based training environments.

The components of SBT are shown in Figure 13.1. Training is supported initially by identifying a master set of tasks or training objectives that describe the complete set of competencies for which the trainer is used. Generally in the military these are task lists (for example, mission essential task list).

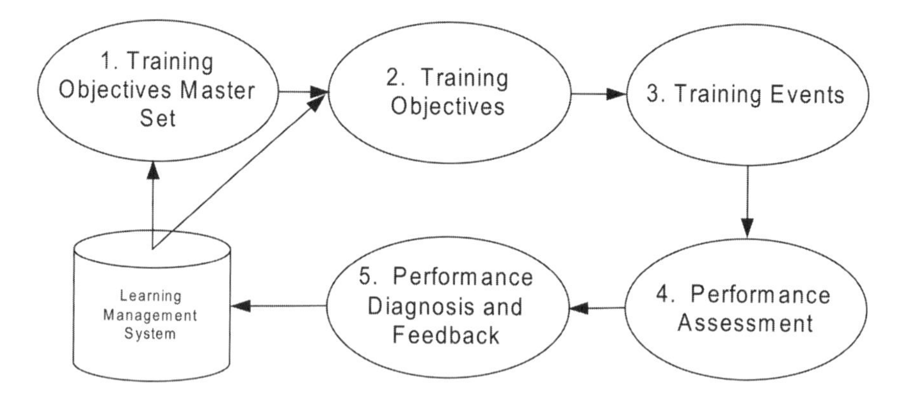

Figure 13.1. Scenario Based Training Framework (Adapted from Cannon-Bowers et al., 1998).

Next, to support a specific training event, a subset of training objectives from the master list is selected. In military training exercises, the subset used would be driven by a variety of factors, including the company commander's or Fire Support Team leader's objectives and the need to prepare for an upcoming deployment. The training objectives selected for a training event in turn drive the development of the scenario events and related products (for example, mission objectives and orders). In some cases, events are deliberately introduced to achieve a specific learning purpose. For example, in teaching a Fire Support Team member the skills needed to control close air support aircraft, events might include the aircraft being on or off the timeline, not carrying the ordnance expected, and having degraded systems. The idea is to produce or encourage situations that allow rich practice opportunities related to the training objectives.

In addition to deliberately introducing events, events can be identified that will occur naturally as a result of the interactions between participants and simulated entities. For these events, timing cannot be known a priori, but instructors can be primed (or prompted) to recognize their occurrences. For example, in an effort assessing army Fire Support Teams, naturally occurring battle events that served as training opportunities included calls for fire, intelligence reports, survivability moves, and battle geometry updates (forward line of own troops and coordinated fire lines) (Fowlkes, Dwyer, Milham, Burns, & Pierce, 1999).

In the SBT model, performance assessment is accomplished by evaluating trainee or team responses to the scenario events. Knowing important events that will occur during training allows instructors to develop a priori expectations for performance assessment and to focus their attention appropriately. These expectations can be used to develop automated performance assessment tools or job aids. In addition, using events as a guide for assessment serves to focus measurement so that not everything has to be observed. This in turn reduces instructor workload and creates a more economical expenditure of time. Feedback provided during after action reviews is also organized around events. Finally, the information collected pertaining to trainee performance (for example, skill inventory and performance history) can be utilized and incorporated into learning management systems to guide future training.

Examples of training topics addressed by SBT include team and tactical training in aviation (Colegrove & Alliger, 2001; Jentsch, Abbott, & Bowers, 1999; Salas, Fowlkes, Stout, Milanovich, & Prince, 1999), team dimensional training (TDT) for navy shipboard teams (Smith-Jentsch, Zeisig, Acton, & McPherson, 1998), emergency management (Schaafstal, Johnston, & Oser, 2001), and advanced skills training for military teams (Salas, Priest, Wilson, & Burke, 2006). We now examine the use of SBT as a framework to support the implementation of advanced learning principles.

ADVANCED LEARNING

Besides the issue of complexity, another reason that it is difficult to achieve pedagogically sound training in discrete multitone modulation systems is that they, either implicitly or explicitly, are focused on supporting advanced learning. Feltovich, Spiro, and Coulson (1993) define advanced learning as "acquiring and retaining a network of concepts and principles about some domain that accurately represents key phenomena and their interrelationships, and that can be engaged flexibly when pertinent to accomplish diverse, sometimes novel objectives" (p. 181). This is a good characterization of the decision making and performance required by combined arms teams. For combined arms, effective coordination of multiple systems and units in a dynamic tactical environment to achieve intended goals requires vast amounts of knowledge (for example, vast systems and organizational, environmental, and enemy knowledge) and a great deal of experience and skill. Further, the challenge associated with conducting this type of warfare is growing as new weapons systems and technologies continue to be introduced and as the enemy becomes increasingly resourceful and skilled at developing plans of aggression that are not easily predicted.

Feltovich et al. (1993) argue that advanced learning is not facilitated very well in any training or education setting. For example, in most educational settings, topics are treated in isolation, whereas in reality they are interdependent; and easily understood examples are provided to illustrate concepts, whereas in the real world, examples are much less clear-cut. In military communities, an additional complication is that operational tempo is so high that it is difficult to get enough time to spend on the higher level curriculum segments. This is a problem with the building-block approach to training. One consequence is that trainees are often unprepared to take advantage of the few advanced skills training opportunities that exist.

For example, predeployment training in the military offers critical opportunities for team members to learn to work with other units and warfighting specialties prior to deploying. However, in these settings, instructors find they have to focus on prerequisite skills rather than on the integration skills that the predeployment opportunities are designed to teach (Rasmussen, 1996). Thus, the trainees have less opportunity to acquire advanced knowledge, and instructors have less opportunity to develop training techniques focused on advanced integration principles.

In the remainder of the chapter we highlight the use of advanced learning strategies for training combined arms teams based on the principles identified by Feltovich et al. (1993) and within the context of the scenario based training framework. The strategies are summarized in Table 13.2.

Training Objectives Master Set

Mission essential task lists provide the core and indisputably important training objectives for military training. But also in support of advanced learning and measurement, other skill decompositions are necessary to identify the complex, higher order skills needed to support complex performance. Examples of these types of decompositions include the U.S. Air Force's mission essential competencies, which define tactical team competencies at various levels of abstraction, from upper military echelons to specific aircrew skills (for example, Colegrove & Alliger, 2001). Fowlkes, Dwyer, Oser, & Salas (1998) used mission-oriented constraints to help decompose complex skills for aviation teams.

From the cognitive task analysis for Fire Support Teams, the authors identified the skills and knowledge needed to support effective performance. Figure 13.2 summarizes the results, as well as the dynamic manner in which we would expect the skills and knowledge to be combined to support effective performance.

In Figure 13.2, the skill categories situation assessment, planning, and plan execution represent key functions performed by Fire Support Team members. Each of the functional areas was associated with exemplars of the specific behaviors performed to implement the functions. The functions are usually performed in parallel so that situation assessment, for example, is performed on an ongoing basis. Fire support planning is also performed throughout a battle and so are plan execution behaviors. These represent the top-tiered functions in Figure 13.2.

Information exchange, adaptability, and team coordination are also skills required by Fire Support Team members. In our view, each of these supports the upper tiered functions and, hence, the arrangement of the pyramid. Information exchange, for example, is critical to situation assessment, planning, and plan execution. In the same way, adaptability enhances the upper level functions as does having an effectively coordinated team. Finally, as illustrated in Figure 13.2, performance of all the skills is facilitated by knowledge associated with expertise. Examples of the types of knowledge used to facilitate performance that was identified from the cognitive task analysis include knowledge of scheme of maneuver, resource capabilities, enemy asset capabilities, enemy tactics, team members, limitations affecting operations, and rules of engagement.

Figure 13.2 is also meant to illustrate via the twisting nature of the pyramid that a given situation requires some context-specific combination of skills and knowledge to form an effective response. Event based techniques can be used

Table 13.2. Advanced Learning Strategies Summarized Using the SBT Framework

SBT Component	Strategy
Training objectives master set	• Use cognitive task analysis and related methods to identify complex skills and knowledge needed for effective performance. • Identify known trainee misunderstandings.
Identify training objectives for training event	• Address experience level of trainees. • Address known trainee misunderstandings.
Design training events	• Insert events (for example, disequilibrium events) that directly challenge trainee misconceptions about the domain. • Provide event sets that can provide opportunities for comparison and contrast to build usable (versus inert) knowledge. • Consider building events that can be compared across scenarios or training events. • Provide situation variations to enhance the identification of regularities.
Performance assessments	• Measures should address qualitative changes in knowledge representation in addition to incremental changes in knowledge accumulation. • Measures should help trainees and instructors compare situations, both potentially within and across scenarios, facilitate understanding, and carry forward important learning themes. • Measures should help alert instructors to important training and assessment opportunities.
Performance diagnosis and debrief/after action review	Facilitate discussions that • enable trainees to assess the linkages between their situation recognition, responses, and the underlying skills and knowledge. • reveal trainee misconceptions with regard to the knowledge they possess or its application to realistic situations (that is, the events in the scenario). • encourage trainees to examine similar cases and link their training experiences to other knowledge they possess.

to create known context for which instructors could anticipate the specific combination of skills and knowledge required for an effective response.

Identify Training Objectives for Training Event

The identification of a subset of training objectives is the first step in the preparation for a specific training event using the SBT model. Considerations for selecting training objectives include the level of experience of the trainees, the

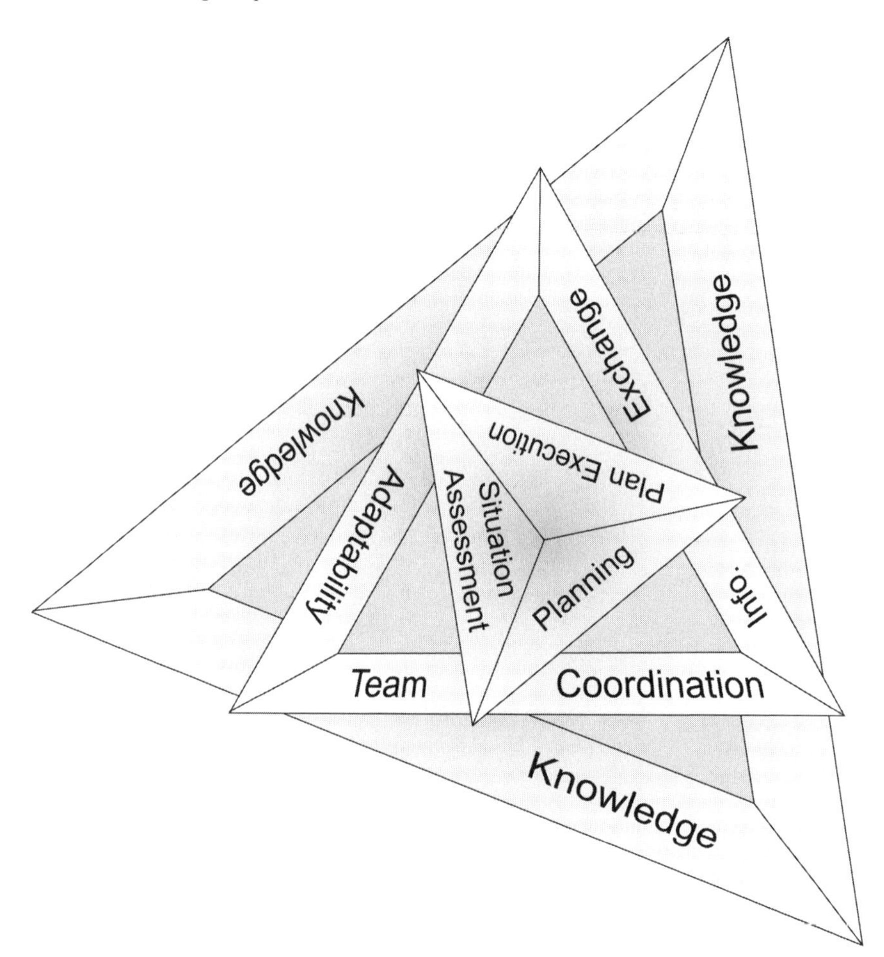

Figure 13.2. Relationship of Skills and Knowledge Associated with Expertise Based on Findings from the Cognitive Task Analysis in the Combined Arms Domain

input of training officers, and the requirements obtained from team leaders (for example, company commander; Stretton & Johnston, 1997).

Another way to select training objectives to support advanced learning is to target known biases and misunderstandings that exist among novices. Feltovich et al. (1993) argue that a simplification bias exists in learners as they acquire information about a domain that may result in faulty mental models. The misunderstandings are often difficult to identify—neither trainees nor instructors may be aware that they exist, and the biases may persist even though information to the contrary is presented. A role of cognitive task analysis is to reveal patterns of misunderstandings among trainees and to use this information to help guide events that can be used to directly challenge the misunderstandings during scenario based training. For the combined arms cognitive task analysis, a number of these examples were identified. For example, synchronizing combined arms is clearly one of the most difficult tasks faced by inexperienced Fire Support Teams. One of the experts interviewed for the cognitive task analysis noted that an inexperienced Fire Support Team leader would be inclined to "bring everything on," referring to artillery, mortars, and close air support, and then would have difficulty with coordination and deconfliction. A scenario could be designed to directly address this novice tendency. Another area related to seeing the "big picture" in terms of the scheme of maneuver and how fire support can be used to support it. One of the experts interviewed said, "If they [trainees] don't understand the concept of how you were moving, and why you were moving, they can't possibly understand what you want as far as fire support."

Develop or Identify Scenarios and Events

Scenario events can be used to directly challenge trainee misunderstandings of combined arms operations, as suggested above, and to more generally train the skills and knowledge that have been linked to effective performance. For example, Fire Support Team activities to be trained and assessed can be identified with the following mission phases: (a) before contact, (b) after contact and during preparation of the package, (c) during package execution, and (d) during attack continuation. In addition to nominal battlefield events, events can also serve as prompts to observe behaviors that occur infrequently or that might not be observable (Fowlkes et al., 1999). An example is a pilot pressuring a forward air controller (part of the Fire Support Team) to run an attack. Instructors would look to the forward air controller to utilize knowledge of the aircraft's time on station, other resources available, and ways the aircraft might have to be supported (for example, the controller might have to talk the aircraft to the target) in forming a response to pressure from the pilot. SBT can also be used to structure other training strategies for facilitating advanced learning.

Pretraining Interventions

Trainees come to a training exercise with knowledge, attitudes, and expectations for what the training will entail. These can dramatically affect training outcomes (Smith-Jentsch, Jentsch, Payne, & Salas, 1996). Our interviews with domain experts revealed that trainees do not fully understand how they will benefit from simulation based training. Thus, a simple intervention would be to familiarize trainees with the purpose of the training and what they are expected to learn. Moreover, if a building-block approach is used, provide trainees an overview of the knowledge and skills they will be acquiring across the scenario based training events to avoid isolating the training topics and to provide a useful framework for future learning—the primary function of an advance organizer (AO; Mayer, 1989). The use of SBT provides an excellent complement to the use of AOs because the framework elicited via AOs will likely be more relevant to the training exercise because SBT permits control over the practice environment. Without this control, the give-and-take of team members may affect the direction the scenario takes, potentially making the framework provided by the AO less relevant to the exercise.

Principles for Advanced Learning

Feltovich et al. (1993) suggest that experiential learning opportunities utilize numerous cases in instruction and emphasize relations among cases and between cases and concepts. The comparison and contrast notion figures prominently in many strategies for training complex skills. For example, Bransford, Franks, Vye, and Sherwood (1989) argue that we can tell trainees about important cues and even allow them practice in recognizing the cues, but still the trainees may not be able to apply their knowledge to new situations. This is known as inert knowledge. Bransford et al. (1989) suggest that perceptual learning may be enhanced by providing perceptual contrasts during experiential learning. These can be used to facilitate "noticing" as well as to enhance the underlying conceptual knowledge. Trainees can be provided with opportunities to assess how new situations are the same or different from situations previously encountered. SBT approaches can help to implement these strategies by engineering situations to be compared.

Scenarios can also be designed to directly challenge trainee misconceptions. That is, instructors and exercise controllers introduce novel events, unexpected events, and *disequilibrium events* (that is, events of a difficulty level with which trainees have not yet gained experience and which cause them to question and reevaluate their assumptions and strategies; Ross & Pierce, 2000). Klein and Baxter (2006) suggest that scenarios can be designed to reveal real world flaws, and then through activities promoting sensemaking and subsequent practice, more accurate mental models can be instantiated.

The domain experts we interviewed described how instructors intervene in ongoing scenarios to make similar interventions. For example, a *what-iffing* strategy (whereby an instructor or controller might opportunistically say, "OK, what if *this* happened. Now what would you do?") is sometimes used. It may be worth noting that a computer based training system would have difficulty generating the wealth of situations that experienced instructors can generate. In addition, instructors introduce events "on the fly" to challenge trainees.

> So the older guys have been there, done that, seen enough things that they can throw that in there. Therefore we're enhancing the learning curve to share that experience. We may not think of everything and all of a sudden, "hey, there's an opportunity to do this and we'll throw that into the mix." The more we can make them think, the more they have to coordinate, the more they learn. . . . we constantly throw these variations into our model, because . . . we find out when someone is good at it and then we try to ratchet it up so they are continuing to learn.

Feltovich et al (1993) also argue that real world complexity should be represented in training experiences. Military education and training often focus on a "building-block" approach to training. A building-block approach is likely to be relatively effective for the USMC and has been used to train thousands of soldiers. However, USMC training may be made more effective by complementing it with additional training strategies. A building-block approach is not necessarily optimal for all tasks or skills (for example, Klein & Pierce, 2001). Notably, it may not be ideal for building more advanced levels of expertise, such as *adaptive expertise* (for example, Holyoak, 1991; Kozlowski, 1998). Hence, other training strategies may be used in concert with the building-block approach to facilitate the acquisition of this critical form of expertise that is quite relevant to the conduct of fire support in a dynamic tactical environment. Examples of such training strategies include those that expose trainees to novel situations that challenge normative skills (Kozlowski, 1998) and unexpected situations and failure (Ross & Pierce, 2000). These training approaches differ from a building-block approach in which skills are gradually developed—not challenged and potentially discarded in the face of completely new situations. Klein and Baxter (2006) argue that what is needed for the training of such complex skills as planning and decision making is cognitive transformation. That is, important learning includes not just the accumulation of facts, but also a transformation of the way causal connections are viewed—a form of insight learning.

Evidence for the importance of complexity in perceptual learning was found in a study by Doane, Alderton, Sohn, and Pellegrino (1996). These researchers found that when training requires difficult discriminations between highly similar stimuli, discriminations of novel stimuli are both faster and more accurate than when training involves less similar discriminations. They concluded that study participants acquired different discrimination strategies depending on whether training discriminations were difficult or easy. Specifically, the harder training on difficult discriminations resulted in the use of more precise and detail-oriented comparison strategies even when easily discriminated stimuli were later presented, whereas the training on easy discriminations resulted in the use of an imprecise or global comparison strategy. They also found the use of these strategies to persist for a long period following their acquisition.

Doane et al. (1996) assert that their findings support the hypothesis that initial training difficulty significantly influences the development of strategic knowledge, a training philosophy that had been advocated by others involved in educational and training research (for example, Schmidt & Bjork, 1992). Similarly, research conducted by Gopher, Weil, and Bareket (1994) demonstrated that certain aspects of expertise acquired using a building-block approach may not transfer to the real and complex task performance environment. Gopher and his colleagues investigated *emphasis-change training,* whereby instructions and feedback are used to focus a trainee's attention on specific different aspects of task performance across the training period. According to Gopher et al., this method teaches trainees alternate ways of coping with a high workload task. Furthermore, evidence obtained in their research suggests that the emphasis-change training technique may lead to the development of attention control strategies that generalize across task performance conditions.

Develop Performance Measures That Capture Responses to Events

When implementing SBT, performance measurement tools are developed around the scenario events to provide links between measurement objectives and diagnosis of performance. This reduces the load on instructors during a training event in that the judgment about elements of acceptable performance has already been accomplished either by the instructor or his or her peers. Johnston, Cannon-Bowers, and Smith-Jentsch (1995) described a variety of individual and team process measures designed to capture responses to events for navy shipboard teams. These included behavioral observation scales, assessment of latencies to events and errors, and ratings. Although the above measures were used to assess performance of navy shipboard teams, they can be adapted for use in a variety of settings. In general, indices of acceptable responses (for example, behaviors and latencies) to events can be developed a priori and incorporated into measurement or job aids.

The foregoing suggests two additional emphases for measurement: measures should address qualitative changes in knowledge representation, not just incremental changes in knowledge accumulation (Klein & Baxter, 2006); measures should help trainees and instructors compare situations, both potentially within and across scenarios, to facilitate understanding and carry forward important learning themes.

Performance Diagnosis and Feedback

Applying the SBT model for diagnosis and feedback, observations and performance assessments are provided to trainees as the events in the scenario are reviewed. Discussions can be facilitated by instructors or by the team members themselves using guided team self-correction (Smith-Jentsch et al., 1998).

Smith-Jentsch et al. (1998) presented the TDT method that can serve as a model for combined arms training focusing on team skills. To implement TDT, a model of team performance is introduced at the outset of training, providing an advance organizer. In response to events, trainee strengths and weaknesses with respect to the team skills or team model are noted by instructors, using handheld data collection tools in either computer based or paper-and-pencil format. After the event, instructors facilitate discussion. Specifically, instructors (a) establish a professional climate and (b) prompt team members to systematically address the team skills introduced at the outset of training. This approach can be used to build a coherent model among trainees of effective task performance.

Similarly, for SBT in the combined arms domain, instructors can prompt trainees to examine how well they utilized the targeted knowledge and skills in response to events in the simulation, whether the events were preplanned, inserted "on the fly" by an instructor, or occurred by chance through the interactions of trainees with the simulated entities. This approach can be used by an instructor or even by other trainees using team self-correction. A procedure for this would be to query trainees regarding the cues they noticed and the considerations and information they used as the basis for their decisions and actions. A structured discussion, resembling knowledge elicitation, can be used to reveal strengths and deficiencies in the trainees' knowledge and skills. The key is to encourage trainees to link key concepts (for example, scheme of maneuver and resources capabilities) to situational cues and to expand the experience base from which they can draw in future situations. For example, consider the following interaction between a trainee and an instructor in response to the trainee's decision to use close air support (CAS).

> *Instructor:* "You decided to use fixed wing CAS. What did you notice about the situation that caused you to go that way?"
>
> *Fire Support Team Leader:* "I keyed in on the ranges of the tanks to my mortar and artillery."
>
> *Instructor:* "Why? What was important about the ranges?"
>
> *Fire Support Team Leader:* "Mortars aren't going to do squat against them. Arty will suppress them, but won't affect the tanks."

To augment the discussions, instructors should be ready to prompt trainees to consider similar situations or cases from the exercise and to relate their training experiences to other experiences they have had. Creating these comparisons and linkages are methods that can be used to facilitate advanced learning and develop flexible knowledge structures, characteristic of dynamic expertise (Feltovich et al., 1993; Kozlowski, 1998). Teams naturally self-correct during task performance during periods of low workload, a tendency that can be taught, encouraged, and even trained and measured.

DISCUSSION

We have argued that SBT combined with advanced learning strategies can be used to enhance experiential learning because together they mitigate some of the challenges to training within complex environments and address training of advanced, higher order skills that simulation based training environments allow the training community to address. For the most part, the strategies discussed are not expensive or difficult to implement, and they have a strong research base. In addition, they are among the types of strategies that have been shown to produce improved retention and transfer (Schmidt & Bjork, 1992).

However, there may be disadvantages as well. Although the combined approach is likely to result in improved transfer and retention, it might also suppress acquisition performance compared to using a building-block approach or to using simpler scenarios. As Schmidt & Bjork (1992) describe, many of the strategies for improving transfer produce difficulties for learners during acquisition. In addition, the strategies (for example, feedback strategies) might take more time than alternative methods. Such factors may affect trainee and instructor perceptions of the training, which are weighted heavily in training evaluation. The disadvantages seem worth confronting, though. There are few guidelines on how to structure simulation based training environments, and this chapter begins to provide such guidance. The strategies reviewed are among those that can be considered in the thoughtful design of complex training environments and are likely to have a high payoff in the development of resilient and high performing teams.

REFERENCES

Bransford, J. D., Franks, J. J., Vye, N. J., & Sherwood, R. D. (1989). New approaches to instruction: Because wisdom can't be told. In S. Vosniadou & A. Ortany (Eds.), *Levels of processing and human memory* (pp. 470–497). Hillsdale, NJ: Erlbaum.

Cannon-Bowers, J. A., Burns, J. J., Salas, E., & Pruitt, J. S. (1998). Advanced technology in scenario-based training. In J. A. Cannon-Bowers & E. Salas (Eds.), *Making decisions under stress: Implications for individual and team training* (pp. 365–374). Washington, DC: American Psychological Association.

Colegrove, C. M., & Alliger, G. M. (2001). *Mission essential competencies: Defining combat mission readiness in a novel way.* Paper presented at the SAS-038 NATO Working Group Meeting, Brussels, Belgium.

Doane, S. M., Alderton, D. L., Sohn, Y. W., & Pellegrino, J. W. (1996). Acquisition and transfer of skilled performance: Are visual discrimination skills stimulus specific? *Journal of Experimental Psychology: Human Perception and Performance, 22,* 218–1248.

Feltovich, P. J., Hoffman, R. R., Woods, D., & Roesler, A. (2004). Keeping it too simple: How the reductive tendency affects cognitive engineering. *IEEE Intelligent Systems, 19* (3), 90–94.

Feltovich, P. J., Spiro, R. J., & Coulson, R. K. (1993). Learning, teaching, and testing for complex conceptual understanding. In N. Frederiksen, R. J. Mislevy, & I. I. Bejar (Eds.), *Test theory for a new generation of tests* (pp. 181–217). Hillsdale, NJ: Lawrence Erlbaum.

Fowlkes J. E., Dwyer, D. J., Milham, I. M., Burns, J. J., & Pierce, L. G. (1999). Team skills assessment: A test and evaluation component for emerging weapons systems. *Proceedings of the 1999 Interservice/Industry Training, Simulation and Education Conference* [CD-ROM]. Arlington, VA: National Training Systems Association.

Fowlkcs, J. E., Dwyer, D. J., Oser, R. L., & Salas, E. (1998). Event-based approach to training (EBAT). *The International Journal of Aviation Psychology, 8,* 209–221.

Fowlkes, J. E., Lane, N. E., Dwyer, D. J., Willis, R. P., & Oser, R. (1995). Team performance measurement issues in DIS-based training environments. *Proceedings of the 17th Interservice/Industry Training Systems and Education Conference* (pp. 272–280). Arlington, VA: American Defense Preparedness Association.

Fowlkes, J., Owens, J., Hughes, C., Johnston, J. H., Stiso, M., Hafich, A., & Bracken, K. (2005). Constraint-directed performance measurement for large tactical teams. *Proceedings of the Human Factors and Ergonomics Society 49th Annual Meeting* (pp. 2125–2129). Santa Monica, CA: Human Factors and Ergonomics Society.

Gopher, D., Weil, M., & Bareket, T. (1994). Transfer of skill from a computer game trainer to flight. *Human Factors, 36,* 387–405.

Holyoak, K. J. (1991). Symbolic connectionism: Toward third-generation theories of expertise. In K. A. Ericsson & J. Smith (Eds.), *Toward a general theory of expertise* (pp. 301–336). Cambridge, England: Cambridge University Press.

Jentsch, F., Abbott, D., & Bowers, C. (1999). Do three easy tasks make one difficult one: Studying the perceived difficulty of simulation scenarios. *Proceedings of the Tenth International Symposium on Aviation Psychology* [CD-ROM]. Columbus: The Ohio State University.

Johnston, J. H., Cannon-Bowers, J. A., & Smith-Jentsch, K. A. (1995). Event-based performance measurement system for shipboard command teams. In *Proceedings of the First International Symposium on Command and Control Research and Technology* (pp. 274–276). Washington, DC: The Center for Advanced Command and Technology.

Klein, G., & Baxter, H. C. (2006). Cognitive transformation theory: Contrasting cognitive and behavioral learning [CD-ROM]. *Proceedings of the Interservice/Industry Training Simulation and Education Conference* [CD-ROM]. Arlington, VA: National Training Systems Association.

Klein, G. & Pierce, L. G. (2001). Adaptive teams. In *Proceedings of the 6th ICCRTS collaboration in the information age track 4: C2 decision-making and cognitive analysis.* Retrieved from http://www.dodccrp.org/6thICCRTS/

Kozlowski, S. W. J. (1998). Training and developing adaptive teams: Theory, principles, and research. In J. A. Cannon-Bowers & E. Salas (Eds.), *Making decisions under stress: Implications for individual and team training* (pp. 115–153). Washington, DC: American Psychological Association.

Lane, N. E., & Alluisi, E. A. (1992). *Fidelity and validity in distributed interactive simulation: Questions and answers* (IDA Document No. 1066). Alexandria, VA: Institute for Defense Analysis.

Mayer, R. E. (1989). Models for understanding. *Review of Educational Research, 59,* 43–64.

Rasmussen, E. (1996). Fallon air wing training curriculum. *Aimpoint, 12,* 38–44.

Ross, K. G., & Pierce, L. G. (2000). Cognitive engineering of training for adaptive battlefield thinking. In *IEA 14th Triennial Congress and HFES 44th Annual Meeting* (Vol. 2, pp. 410–413). Santa Monica, CA: Human Factors.

Salas, E., Fowlkes, J., Stout, R., Milanovich, D., & Prince, C. (1999). Does CRM training improve teamwork skills in the cockpit? Two evaluation studies. *Human Factors, 41,* 326–343.

Salas, E., Priest, H. A., Wilson, K. A., & Burke, C. S. (2006). Scenario-based training: Improving military mission performance and adaptability. In A. B. Adler, C. A. Castro, & T. W. Britt (Eds.), *Military life: The psychology of serving in peace and combat: Vol. 2. Operational stress* (pp. 32–53). Westport, CT: Praeger Security International.

Schaafstal, A. M., Johnston, J. H., & Oser, R. L. (2001). Training teams for emergency management. *Computers in Human Behavior, 17,* 615–626.

Schmidt, R. A., & Bjork, R. A. (1992). New conceptualizations of practice: Common principles in three paradigms suggest new concepts for training. *Psychological Science, 3,* 207–217.

Smith-Jentsch, K. A., Jentsch, F. G., Payne, S. C., & Salas, E. (1996). Can pre-training experiences explain individual differences in learning? *Journal of Applied Psychology, 81,* 110–116.

Smith-Jentsch, K. A., Zeisig, R. L., Acton, B., & McPherson, J. A. (1998). Team dimensional training: A strategy for guided team self-correction. In J. A. Cannon-Bowers & E. Salas (Eds.), *Making decisions under stress: Implications for individual and team training* (pp. 271–297). Washington, DC: American Psychological Association.

Stretton, M. L., & Johnston, J. H. (1997). Scenario-based training: An architecture for intelligent event selection. *Proceedings of the 19th Interservice/Industry Training Simulation and Education Conference* (pp. 108–117). Arlington, VA: National Training Systems Association.

Chapter 14

EXAMINING MEASURES OF TEAM COGNITION IN VIRTUAL TEAMS

C. Shawn Burke, Heather Lum, Shannon Scielzo, Kimberly Smith-Jentsch, and Eduardo Salas

The use of work teams in organizations is no longer a distinct competitive advantage used by only the most successful companies, but a common practice driven by the complexity of the work environment. As the use of work teams has increased, teams have taken many different forms in order to meet the needs of a dynamic environment. One form that has become prevalent is virtual teams, with 60 percent of professional employees reporting working in virtual teams (Kanawattanachai & Yoo, 2002). Virtual teams have most recently been defined as "teams whose members use technology to varying degrees in working across locational, temporal, and relational boundaries to accomplish an interdependent task" (Martins, Gilson, & Maynard, 2004, p. 808). Virtual teams have been argued to be a mechanism that can reduce travel time and costs associated with bringing together distributed members working on a common task. While virtual teams offer organizations flexibility, they also create challenges. Difficulties have been identified in the areas of planning and coordination across time zones, cultural differences (Kayworth & Leidner, 2000), effective communication (Sproull & Kiesler, 1986), team monitoring, and backup behavior (Martins et al., 2004).

Underlying many of these challenges are differences in the cognitive processes and states that emerge as a result of individuals enacting their respective roles. Although individuals assigned to virtual teams are often experts in their individual roles, the processes and states that emerge do not always serve to promote effective team performance. While progress has been made in understanding the knowledge structures and cognitive processes that promote effective coordination within conventional teams, this has been a relatively neglected area within virtual teams (Martins et al., 2004).

Many of the challenges mentioned with regard to teamwork within virtual teams have at their root problems in building compatible knowledge structures that allow members to be anticipatory in their prediction of member needs in the face of degraded social cues. While examining team cognition is often not easy, much progress has been made in this area (for example, Cooke, Salas,

Kiekel, & Bell, 2004; Lewis, 2003). However, organizations often fail to leverage what is known.

In light of the above, the purpose of the current chapter is to create a frame within which measures of team performance, specifically team cognition, can be assessed with regard to their applicability to virtual teams. In building the requisite framework virtual teams and the role that team cognition occupies in their effectiveness will be defined. Next, the basic components of team performance measurement systems will be described. Finally, performance measurement characteristics will be used to review how current measures of team cognition may apply within virtual teams, culminating in a set of guidelines.

WHAT ARE VIRTUAL TEAMS?

Since their inception, virtual teams have been defined in several ways. Driskell, Radtke, and Salas (2003) define virtual teams as those teams "whose members are mediated by time, distance, or technology" (p. 297). While there are variations across definitions (see Priest, Stagl, Klein, & Salas, 2005, for a review), a fair amount of consistency exists in how the boundaries between virtual and traditional teams have been described. Bell and Kozlowski (2002) argue for two primary characteristics that distinguish virtual and conventional teams: spatial distance and mode of communication. Contrary to conventional teams, virtual teams are not colocated, but are geographically and, often, temporally distributed. The second boundary condition, communication mode, refers to the fact that while conventional teams may augment face-to-face communication with other more technologically enabled forms, communication within virtual teams must be technologically mediated. Thereby, it is not the task itself that distinguishes virtual from conventional teams, but the manner in which tasks are accomplished based on the configural properties of virtual teams.

Researchers have recently begun to argue that virtual teams lie along a continuum and vary on their virtualness (see Bell & Kozlowski, 2002; Priest et al., 2005). Four properties have been identified that, when combined, result in virtual team types that vary in workflow patterns and task complexity: member roles, boundaries, lifecycle, and temporal distribution (Bell & Kozlowski, 2002). The first characteristic is the degree to which team members hold singular or multiple roles. As the number of roles team members hold increases, so does the potential for role conflict and ambiguity. A second distinguishing characteristic, but at the team level, is the team's boundaries. Virtual teams can contain cross-functional, organizational, and cultural boundaries or be more similar to conventional teams and be bounded within a single organization, cultural, or functional boundary (Bell & Kozlowski, 2002). As the degree to which the team crosses different boundaries increases, it becomes more difficult to establish and maintain a team identity, cohesion, and leadership. The third characteristic that has been argued to distinguish between types of virtual teams is their lifecycles. Within virtual teams, members often rotate in and out, disrupting team development and engendering a shorter lifecycle than is typical in most conventional teams (Bell &

Kozlowski, 2002). Finally, the temporal distribution may range from operation in real time due to tightly coupled interdependencies to more sequential and asynchronous interaction for those teams that are more loosely coupled. Given this brief examination of the distinguishing features of virtual teams, the next logical question arises: What are the competencies needed to facilitate successful navigation of the virtual team terrain?

TEAM COGNITION WITHIN VIRTUAL TEAMS

While the research literature on virtual teams is relatively young, it is reasonable to expect that the core competencies (that is, teamwork and taskwork, see also Marks, Mathieu, & Zaccaro, 2001) identified within conventional teams are a necessary, but not sufficient, condition for success within virtual teams. While taskwork knowledge, skills, abilities, and other characteristics (KSAOs) provide the initial foundation for performance, teamwork KSAOs provide the mechanism by which members are able to coordinate to accomplish the task. It is often not taskwork that poses the greatest challenge for virtual teams, as members are often purposely selected based on their taskwork capabilities, but teamwork.

There have been a number of processes and states identified as necessary for virtual team effectiveness (Martins et al., 2004); however, the focus here will be on those that theoretically underlie a team's ability to implicitly coordinate their actions. This focus was chosen as the environment within which virtual teams operate often produces decrements in the quality and number of nonverbal cues guiding coordination within traditional teams; thereby coordination is often most similar to the notion of implicit coordination as seen within traditional teams. Therefore, a focus on measures of team cognition and how they might apply within virtual teams is warranted.

Team Cognition Defined

Team cognition has been defined as the interaction of internalized and externalized processes, which emerge from individual cognition, team interactions, and process behaviors (Fiore & Schooler, 2004). It has been characterized as a type of awareness used to bind a team's actions (Gutwin & Greenberg, 2004) and communication. Recently, another term, macrocognition, has begun to emerge to describe many of the cognitive processes and states that comprise team cognition. Macrocognition has been defined as "the internalized and externalized high level mental processes employed by teams to create new knowledge during complex, one of a kind, collaborative problem solving" (Letsky, Warner, Fiore, Rosen, & Salas, 2007, p. 7). Major processes have been argued to include the following: individual knowledge building, team knowledge building, developing shared problem conceptualizations, team consensus development, and outcome appraisal (Warner, Letsky, & Cowen, 2005).

Within the current chapter, we focus on the intersection of macrocognition and team cognition as traditionally defined. Components of team knowledge building and developing shared problem conceptualizations will be focused on as they move beyond individual knowledge building, thereby providing the foundation for coordination within virtual teams. Within team knowledge building measurement, developments related to transactive memory and shared mental models will be examined, while team situation awareness will be examined within developing shared problem conceptualizations.

Shared Mental Models

Shared mental models (SMMs) have been defined as organized knowledge structures that are held by more than one team member and involve the integration of information and the comprehension of a given phenomenon (Johnson-Laird, 1983; Cannon-Bowers, Salas, & Converse, 1993). These structures do not have to be shared in the truest sense, but instead represent compatible knowledge structures.

It has been argued that compatible mental models facilitate the effective coordination of action, as well as promote a similar method for processing new information within the team (Klimoski & Mohammed, 1994). SMMs have been found to increase members' abilities to recognize causal relationships, make sound inferences, and create better explanations regarding the task and team member actions (see Mathieu, Heffner, Goodwin, Salas, & Cannon-Bowers, 2000).

Transactive Memory Systems

The construct of transactive memory is an expansion on most conceptualizations of SMM in that it speaks to the storage of information, not the interrelationships among the stored items. Wegner (1986) argues that a transactive memory system (TMS) is composed of each individual's and a collective awareness of with whom that knowledge resides. Others argue that team members' metaknowledge, consensus/agreement, and accuracy are necessary components (Yoo & Kanawattanachai, 2001; Austin, 2003).

The TMS is founded on the idea that, due to complexity, individuals know only part of what the team as a whole knows, and this team knowledge is distributed unequally among members (Moreland, Argote, & Krishnan, 1998). Initial research indicates that a TMS enhances a team's performance, particularly when the task is complex and requires a considerable contribution of knowledge from individual team members (Faraj & Sproull, 2000).

Team Situation Awareness

The importance of situational awareness in complex tasks has received extensive empirical support, both at the individual and team levels (for example, Endsley, 1995; Salas, Prince, Baker, & Shrestha, 1995). Situation awareness has been defined as the "perception of the elements in the environment within a volume of time and space, the comprehension of their meaning, and the projection of their status in the near future" (Endsley, 1995, p. 36). Although often analyzed at an individual level, researchers have begun to conceptualize team situation awareness (TSA). Salas et al. (1995) suggest that TSA is more than the sum of the individual members' situation awareness, but includes team process behaviors as well.

So given the argued importance of team cognition, how do those charged with developing/maintaining virtual team effectiveness diagnose the aforementioned aspects of team cognition? Team performance measurement systems are a proven method that can guide instructors, practitioners, and team members themselves in the diagnosis process.

PROPERTIES OF TEAM PERFORMANCE MEASUREMENT SYSTEMS

Team performance measurement is by no means new; however, it is often one of the most overlooked and misunderstood components within team development. The term is somewhat of a misnomer as it does not only refer to the measurement of team performance as an outcome, but also to the processes and states that comprise such performance. Team performance measurement systems serve multiple purposes within organizations, not the least of which is as a diagnostic aid. Quality team performance measurement systems are the only method by which teams can be systematically evaluated to catch problems early before they become ingrained in members' thought processes and actions. The information gained as a result of systematically designed and soundly implemented team performance measurement systems can also serve as a basis for the design, delivery, and choice of interventions to further team development. For a review of the basic properties of team performance measurement system, see Figure 14.1.

When examining team performance measures to assess fit for a particular purpose, it is important to realize that each measure can be delineated into its component parts. The content, the elicitation source, the elicitation method, and the indexing/aggregation method combine to create a single measure. It is not uncommon for these four components to be spoken of as one entity; however, they are distinct components that should each be taken into account when determining the best measure for a situation. The content of a measure refers to what the measure is actually attempting to capture (that is, knowledge, behavior, or attitudes). This can be further subdivided into the specific content of that knowledge, behavior, or attitude. For example, when talking about shared cognition, the content may be knowledge of the equipment, task, team member roles/responsibilities, expertise, or situation. The second component of a performance measure is the source from which the response is elicited (for example, subjective source—peer, participant/self-report, or observer; objective source—equipment). The third component, method, refers to the manner in which the information is extracted (for example, interviews, checklists, card sorts, Likert scales, vignettes,

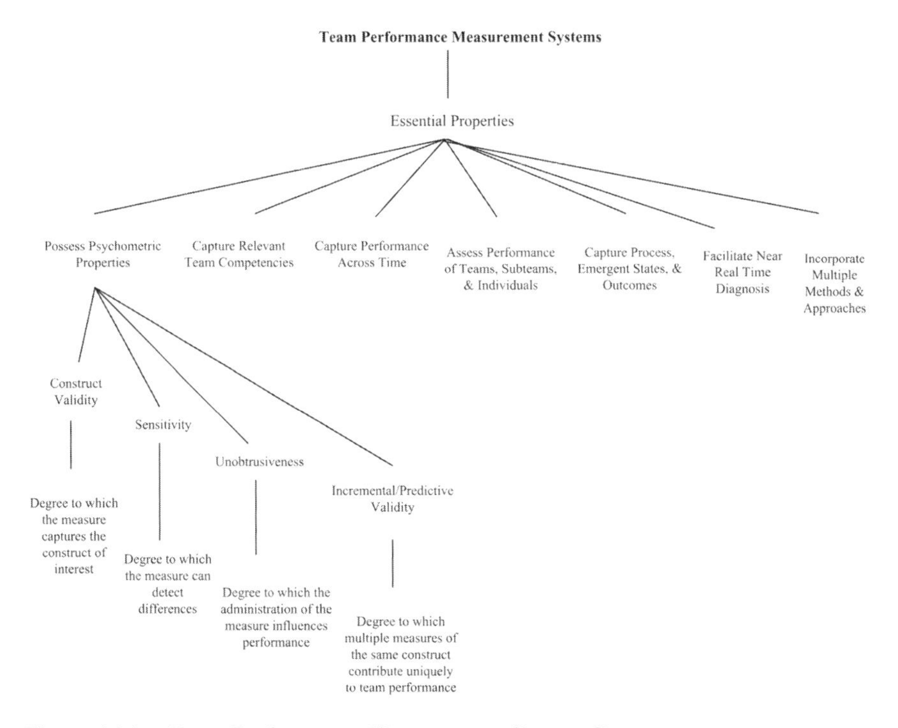

Figure 14.1. Team Performance Measurement System Components

or think-alouds). Finally, when dealing with teams, the method of aggregation becomes an important component. This refers to the manner in which the information is scored and compiled (for example, percentage, mean, sum, distance, or correlation). These four components form the basis for understanding any diagnostic measure.

TEAM PERFORMANCE MEASUREMENT IN VIRTUAL TEAMS

While there are differences between conventional and virtual teams, the components of a diagnostic metric and the associated decisions to be made remain the same. What differs is not the manner in which metrics for virtual teams are designed, but the task and team characteristics that drive the information provided to the metrics. Next, the high level questions to be asked are identified, and prescriptive guidance as to how metrics may look dependent on the characteristics of virtual teams is offered.

What to Measure?

The first question that must be answered within any diagnostic endeavor is what the metric should capture. The answer is driven by two factors: the construct being diagnosed and the important components within that construct. As we have argued that team cognition is especially important due to its role in promoting coordination within virtual teams, we first briefly review the various manner in which content is conceptualized with regard to SMM, TMS, and TSA. Next, how content is driven by the defining characteristics of virtual teams will be specified.

Shared Mental Models—Content

With regard to SMM, the literature has primarily argued for knowledge structures to be compatible around four foci or content areas: the equipment, task, team (Rouse, Cannon-Bowers, & Salas, 1992), and team interaction (Cannon-Bowers et al., 1993). Metrics whose content is equipment knowledge focus on how the equipment that the team is interacting with works. This knowledge allows team members to predict what the equipment is likely to do and when to make a response. Conversely, metrics that focus on diagnosing task knowledge structures query knowledge relating to the basic attributes of the task and how to accomplish it. Content within such metrics focuses on diagnosing knowledge and beliefs regarding task procedures, goals, strategies, and the interrelationships among this content. Compatibility in terms of task knowledge has been argued to allow members to describe why task performance is important, what situations may occur, explain task procedures, and predict consequences of performance (Rouse et al., 1992).

The last two types of content contained within metrics of SMM are team and team interaction (see Cannon-Bowers et al., 1993). The team mental model (TMM) contains knowledge about team member characteristics, including their task knowledge, skills, abilities, and preferences. The team interaction model (TIM) contains information about the team in relation to the individual and collective requirements needed for effective team interaction. While the TMM allows members to form expectations and predict future performance, the TIM allows members to anticipate and sequence their collective actions. Compatibility between the last two types of content (that is, team and team interaction) has been argued to be most important for coordinated team action (see Cannon-Bowers et al., 1993). As these knowledge structures have been argued to be hierarchical in nature (Rentsch & Hall, 1994), metrics have tended to focus on the content indicative of knowledge structures higher in the hierarchy (for example, task, team, and team interaction).

Transactive Memory Systems—Content

TMSs are a relatively new development within the team literature. Existing metrics either directly (for example, Rau, 2006) or indirectly (for example, Lewis, 2003) diagnose TMS. For example, Austin (2003) created a questionnaire that directly assessed the team's knowledge content by measuring the team's collective knowledge, as well as specialization of, agreement about the location of, and accurate perceptions of who possesses said knowledge, in relation to a specific topic. Other measures of TMS have included content that is indirectly related to this construct (that is, memory differentiation, task coordination, and task credibility; see, for example, Moreland et al., 1998).

Team Situation Awareness—Content

Situation awareness has been argued to be comprised of three levels: perceiving components in the environment, comprehending the situation, and predicting future scenarios (Endsley, 1995). Perceiving requires one to be familiar with environmental features and the changes that occur. Comprehending, the second level, engages working memory to assist in comprehending the significance of complex environmental cues. The third level, predicting future actions, requires team members to not only perceive and comprehend the environment, but to implement a mental model of the surroundings. Building from this conceptualization, the content within TSA measures revolves around the perception of key elements within the environment, comprehension of their meaning, and projection of their status in the near future. Most measures focus more heavily on the perception of elements as compared to meaning and projection.

Application to Virtual Teams

While the preceding paragraphs have afforded a brief review of the content contained within measures of team cognition, the exact content that should be included is dependent on the task and team characteristics that drive performance. In this regard there are four primary characteristics that may impact the content to be assessed: variations in roles, lifecycle, temporal distribution, and interdependence.

Beginning with differences in role configurations, Bell and Kozlowski (2002) argued that as teams move along the virtual continuum, the degree to which members occupy multiple roles is likely to vary. As the number of roles increases, there is a greater propensity for role ambiguity and corresponding role conflict. This role conflict is likely to translate to other team members. Specifically, it becomes increasingly likely that members will have different conceptualizations concerning the true nature of particular member roles and responsibilities. This ambiguity often results in less compatibility among the knowledge structures that serve to guide coordination. Given the above, measures designed for virtual teams should, at a minimum, be focused on capturing the content within TMS, as well as team models.

Guideline 1: Assess the degree to which members of virtual teams have singular or multiple roles within and across virtual teams.

Guideline 1a: Design measures to directly assess a collective awareness of who knows what and areas of expertise (that is, TMS); this becomes increasingly important as role complexity increases.

Guideline 1b: Design measures to assess knowledge about team member characteristics and responsibilities (that is, TMM); this becomes increasingly important as role complexity increases.

A second characteristic that may differentiate virtual teams is the team's lifecycle. Within virtual teams there is a tendency for rotating members resulting in a shorter lifecycle (Bell & Kozlowski, 2002). Thereby, it becomes more difficult to maintain the compatible knowledge and affective structures that guide behavior. While mental models at the higher levels (that is, team and team interaction) have been shown to be primarily responsible for the seamless coordination and adaptation within effective teams, these mental models often take more time to fully develop than equipment or task models. Therefore, those mental models that play the heaviest role in implicit coordination and adaptation (skills that are often more challenging for virtual teams due to degraded social cues, loss of face-to-face contact, distribution, and added role complexity) may not be fully developed as lifecycles become shorter or membership is changing.

Guideline 2: Assess the degree to which virtual team members rotate in and out of the team (that is, a short versus a long lifecycle).

Guideline 2a: Design measures to include content related to knowledge of the team and team interaction and within team compatibility of this knowledge; this information becomes more diagnostic within short lifecycles.

Guideline 2b: Target the entire content within TMS to better diagnose virtual teams with short lifecycles.

The distributed nature of many virtual teams also impacts the content that should be included within measures of team cognition. As members become further distributed in time and space, it becomes more difficult to maintain a common awareness of environmental elements that impact team action. Virtual team members may see very different elements of the situation; consequently "ground truth" regarding TSA must be determined. This argues for a need for measures of content related to both individual and team level situation awareness. Content at the individual level will assist in diagnosing where the breakdown actually occurred (for example, perception, meaning assignment, or communication).

Research has indicated that shared mental models allow team members to predict the needs of their teammates (Mathieu et al., 2000), oftentimes without explicit communication (Entin & Serfaty, 1999). This becomes important for as distribution increases, it becomes more difficult to explicitly coordinate, thereby creating an increased reliance on coordination resembling implicit coordination.

Guideline 3: Assess the degree of temporal and physical distribution within virtual teams.

Guideline 3a: Design content to capture the knowledge contained within team and team interaction models; these models become more difficult to maintain as distribution increases.

Guideline 3b: Design content to capture both individual and team level situation awareness; it will assist in diagnosing whether decrements are in terms of failures of perception and meaning at the individual level or communication at the team level, each of which become more challenging as distribution increases.

Finally, virtual teams may vary in the level of required task interdependence. As task interdependency increases from pooled to team (see Saavedra, Earley,

& Van Dyne, 1993), there is a corresponding need to coordinate and synchronize member actions. Additionally, there is a tendency for tasks to become more complex as teams move up the task interdependence hierarchy. Teams that are operating under higher levels of interdependency have more freedom in how the task and member roles are structured.

Guideline 4: Assess the degree of task interdependence required within the virtual team prior to designing or choosing measures.

Guideline 4a: Design measures to include content related to team and team interaction knowledge for virtual teams that require moderate to high degrees of task interdependence.

Where to Collect Information?

Once the content of measures is decided upon, a second decision that often serves to categorize a metric is the source of the content. Within the larger team performance measurement literature, the predominant source of elicitation is the individual being diagnosed. Although other elicitation sources are used (for example, supervisors, peers, and trained raters), self-report is the most predominant. While it may be argued that no one person has a better understanding of an individual's cognition than the individual in question, even experts may not have sufficient levels of insight to successfully verbalize the information, and self-reports have repeatedly been criticized for biases.

Application to Virtual Teams

The determination of whom to elicit the data from should be driven by the research question, as well as an awareness of who has opportunities to observe the behavior being diagnosed. Recently researchers have begun to argue that team cognition manifests itself in the behavioral actions that teams enact (Cooke et al., 2004). Thereby, information can be collected from other sources in addition to the targeted individual. Within virtual teams it is especially important to use a multisource approach to knowledge elicitation as often individual members who are distributed may have very different perceptions serving as input to their knowledge structures. By gathering information from multiple personnel and from objective indices it becomes possible to assess where breakdowns in team cognition are occurring. For example, with regard to TSA, are breakdowns originating due to individual members misperceiving their unique perspectives or due to unsuccessful sharing of critical information?

Guideline 5: Collect information from a variety of sources as the triangulation of information will provide a fuller picture of the state of team cognition, especially as distribution increases.

How to Elicit Information?

Elicitation methods span numerous dimensions: qualitative to quantitative, subjective to objective, and explicit to implicit. Looking across the three cognitive constructs discussed within the current chapter, elicitation methods are primarily subjective and explicit. In particular, methods used to elicit SMM include questionnaires, network scaling methods, concept mapping, causal mapping, card sort, content analysis of communication, and observation. Questionnaires, concept mapping, and card sorts all tend to be explicit/intrusive, rely heavily on self-report data, and use paper-and-pencil instruments. Several programs have emerged that allow the electronic delivery of card sorts and concept mapping (see Hoeft et al., 2003). More indirect methods include network scaling, communication content analysis, and observation. While the former tends to use computer algorithms to score content relationships and network structures, the basis for the input is normally team member relatedness ratings of concepts that tap equipment, task, team, or team interaction knowledge. Conversely, content analysis of communication and observations are normally conducted post hoc or near real time via the use of trained observers.

As some have argued that TMS are a subset of shared mental models, it is not surprising that the methods of extraction appear very similar, albeit slightly more limited. Specifically, the predominant methods used to assess TMS are questionnaires, observations, and communication analysis.

Finally, with regard to TSA, the most common forms of elicitation are query methods, followed by self/peer ratings, event based observation, and communication analysis. Query methods are programs that ask specific questions about the situation while a participant is performing a task with methods varying as to the query's obtrusiveness (see Situation Awareness Global Assessment Technique [SAGAT], Endsley, 2000; Situation-Present Assessment Method [SPAM], Durso, Hackworth, Truitt, Crutchfield, & Nikolic, 1999). The SAGAT method freezes the display to ask the question, whereas SPAM presents queries in the task allowing the user to obtain information from the task environment (Cooke, Stout, & Salas, 2001). Often using the same referent as query methods are rating scales in which perceived levels of TSA are assessed. For example, the Situational Awareness Rating Scale (SARS; Bell & Waag, 1995; Waag & Houck, 1994) and the Situational Awareness Rating Technique (SART; Taylor, 1989) have been used to obtain self-report data. The SARS has also been used to collect peer ratings.

Less obtrusive methods include event based observation and post hoc communication analysis. Event based observation methods use trained raters to assess the presence of behavioral markers of TSA. This method has been used extensively within a variety of environments and specifies a priori defined events within which markers are created (see Dwyer, Fowlkes, Oser, & Lane, 1997). Finally, communication analysis may overcome limitations associated with query methods and event based ratings. The latter may pose challenges within field environments where there is much ambiguity. The use of communication analysis, whereby the content and pattern of a team's communication is analyzed, can provide an assessment of TSA.

Application to Virtual Teams

The heavy reliance on self-report and paper-and-pencil measures that can be cumbersome in virtual distributed teams argues for a need to move elicitation methods beyond the sole use of traditional methods. Moreover, the distributed nature of virtual teams makes it difficult to conduct observations within context of the entire team. Thus, it is recommended that, in addition to broadening our toolbox, the technology present in such teams is leveraged to translate traditional measures to electronic formats.

Guideline 6: Design measures to reduce the additional burdens put on those assessing teams distributed across time and space.

Guideline 6a: Take advantage of the technology embedded within virtual teams to translate paper-and-pencil measures to electronic formats.

Guideline 6b: Use embedded measurement and design integration mechanisms within the system to further reduce the burden of assessors.

While there is a wide variety of methods, there is also a need to move away from existing methods in that they are subjective and self-reporting. The technology present within virtual teams should be leveraged to not only reduce the obtrusiveness and cumbersome nature of existing measures, but also to incorporate techniques in other domains. One such nontraditional, emerging tool is the use of psychophysiological measurement techniques. The term "psychophysiological measurement" is a blanket expression for measures that examine changes in physiological data and how that may translate into differences in psychological states. This provides unique information that can be streamlined into simulation experimental designs (Cacioppo, Bernston, Sheridan, & McClintock, 2000). One downfall is that there is no direct link between psychological and physiological processes, so a certain level of inference must be used to analyze data. The following psychophysiological tools may prove the most beneficial in team cognition assessment: eye tracking, electroencephalogram, and vocal characteristics. As the first two have the most direct connection to team cognition, these will be briefly discussed.

An eye tracker measures eye movements, pupil size, focus, and other characteristics of one or both eyes while engaged in a task. An eye tracker can capture certain metrics that are important indicators of cognitive and social processes (Lum, Feldman, Sims, & Salas, 2007). This device provides information regarding an individual's gaze allowing a researcher to identify what a person is looking at any given time and eye movement patterns across an entire task (Poole & Ball, 2006). Previous studies have measured information exchange by indexing when, and how often, a piece of information was passed from one team member to another, yet this may not indicate that this information was received and used by the member for whom it was intended. Employing an eye tracker, along with traditional measures, might more definitively determine if the information was received and what was done with it. This would be especially useful in diagnosing virtual team performance, as often information gets lost due to the temporal and physical distribution inherent in the medium. Another potential application includes an examination of individual and group differences in shared mental model generation in a virtual environment. For example, eye-tracking data could be collected while participants perform a simulation, and eye patterns while performing the simulation could be used (for example, some people may look more at certain events or information) to predict how team members interacted in the simulation.

Electroencephalogram (EEG) is another possible objective method, albeit an intrusive one. EEG measures the electrical activity of the brain using electrodes strategically placed on the scalp that send electrical impulses occurring at the millivolt level to a digital or analog acquisition device. EEG is a more proximal indicator of cognitive activity than some of the other psychophysiological metrics because it measures actual cortical activation (Davidson, Jackson, & Larson, 2000). Certain metrics within EEG, such as bandwidth, are able to discriminate between general arousal and focused attention (Klimesch, 1999). This may have implications for diagnosing virtual teams with regard to TSA, as well as changes in arousal levels during stressful situations and related performance decrements.

While these methods have not been validated as measures of team cognition, comparison of validated traditional measures with these newer methods would increase confidence. For example, team processes and macrocognition may be aided by the use of psychophysiological measurement in detection of how often and exactly when teammates look at stimuli in their environment during interaction of a team task. Studies that have used team indices employ measures of physiological compliance or the similarity of physiological activity between team members and show that compliance positively predicts team coordination efforts and team performance (Henning & Korbelak, 2005; Henning, Boucsein, & Gil, 2001). These alternative methods may not only pick up on some of the truly challenging aspects of virtual teams, but also provide a fuller picture of team cognition.

Guideline 7: Think outside the box in deciding on methods to employ, consider psychophysiological methods as augmenting traditional methods of gathering information pertaining to team cognition.

Guideline 7a: Use eye tracking to obtain objective indices of perception of environmental elements and, in turn, diagnose the components of TSA.

Guideline 7b: Pair psychophysiological measures with traditional measures to begin to establish convergent and/or divergent validity among methods of assessing team cognition.

How to Index and Aggregate Elicited Information?

A final component of every measure of team cognition is the level (that is, individual or team) at which the construct of interest is captured. The nature of the specific construct of interest should be what drives the decision as to the level of measurement to target. Within measures of team cognition, it is most common to collect information at the individual level and then aggregate to the team level. Typically indexing has been done through averaging. Recently there has been an increased focus on the manner in which team level constructs when measured at the individual level should be aggregated. Kozlowski and Klein (2000) have argued that there are two primary ways in which individual constructs may manifest themselves at the team level: compilation and composition. In the current situation, composition describes a process whereby an individual construct (that is, situation awareness) emerges upward to the team level (that is, team situation awareness), but essentially remains the same. When this happens and within-unit variance is demonstrated, aggregation to the team level can be represented by the mean or sum. Conversely, compilation is the process whereby similar but distinctively different lower level properties combine into a higher level (for example, team) property that is related to but different from its diverse lower level constituent parts (Kozlowski & Klein, 2000). Constructs that emerge through compilation do not represent shared properties across levels, but rather are qualitatively different (that is, constructs are characterized by patterns). Thereby, constructs that emerge in this manner are best represented by the minimum or maximum, indices of variation, profile similarity, multidimensional scaling (Kozlowski & Klein, 2000, p. 34) along with a number of other related techniques.

The specific manner in which team cognition materializes, and how that materialization is operationalized, is contingent upon organizational context, work-flow interdependencies, and other situational factors (Klein & Kozlowski, 2000). Given this, some researchers have argued that capturing individual level cognition and aggregating it to the team level through a mean index may not always be the most appropriate. Cooke, Kiekel, Bell, and Salas (2002) argue that as role specialization increases within a team, it is no longer appropriate to use the mean as an aggregating index. Thereby, Cooke et al. (2002) have proposed a more holistic assessment of team cognition that results from "the interplay of the individual cognition of each team member and team process behaviors" (p. 85). Although this may be more difficult to develop, it may ultimately be a better way of determining certain aspects of team knowledge. Within this framework content is directly assessed at the team level.

Guideline 8: Recognize that there is not one correct manner in which to index measures of team cognition—it depends on task and team characteristics.

Guideline 8a: Use task interdependence and role structure to assist in guiding choice of aggregation and indexing method.

CONCLUDING COMMENTS

As organizations continue to invest in technology and location becomes less of an issue in selecting team members, the use of virtual teams will continue to rise. While many of the lessons that have been learned concerning effectiveness in conventional teams are expected to hold within virtual teams, virtual teams possess some unique challenges. Due to their distributed nature breakdowns in team process, and the knowledge structures that guide such a process, may propagate over time with members not being aware until these have become rather large and ingrained. While systematic, frequently delivered diagnosis with corresponding feedback has been argued to be essential in order for teams to continually evolve and adapt, it is even more important in situations where errors can easily propagate, as is the case with virtual teams.

Within the current chapter we have identified several characteristics that distinguish virtual from conventional teams, as well as distinguish among varieties of virtual teams. Knowledge about performance measurement within conventional teams was then leveraged against these characteristics to identify a set of guidelines for metrics of team cognition within virtual teams. While we acknowledge that there is much to learn about virtual teams, we hope that what is offered within the current chapter will begin to foster thinking concerning measurement within such teams. Finally, we hope that the current chapter encourages those responsible for team performance measurement to remember that in the creation and implementation of diagnostic instruments, one must not only consider psychometric properties, but the series of decisions related to content, elicitation source, elicitation method, and the indexing/aggregation.

ACKNOWLEDGMENTS

The views expressed in this chapter are those of the authors and do not necessarily reflect official U.S. Navy policy. This work was supported in part by an ONR MURI Grant No. N000140610446 (Dr. Michael Letsky, Program Manager).

REFERENCES

Austin, J. R. (2003). Transactive memory in organizational groups: The effects of content, consensus, specializations, and accuracy on group performance. *Journal of Applied Psychology, 88,* 866–878.

Bell, B. S., & Kozlowski, S. W. J. (2002). A typology of virtual teams: Implications for effective leadership. *Group and Organization Studies, 27*(1), 14–19.

Bell, H. H., & Waag, W. L. (1995). Using observer ratings to assess situational awareness in tactical air environments. In D. J. Garland & M. R. Endsley (Eds.), *Experimental analysis and measurement of situation awareness* (pp. 93–99). Daytona Beach, FL: Embry-Riddle Aeronautical University Press.

Cacioppo, J. T., Bernston, G. G., Sheridan, J. F., & McClintock, M. K. (2000). Multilevel integrative analyses of human behavior: Social neuroscience and the complementing nature of social and biological approaches. *Psychological Bulletin, 6,* 829–843.

Cannon-Bowers, J. A., Salas, E., & Converse, S. A. (1993). Shared mental models in expert team decision making. In N. J. Castellan, Jr. (Ed.), *Current issues in individual and group decision making* (pp. 221–246). Hillsdale, NJ: Lawrence Erlbaum.

Cooke, N. J., Kiekel, P. A., Bell, B., & Salas, E. (2002, October). Addressing limitations of the measurement of team cognition. *Proceedings of the 46th Annual meeting of the Human Factors and Ergonomics Society* (pp. 403–407). Santa Monica, CA: Human Factors and Ergonomics Society.

Cooke, N. J., Salas, E., Kiekel, P. A., & Bell, B. (2004). Advances in measuring team cognition. In E. Salas & S. M. Fiore (Eds.), *Team cognition* (pp. 83–106). Washington, DC: American Psychological Association.

Cooke, N. J., Stout, R. J., & Salas, E. (2001). A knowledge elicitation approach to the measurement of team situation awareness. In M. McNeese, E. Salas, & M. Endsley (Eds.), *New trends in cooperative activities: Understanding system dynamics in complex environments* (pp. 114–139). Santa Monica, CA: Human Factors and Ergonomics Society.

Davidson, R. J., Jackson, D. C., & Larson, C. L. (2000). Human electroencephalography. In J. T. Cacciopo, L. G. Tassinary, & G. G. Berntson, (Eds.), *Handbook of psychophysiology* (pp. 27–52). Cambridge, MA: Cambridge University Press.

Driskell, J. E., Radtke, P. H., & Salas, E. (2003). Virtual teams: Effects of technological mediation on team performance. *Group Dynamics: Theory, Research, and Practice, 7* (4), 297–323.

Durso, F. T., Hackworth, C. A., Truitt, T. R., Crutchfield, J., & Nikolic, D. (1999). *Situation awareness as a predictor of performance in en route air traffic controllers* (Rep. No. DOT/FAA/AM-99/3). Washington, DC: Office of Aviation Medicine.

Dwyer, D. J., Fowlkes, J. E., Oser, R. L., & Lane, N. E. (1997). Team performance measurement in distributed environments: The TARGETs methodology. In M. T. Brannick, E. Salas, & C. Prince (Eds.), *Team performance assessment and measurement: Theory, methods, and applications* (pp. 137–153). Mahwah, NJ: Lawrence Erlbaum.

Endsley, M. R. (1995). Toward a theory of situation awareness in dynamic systems. *Human Factors, 37,* 32–64.

Endsley, M. R. (2000). Direct measurement of situation awareness: Validity and use of SAGAT. In M. R. Endsley & D. J. Garland (Eds.), *Situation awareness analysis and measurement* (pp. 147–174). Mahwah, NJ: Lawrence Erlbaum.

Entin, E. E., & Serfaty, D. (1999). Adaptive team coordination. *Human Factors, 41*(2), 312–325.

Faraj, S., & Sproull, L. (2000). Coordinating expertise in software development teams. *Management Science, 46,* 1544–1568.

Fiore, S. M., & Schooler, J. W. (2004). Process mapping and shared cognition: Teamwork and the development of shared problem models. In E. Salas & S. M. Fiore (Eds.), *Team cognition: Understanding the factors that drive process and performance* (pp. 133–152). Washington, DC: American Psychological Association.

Gutwin, C., & Greenberg, S. (2004). The importance of awareness for team cognition in distributed collaboration. In E. Salas & S. M. Fiore (Eds.), *Team cognition: Understanding the factors that drive process and performance* (pp. 177–201). Washington, DC: American Psychological Association.

Henning, R., & Korbelak, K. T. (2005). Social-psychological compliance as a predictor of future team performance. *Psychologia, 45*(2), 84–92.

Henning, R. A., Bouscein, W., & Gil, M. C. (2001). Social-physiological compliance as a determinant of team performance. *International Journal of Psychophysiology, 40,* 221–232.

Hoeft, R. M., Jentsch, F. G., Harper, M. E., Evans, A. W., Bowers, C. A., & Salas, E. (2003). TPL-KATS—concept map: A computerized knowledge assessment tool. *Computers in Human Behavior, 19* (6), 653–657.

Johnson-Laird, P. (1983). *Mental models.* Cambridge, MA: Harvard University Press.

Kanawattanachai, P., & Yoo, Y. (2002). Dynamic nature of trust in virtual teams. *Strategic Information Systems, 11,* 187–213.

Kayworth, T. & Leidner, D. (2000). The global virtual manager: A prescription for success. *European Management Journal, 18*(2), 183–194

Klimesch, W. (1999). EEG alpha and theta oscillations reflect cognitive and memory performance: A review and analysis. *Brain Research Reviews, 29,* 169–195.

Klimoski, R., & Mohammed, S. (1994). Team mental model: Construct or metaphor? *Journal of Management, 20,* 403–437.

Kozlowski, S. W. J., & Klein, K. (2000). A multilevel approach to theory and research in organizations: Contextual, temporal, and emergent processes. In K. J. Klein & S. W. J. Kozlowski (Eds.), *Multilevel theory, research, and methods in organizations: Foundations, extensions, and new directions* (pp. 3–90). San Francisco: Jossey-Bass, Inc.

Letsky, M., Warner, N., Fiore, S. M., Rosen, M. A., & Salas, E. (2007). *Macrocognition in complex team problem solving.* Paper presented at the 11th International Command and Control Research and Technology Symposium (ICCRTS), Cambridge, United Kingdom.

Lewis, K. (2003). Measuring transactive memory systems in the field: Scale development and validation. *Journal of Applied Psychology, 88,* 587–604.

Lum, H., Feldman, M., Sims, V., & Salas, E. (2007). Eye tracking as a viable means to study augmented team cognition. In D. D. Schmorrow, D. M. Nicholson, J. M. Drexler, & L. M. Reeves (Eds.), *Foundations of Augmented Cognition* (4th ed., pp. 190–196). Arlington, VA: Strategic Analysis, Inc.

Marks, M. A., Mathieu, J. E., & Zaccaro, S. J. (2001). A temporally based framework and taxonomy of team processes. *Academy of Management Review, 26*(3), 356–376.

Martins, L. L., Gilson, L. L., & Maynard, M. T. (2004). Virtual teams: What do we know and where do we go from here? *Journal of Management, 30,* 805–835.

Mathieu, J. E., Heffner, T. S., Goodwin, G. F., Salas, E., & Cannon-Bowers, J. A. (2000). The influence of shared mental models on team process and performance. *Journal of Applied Psychology, 85,* 273–283.

Moreland, R. L., Argote, L., & Krishnan, R. (1998). Training people to work in groups. In R. S. Tindale, L. Heath, J. Edwards, E. J. Posavac, F. B. Bryant, Y. Suarez-Balcazar, E. Henderson-King, & J. Myers (Eds.), *Theory and research on small groups* (pp. 37–60). New York: Plenum.

Poole, A., & Ball, L. J. (2006). Eye tracking in HCI and usability research. In C. Ghaoui (Ed.), *Encyclopedia of human-computer interaction* (pp. 211–219). Hershey, PA: Idea Group, Inc.

Priest, H. A., Stagl, K. C., Klein, C., & Salas, E. (2005). Creating context for distributed teams via virtual teamwork. In C. A. Bowers, E. Salas, & F. Jentsch (Eds.), *Creating high tech teams* (pp. 185–212). Washington, DC: American Psychological Association.

Rau, D. (2006). Top management team transactive memory, information gathering, and perceptual accuracy. *Journal of Business Research, 59,* 416–424.

Rentsch, J. R., & Hall, R. J. (1994). Members of great teams think alike: A model of team effectiveness and schema similarity among team members. *Advances in Interdisciplinary Studies of Work Teams, 1,* 223–261.

Rouse, W. B., Cannon-Bowers, J. A., & Salas, E. (1992). The role of mental models in team performance in complex systems. *IEEE Transactions on Systems, Man, and Cybernetics, 22,* 1296–1308.

Saavedra, R., Earley, P. C., & Van Dyne, L. (1993). Complex interdependence in task-performing groups. *Journal of Applied Psychology, 78*(1), 61–72.

Salas, E., Prince, C., Baker, D. P., & Shrestha, L. (1995). Situation awareness in team performance: Implications for measurement and training. *Human Factors, 37*(1), 123–136.

Sproull, L., & Kiesler, S. (1986). Reducing social context cues: Electronic mail in organizational communications. *Management Science, 32*(11), 1492–1512.

Taylor, R. M. (1989). Situational awareness rating technique (SART): The development of a tool for aircrew system design. In *Situational Awareness in Aerospace Operations* (AGARD-CP-478; pp. 3.1–3.17). Neuilly Sure Seine, France: NATO-AGARD.

Waag, W. L., & Houck, M. R. (1994). Tools for assessing situational awareness in an operational fighter environment. *Aviation, Space and Environmental Medicine, 65*(5), A13–A19.

Warner, N., Letsky, M., & Cowen, M. (2005). Cognitive model of team collaboration: Macro-cognitive focus. *Proceedings of the Human Factors and Ergonomics Society 49th Annual Meeting* (pp. 269–273). Santa Monica, CA: Human Factors and Ergonomics Society.

Wegner, D. M. (1986). Transactive memory: A contemporary analysis of the group mind. In B. Mullen & G. R. Goethals (Eds.), *Theories of group behavior* (pp. 185–208). New York: Springer-Verlag.

Yoo, Y., & Kanawattanachai, P. (2001). Developments of transactive memory systems and collective mind in virtual teams. *The International Journal of Organizational Analysis, 9*(2), 187–208.

Chapter 15

VIRTUAL ENVIRONMENT PERFORMANCE ASSESSMENT: ORGANIZATIONAL LEVEL CONSIDERATIONS

Robert D. Pritchard, Deborah DiazGranados, Sallie J. Weaver, Wendy L. Bedwell, and Melissa M. Harrell

Measuring performance in virtual environments (VEs) is a critical part of VE training design and application. The premise of our chapter is that while individual and team level measurement issues are important (covered in Fowlkes, Neville, Nayeem, and Eitelman, Volume 1, Section 3, Chapter 13; Burke, Lum, Scielzo, Smith-Jentsch, and Salas, Volume 1, Section 3, Chapter 14), there are also several organizational level issues that must be considered for optimal VE training performance measurement. This level of analysis issue is an important topic (Klein & Kozlowski, 2000) because individuals and teams should contribute to the broader organization. Without considering the broader organization, important aspects of measuring performance are missed.

Organization level measurement is defined as levels of measurement beyond individuals and teams. It would include multiple team coordination (for example, teams of teams; see Marks, DeChurch, Mathieu, Panzer, & Alonso, 2005), collections of units, broader departments or divisions, and ultimately the entire organization.

In this chapter, we first discuss general topics about performance and performance measurement, highlighting problems especially relevant to the organizational level of analysis. We next present a series of specific organizational level issues to be considered in designing any performance measurement system, including VE when used for training. We conclude with some ideas for implementing these suggestions.

PERFORMANCE, PERFORMANCE MEASUREMENT, AND MANAGEMENT

The primary reason for measuring performance is to maximize performance. One approach is utilizing performance measures to provide feedback to trainees, evaluate effects of an intervention, make decisions about resource allocations, or assess contributions of a large department in the organization. As behavioral scientists, our focus is on performance of people in specific situations. The goal is to encourage people to behave in a manner that produces outputs or results of maximal organization.

The organization's performance management system is designed to do exactly that: manage people so they generate results of maximal value (DeNisi, 2000). This occurs through three conceptually and operationally distinct organizational systems: the measurement, evaluation, and reward systems. The *measurement system* is what the organization chooses to measure, containing descriptive information about how much of which results are being produced (that is, number of trainees trained, trainee criterion test scores, or mean cost per trainee). The measurement system is important because it defines what evaluators, such as supervisors or training directors, believe are important organizational results. The *evaluation system* utilizes measures from the measurement system and places them on a continuum from good to bad. Stating 78 percent of trainees are meeting a criterion is a measurement; noting this is unsatisfactory is an evaluation. The measurement system indicates how much was done; the evaluation system indicates how good that amount is. This continuum is ideally a translation of how much is done into how valuable it is for the organization. Finally, the *reward system* is the process or set of rules by which outcomes are tied to evaluations. If outcomes of value are tied to evaluations that accurately reflect level of output, greater work motivation can result (Latham, 2006; Pritchard & Ashwood, 2008).

PERFORMANCE MEASUREMENT: ORGANIZATIONAL LEVEL ISSUES

With this background in mind, we now turn to specific organization level issues to be considered in VE training performance measurement. The well-known Kirkpatrick (1998) framework for evaluating training focuses on four levels of evaluation: reactions, learning, behavior, and results. Our discussion of organizational level issues is not inconsistent with that framework; our arguments can apply to any of the four levels of evaluation.

Alignment with Organizational Objectives

The single most important organizational level factor for performance measurement is alignment of performance measures with the organization's overall strategic objectives. This means that what and how things are measured in VE training should be consistent with what is truly of value to the organization, that is, what will lead to meeting an organization's strategic objectives.

While this point may seem obvious, it is more difficult than it appears. One example comes from a military maintenance unit's development of a performance measurement system (Pritchard, Jones, Roth, Stuebing & Ekeberg, 1989). In this example, a key measure was average time to complete repairs. This implied that taking less time to do repairs added more value to the overall organization. However, after careful analysis, unit personnel realized that meeting repair item demand, not average repair time, was most important for the organization. If demand was low, it was better to do a more thorough repair, including preventative maintenance. If things were busy, it was better to do the minimum required to get the repaired item operational. This led to changing the average repair time measure to percentage of demand met. In a training context, training should focus not only on how to do timely repairs, but also how to balance rapid repairs with preventative maintenance. Understanding this difference would help align learning from the training with organizational goals.

It is important to realize the critical issues to performance measurement in VEs are very similar to issues in any other training situation; VE systems are not "paradigmatically" different from other training systems (Caird, 1996). Training content should focus on the knowledge, skills, and attitudes critical to achieving organizational objectives (Caird, 1996). Therefore, just as in other types of training, the alignment with organizational goals should drive training design.

Assessing Alignment

While scholars have argued the importance of aligning employee contributions with the organization's strategic objectives (for example, Boswell & Boudreau, 2001; Latham, 2006), there is no clear, objective method of assessing alignment of performance measures with organizational goals. This assessment requires careful, logical analysis of each measure. One approach involves examining what would happen if the measure were maximized. Specifically, what would the person/team do to maximize his or her/its score on the measure and how would that impact the broader organization? Consider the maintenance unit repair example. If the measure is average time to complete repairs, to maximize the measure, unit personnel should minimize work done on each repair, avoid any extra maintenance, avoid working on repairs known to take longer to complete, and when workload was low, relax until more items come in for repair. These work strategies would not be consistent with organizational goals.

Another approach to assessing alignment is to ensure upper management supports the use of the measure. In other words, are members of higher management committed to maximizing that measure, and do they agree that the measure will be used to evaluate the individual or unit? Such discussions help identify areas lacking alignment.

Measurement Characteristics for Alignment: Identifying and Communicating Value

This discussion suggests measures must be aligned, that is, accurately reflect organizational value, and this organizational value must be communicated accurately to managers, supervisors, and unit personnel/trainees.

Identifying value means identifying value of different levels of each measure. That is, how much value is being added when trainees' final score on a performance test is 80 percent? How much value is created if 50 or 100 percent of a field unit is trained? To accurately communicate value to trainees and training managers, the value of different levels of performance on each measure must be known. This issue reflects the evaluation system: the system that places each level of a measure on a good-bad continuum by identifying how much value that level of performance has for the overall organization. This identification of value to the organization is critical to alignment. In fact, specifying this value actually operationalizes organizational value, allowing it to be communicated to personnel at all levels.

This identified value must accurately describe value to the organization. If the evaluation system does not match what is of value to the organization, the employee's behavior will be consistent with the evaluation system, not with what is optimal for the organization (for example, DeNisi, 2000). For example, suppose what is measured and evaluated in VE training is the number of tasks completed. However, if the communication and backup behaviors by which the team accomplishes those tasks are actually most valuable, the evaluation system is not consistent with what is of value to the organization.

We now turn to characteristics needed in performance measurement systems to lead to alignment and accurate identification and communication of value.

Measure All Important Aspects of Performance

For alignment to be present, the measures as a set must cover all important aspects of performance valuable to the organization. A typical approach is to use easy-to-collect measures and ignore important, but difficult to measure, aspects of performance (Borman, 1991). For example, it is often easier to measure quantity than quality, but both are valuable to organizations. A VE training situation might measure performance of a physician treating a virtual cardiac arrest patient by whether the correct medications were given, at the correct time, and in the correct dosage, but not measure how well the physician coordinated activities of other medical personnel team members because this is more difficult to measure. There is no objective methodology for assessing whether all aspects of value to the organization are included in the set of measures. This assessment requires careful, logical analysis by people who do the work, immediate supervisors, management, and key customers. In VEs, this is particularly challenging as VE tasks are often complex. For example, potential tasks that can be trained utilizing VEs are nonroutine procedures, planning and coordination, decision making, and dealing with hazardous situations to trainee health (Caird, 1996). Regardless of difficulty, each aspect of performance should be measured in these situations to accurately reflect the value of VE training to the organization.

Include Descriptive and Evaluative Information

Any assessment of training performance will include descriptive measures (that is, scores on performance tests, time to reach criterion performance, or percentage of failed trainees). These descriptive measures are part of the measurement system. They are result measures, which indicate how much was done. To identify, assess, and communicate organizational value, descriptive measures should be translated into evaluative measures identifying how good the output is and the organizational value.

One approach, in general terms, is to define what is considered poor, adequate, and excellent levels of results by the broader organization. Most training managers have a sense of this level of evaluation. However, it is feasible to get more precise in determining value using quantitative measures to precisely identify the value of each level of each measure. We will discuss techniques for doing this later in this chapter, but for now we want to make the point that quantification has important, practical advantages for performance assessment.

An Overall Index of Performance

Quantifying value allows for an overall index of performance. Performance in VE training of any complexity will be assessed with multiple, qualitatively different measures. For example, measures of (1) control of a virtual military aircraft in space, (2) correct identification of friends and foes, (3) use of offensive weapons, and (4) use of defensive weapons will produce a series of measures not easily combined. It is important that scores on separate components of performance be measured and used for a variety of purposes (that is, feedback and training). It is also valuable to have an overall index of performance (Pritchard, 1990; Salas, Rosen, Burke, Nicholson, & Howse, 2007). An overall index can be used for monitoring training performance over time, providing ongoing feedback, and evaluating overall training effectiveness (for example, mean across trainees). Additionally, if developed correctly, it is an index of the overall value of the training created for the organization.

Suppose levels of output, such as correctly identifying friends 80 percent of the time, are accurately placed on a scale of value to the organization, and this is done for each output measure. If these conditions were met, it would be easy to convert scores on each measure to their corresponding value score and sum value scores for an overall performance index. As a hypothetical example, if correctly identifying friends 80 percent of the time gets a value score of 10, correctly identifying foes 97 percent of the time receives a value score of 75, and having 12 hits with offensive weapons results in a value score of 35, the overall score is the sum of the three, 120. We will discuss specific techniques for doing this later in the chapter.

Identifying Relative Importance

Not all measures of performance are equally important. In other words, different measures do not contribute equal value to the organization. A good assessment system will identify and incorporate this differential importance. In VE systems, the greatest importance might be placed on tasks that are not easily trained via traditional methods, yet are of greatest value to the organization (Caird, 1996). One approach to differential importance is weighting measures by their relative importance. For example, each measure could be standardized and multiplied by an importance weight; the resulting products are summed to produce an overall evaluation. This standardization allows for adding measures (now on a common scale), and the weight applies the differential importance. Using this approach, weights are determined by organizational personnel who identify how much value is added to the organization.

Identifying Nonlinearities

There are limitations to this weighting approach. It assumes that the importance of each measure is always the same, no matter how good or poor performance is. Consider the example of VE training in the operation of an electrical power plant where one measure is time to effectively respond to system-generated warning alarms. If the alarm response measure is time, it may be important that response be done within 5 minutes, and critical that response be done within 15 minutes. If response takes more than 15 minutes, damage has been done, so responses slower than 15 minutes are no worse than a response time of 15 minutes. If response is less than 5 minutes, this is no better than 5 minutes because for the first 5 minutes of the warning, nothing serious occurs.

The point is there is a nonlinear relationship between amount, minutes before response, and value. In this example, value to the organization is equal for performance between 1 and 5 minutes and for performance of 15 minutes or longer, but it varies greatly between 5 and 15 minutes. A simple weighting approach will not capture this nonlinearity. Simple weighting assumes a change in the measure of any given size will be equally valuable at all points along the scale. A number of scholars have argued for the importance of incorporating nonlinearities (Campbell & Campbell, 1988; Pritchard, 1992). Research on performance assessment systems by Pritchard and his colleagues has shown the vast majority of performance measures have nonlinear relationships with value (Pritchard, Paquin, DeCuir, McCormick, & Bly, 2002). For optimal alignment with organizational values, the system must account for nonlinearities.

Agreement across Evaluators

There are multiple important evaluators for every person within an organization. These include supervisors, peers, subordinates, higher management, internal and external customers, and the employee himself or herself. Using our conceptualization, these diverse evaluators value results differently. Subordinates value different things than peers, one manager values different things than another manager, and the employee may value different things than the supervisor. In fact, role conflict is exactly this—different evaluators of importance valuing diverse things from the individual. Role conflict has several negative effects (Jackson & Schuler, 1985; Tubre & Collins, 2000). A necessary condition for presence of alignment is that different evaluators agree on what is important.

Perfect agreement of all evaluators is unrealistic, but it is important that supervisors, higher management, and critical customers largely agree. If important evaluators knowingly or unknowingly place different value on the same levels of output, this sends a conflicting message, making optimal performance difficult. In VE training, subject matter experts (SMEs), trainers, the training supervisor, higher management, and important customers of the trainees must agree on (1) measures used to assess performance and (2) the organizational value of different measures.

In VE, measures typically center on accuracy, task completion time, or both (Nash, Edwards, Thompson, & Barfield, 2000). Evaluators should agree on the relative importance of these measures. If different evaluators place different importance on training measures, when trainees return to the job they will most likely behave according to the importance communicated by their most influential evaluator, for example, the person completing their performance appraisals. Therefore, it is critical that managers, supervisors, and trainers agree on the importance of different training performance measures.

The first test of agreement is whether there is consensus on measures used to evaluate performance. All SMEs should agree that each measure is a good one and the set as a whole covers all the important aspects of the work or training. While time consuming, this is not typically difficult for personnel to do once measures are identified. The second test is that SMEs agree on the organizational value of each level of performance for each measure.

Identifying Improvement Priorities

One characteristic needed for maximizing performance is the identification of improvement priorities. Performance is multidimensional; therefore, it is difficult to improve all aspects of performance at once in training or in normal job performance (Brannick, Prince, & Salas, 1993; Smith-Jentsch, Zeisig, Acton, & McPherson, 1998). At any single point in time, trainees or job incumbents focus improvement efforts on a subset of the task. To do this effectively, the value of the subtasks to the organization must be known. Supervisors and incumbents or trainees need to know how much value is placed on different improvements in order to focus improvement efforts on areas of maximum value to the organization. If the value of different levels of performance on each measure has been identified, this information can be used to identify improvement priorities. Specifically, if value of each level of performance has been accurately quantified, the value of improvement on each measure can be calculated and communicated to the incumbent or trainee.

Sensitivity to Changing Objectives or Mission

While it is important that measurement and evaluation systems be stable over reasonable periods of time, it is also important that they be sensitive to changes in objectives/missions. For example, in the simulation based training system GUARD FIST II the trainee can be trained on (a) locating targets or (b) calling for and adjusting indirect fire support. The objective of this training system can change to emphasize one or both tasks.

If mission flexibility is important to the job, the VE needs to train for the changing mission and accurately assess performance in different missions. It is sometimes the case that measures change when missions change, and, therefore, new measures must be developed. Often, however, the measures do not change; it is the organizational value placed on them. For example, in VE training of military flight crews, the measures may consist of aircraft control and navigation, identification of friend or foe, and use of offensive and defensive weapons. In most missions, all factors have some importance, but importance varies from mission to mission. For one type of mission, effective use of offensive weapons may be essential, but for a reconnaissance mission it may be less important. Navigation is always important, but if the mission is to run a search grid to locate survivors in a small raft, it is especially critical.

When missions change, the ability to quickly change organizational value placed on the measures is necessary. If the set of possible missions can be identified in advance, different sets of quantitative indices of value can be developed in advance. If changing missions are unpredictable, this requires a cost-effective way to identify value that can quickly be customized for unique missions.

Other Measurement Characteristics

In the prior section we discussed organizational level measurement characteristics around the issue of alignment with organizational value. We now turn to other measurement characteristics at the organizational level.

The Organizational Reward System

We discussed the issue of making the evaluation system consistent with what is actually of value to the broader organization. It is also important that the organization's reward system match what is of value to the organization. For example, if the reward system leads to formal or informal rewards for speed but not for quality, a mismatch between value and reward occurs. In such situations, employee behavior is usually consistent with the evaluation and reward systems, not with what is of value to the organization (Johnston, Brignall, & Fitzgerald, 2002). Furthermore, the closer the reward system is tied to these evaluations, the more the person will behave consistently with the evaluation and reward systems. It is ironic that with a mismatch between evaluations and actual organizational value, the stronger the reward system, the worse things become for the organization. This is an especially salient point in the VE. Utilizing a VE is a costly endeavor (Caird, 1996). If the focus of VE training is not on behaviors consistent with valued organizational goals appropriately tied to the reward system, the training will be less effective.

Controllability

Measures used to evaluate training performance must be largely under the control of the trainees. We define controllability as the degree to which trainees or incumbents can control the level of their performance measures by varying the amount of effort allocated to the tasks that lead to those performance measures (Pritchard et al., 2008). As Pritchard and his colleagues noted, lack of controllability leads to a variety of negative outcomes and ultimately influences motivation and performance. Often, measures with good face validity are used, but people in the training program or job have limited control over their performances on the measure. For example, the rate of medication errors is a highly face valid measure of physician performance. However, in reality the physician may have a low level of control over this measure because the errors may be due to mistakes by pharmacists or nurses. To improve physician control, changes would need to occur at the organization level such that variance due to other people and processes in the medication system itself is removed, leaving variance only due to the physician's own actions. This is especially relevant to VE situations. If trainees do not feel they can affect their levels of performance within a VE training situation, they will not fully engage in training. This provides implications for VE training design as well as assessment. The target training behavior should be measured in a manner reflective of varying level of effort from the trainee.

Buy-In

One notable problem with VE training in particular is buy-in. Trainees sometimes do not perceive measures as valid due in part to negative transfer or hindered, correct, real world performance (Rose, Attree, Brooks, Parslow, & Penn, 2000). Rose and colleagues (2000) provide an example of teaching children to cross a street using VE. While training may provide valuable skills, such as looking both ways and timing involved in actually crossing a street with traffic, it is nearly impossible to include the possibility of injury from oncoming traffic. One way to improve buy-in is to include measures of the all-important aspects of performance. In the street crossing example this suggests simulating what would happen to the child if hit by a vehicle and measuring this knowledge.

IMPLEMENTING ORGANIZATION LEVEL FACTORS

In the section above, we identified a series of important organizational level considerations in the development and design of measuring performance in VE training, or measuring performance in general. The final section of this chapter discusses specific ways to deal with these issues.

A number of the factors are fairly straightforward, and implementation issues are discussed. Assessing alignment of measures, measuring all important aspects of performance, matching evaluation and reward systems with what is of value to the organization, and controllability of measures can all be handled through logical analysis of the setting. Some specific tests are presented; however, there is no objective methodology. Each requires discussion and judgment by relevant people, especially those doing the work and SMEs.

All other factors have one thing in common: they require the determination of how much value to the organization is produced for each level of performance on each performance measure. We now turn to ways to operationalize this value.

Identifying and Communicating Value: Specific Techniques

The question remains of how to accurately attach value to different levels of performance on performance measures. There are multiple ways to do this. The simplest is to attach some number of points to each level of performance on each measure. For example, five errors are worth 0 points, 4 errors are 5 points, or 3 errors are 10 points. If points are determined for each level of each measure, and these points reflect true value to the organization, a measure of value is generated. If we wanted an overall measure of performance, we could simply sum points an individual or team earned for a given training trial or performance period. This is essentially how the Balanced Scorecard works (Kaplan & Norton, 1996).

This approach has the advantage of simplicity, but it has disadvantages as well. What we really need is a value scale that is comparable across performance measures. We want a value score of 20 to lead to the same value to the organization from all measures. So 20 points on an error measure has the same value as 20 points on a speed measure. This means a scale that is the same scale for all performance measures is needed. Giving points to each level of performance separately, the way it is typically done with approaches like the Balanced Scorecard, does not ensure the value scale is comparable across measures. The second disadvantage is that this approach does not formally consider nonlinearities.

ProMES Contingencies

An approach that overcomes these limitations is that used by the first author and his associates in the productivity measurement and enhancement system (ProMES) intervention (Pritchard, 1990; Pritchard, Harrell, DiazGranados, & Guzman, 2008; Pritchard et al., 1989). ProMES is designed as an intervention to measure performance and to provide feedback used to improve performance through a continuous improvement model. Once the set of performance measures is developed in ProMES, the next step is to develop contingencies. These contingencies capture value to the organization.

ProMES contingencies are derived from theory; they operationalize the results-to-evaluation contingencies in the Naylor, Pritchard, and Ilgen (1980) and Pritchard and Ashwood (2008) theories of motivation. A ProMES contingency is a type of graphic utility function that relates the amount of each performance measure to organizational value. Example contingencies from a hypothetical VE training of combat aircraft teams are shown in Figure 15.1. The contingency in the upper left is for a measure of aircraft control, the distance in miles the aircraft is from its intended location. The horizontal axis shows levels of the measure ranging from a low performance level of 10 miles away to a high performance level of 0 miles away from the intended location. The vertical axis is

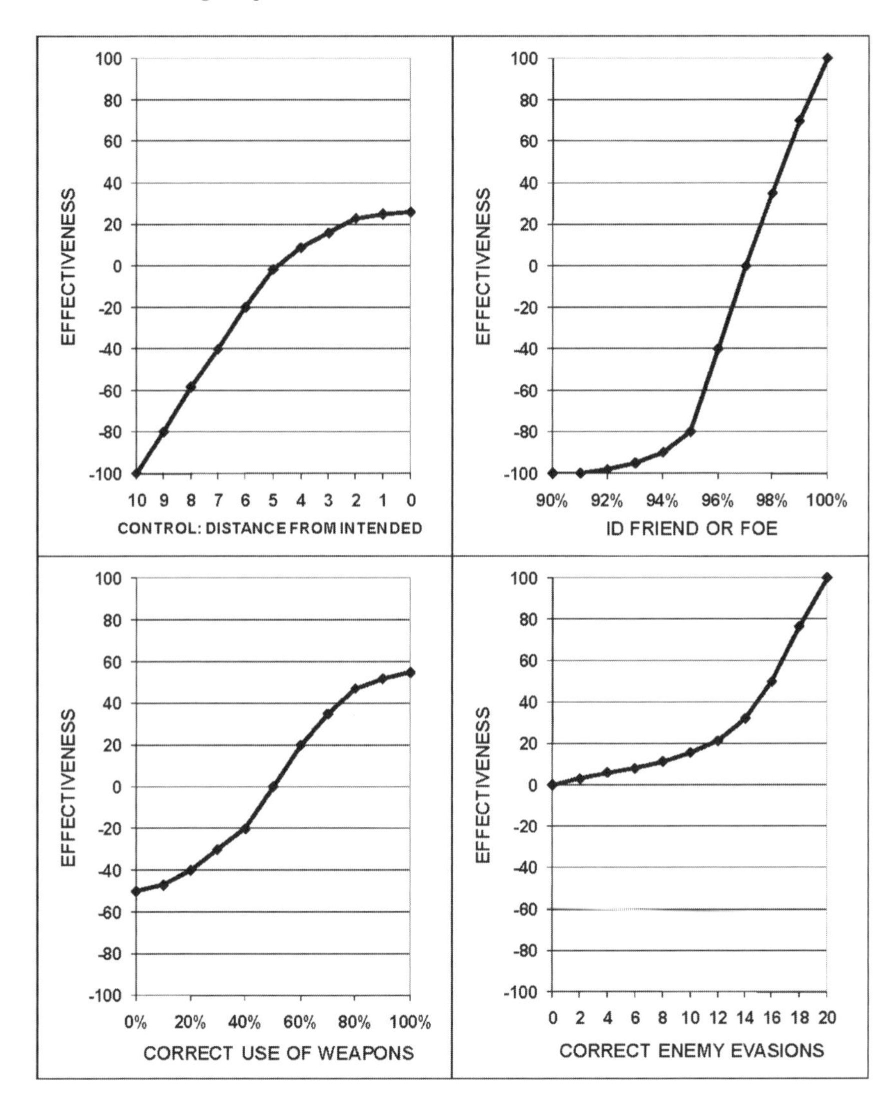

Figure 15.1. Example Contingencies

called effectiveness, defined as the amount of contribution being made to the organization. The effectiveness scale ranges from –100, which is highly negative, through zero, which is meeting minimum expectations, to +100, which is highly positive. The function itself defines how each level of the measure is related to effectiveness. As depicted in Figure 15.1, a contingency is generated for each performance measure.

Contingencies are developed using a discussion to consensus process. A design team is formed composed of job incumbents, at least one level of supervision, and a facilitator. In the case of VE training, the design team would be composed of subject matter experts, the training manager, a facilitator, and possibly some key customers. The basic idea is for a facilitator to break development of contingencies into a series of steps that the design team can do. For example, the first step is to identify the maximum and minimum realistic levels for each measure. In the example aircraft control contingency, the design team decided that the minimum realistic value was 10 miles from the intended location; the maximum possible value was 0 miles. Next, the design team decides a minimum level of acceptable performance, defined as the point of just meeting minimum expectations. Members of the design team discuss this value until a consensus is reached. This point becomes the point of zero effectiveness in the contingency, 5 miles in the aircraft control example. The design team then continues through a set of steps that ultimately lead to the creation of the set of contingencies. More detail on how contingencies are done can be found in Pritchard (1990) and Pritchard, van Tuijl et al. (2008).

Advantages of Contingencies

Contingencies have several important features that meet criteria for value assessment as identified above. The *relative importance* of each measure is captured by the overall range in effectiveness score. Those measures with larger ranges, such as identifying friend or foe, see Figure 15.1, can contribute to or detract from the organization in greater amounts and are thus more important than those with smaller ranges. Contingencies also *translate measurement into evaluation* by identifying how each level of each measure contributes to effectiveness. For example, if the unit had a use of weapons score of 80 percent, this translates into an effectiveness score of +50, quite positive and well above minimum expectations. Contingencies also identify *nonlinearities.* Figure 15.1 shows several forms of nonlinearity. The control contingency indicates there is a point of diminishing returns where the slope decreases once the aircraft is within 2 miles, indicating improved location accuracy above 2 miles decreases in value. The identify (ID) friend or foe contingency is a critical mass type contingency, where organizational value is very low until the performance reaches a certain point (at least 96 percent) and then increases rapidly.

Contingencies also help *identify priorities for improvement.* One can readily calculate the gain in effectiveness if the unit improved on each measure. For example, suppose the flight crew has a score of 3 miles from the intended location and an ID friend or foe score of 96 percent. The contingency indicates going from 3 miles from the intended location to 2 miles produces a gain in effectiveness of 10 effectiveness points; going from an ID score of 96 percent to a score of 97 percent shows a gain of 20 effectiveness points. Improvement in ID is worth twice as much to the organization as improvement in location. This can be done for each measure. Effectiveness gain scores are a quantification of how valuable each improvement is. Finally, contingencies rescale each measure to the common metric of effectiveness so a single, *overall effectiveness score* can be formed by summing each indicator's effectiveness scores. For example, if the effectiveness

Table 15.1. Organizational Levels Performance Measurement Guidelines

Performance Measures Guidelines	Definition	Key Points for VE
1. Alignment with organizational objectives	What and how things are measured should be consistent with what is truly valued to the organization	• Identify value for different levels of each measure. • Ensure value accurately reflects true value to the organization. • Design VE training to focus on knowledge, skills, and attitudes (KSAs) that reflect the most value to the organization.
2. Measure all important aspects of performance	Set of measures cover all dimensions of performance that are of value to the organization	• VE training is often more complex (that is, situations therein are too hazardous to train in the "real world"). • Measures must capture all aspects of performance within complex tasks that are of value to the organization.
3. Descriptive and evaluative information	Result measures designed to indicate how much was done (descriptive) and how good it was (evaluative)	• The goal of any training should be objective measures of performance that provide feedback to the trainee, especially in a VE situation. • Feedback regarding how much was accomplished and how good it was relative to the organizational goals is important.
4. Overall index of performance	Summation of all quantified indices of performance	• Provides meaningful information on effectiveness of VE training • Provides meaningful information on value of training to the organization
5. Identify relative importance	Differentiating contribution of each measure to the value of the organization	• Provides manner in which to prioritize training in VE systems • Greatest weight should be placed on tasks that are of highest value to the organization, but not easily trained utilizing training methods other than VE.
6. Agreement across evaluators	Shared mental models regarding value of output	• Evaluators must agree on most valued levels and utilize measures that accurately reflect VE training performance based on desired levels.

7. Identify improvement priorities	Determination of which aspect of performance is most critical for improvement at any given moment	• Utilizing values placed on different levels of performance, it is easy to determine which area requires work. • VE systems can therefore be utilized more effectively if they can target the required KSAs.
8. Sensitive to changing objectives or missions	Measurement system should be stable (when considering evaluation system), yet flexible to changes in objectives/mission	• Should be able to utilize VE system for just-in-time training to maximize value of training to the organization
9. Organizational reward system tie	Measures should tie to the organizational reward system	• For effective transfer of VE training, evaluation of resulting behaviors should be directly tied to the organizational reward system.
10. Controllability	Degree to which trainees can control level of performance measures by varying amount of effort	• If trainees do not feel they can affect their levels of performance within a VE training situation, they will not engage fully in training. This provides implications for VE training design as well as assessment.
11. Buy-in	Acceptance (or lack thereof) of applicability of training to the real world	• All important aspects of work must be measured to ensure buy-in, even those aspects that are difficult to train or measure (for example, bodily injury that can result from children not paying attention when crossing the street is not easy to demonstrate in a VE training system).

score for the location indicator was +60, it would be added to effectiveness scores from other indicators. If there were 12 indicators, there would be 12 effectiveness scores summed to create the overall effectiveness score. This overall effectiveness score provides a single index of overall productivity. Because contingency development includes inputs from a variety of different subject matter experts, levels of supervision and management, and key customers, it maximizes *agreement across evaluators*.

Finally, contingencies allow for *changing objectives or missions*. If the mission of the aircraft crew changed from search and destroy to recovery, the contingency for location control might become steeper and the expected minimum performance might change. Once a design team has created contingencies, it takes little time to make such changes when missions adjust.

CONCLUSION

While it is essential to consider a variety of issues at the individual and team levels in developing measures of performance in VE training settings, there are also issues of importance at the organizational level (refer to Table 15.1 for summary).

The most difficult is determination of organizational value for different levels of performance on each measure. If value can be accurately determined, a series of advantages can be realized. Utilizing these in VE training design will allow for more effective VE training.

REFERENCES

Borman, W. C. (1991). Job behavior, performance, and effectiveness. In M. D. Dunnette & L. Hough (Eds.), *Handbook of industrial and organizational psychology* (2nd ed., Vol. 2, pp. 271–326). Palo Alto, CA: Consulting Psychologists Press.

Boswell, W. R., & Boudreau, J. W. (2001). How leading companies create, measure, and achieve strategic results through "line of sight." *Management Decision, 39,* 851–859.

Brannick, M. T., Prince, A., & Salas, E. (1993). Understanding team performance: A multimethod study. *Human Performance, 6*(4), 287–308.

Caird, J. K. (1996). Persistent issues in the application of virtual environment systems to training. *Proceedings of the 3rd Symposium on Human Interaction with Complex Systems—HICS* (pp. 124–132). Dayton, OH: IEEE.

Campbell, J. C., & Campbell, R. J. (1988). Industrial-organizational psychology and productivity: The goodness of fit. In J. C. Campbell & R. J. Campbell (Eds.), *Productivity in organizations* (pp. 82–94). San Francisco: Jossey-Bass.

DeNisi, A. S. (2000). Performance appraisal and performance management. In K. J. Klein & S. W. J. Kozlowski (Eds.), *Multilevel theory, research, and methods in organizations* (pp. 121–156). San Francisco: Jossey-Bass.

Jackson, S. E., & Schuler, R. S. (1985). A meta-analysis and conceptual critique of research on role ambiguity and role conflict in work settings. *Organizational Behavior and Human Decision Processes, 33,* 1–21.

Johnston, R., Brignall, S., & Fitzgerald, L. (2002). 'Good enough' performance measurement: A trade-off between activity and action. *Journal of the Operational Research Society, 53,* 256–262.

Kaplan, R. S., & Norton, D. P. (1996). *Translating strategy into action: The balanced scorecard.* Boston: Harvard Business School Press.

Kirkpatrick, D. L. (1998). *Evaluating training programs.* San Francisco: Berrett-Koehler Publishers, Inc.

Klein, K. J., & Kozlowski, S. W. J. (Eds.). (2000). *Multilevel theory, research, and methods in organizations.* San Francisco: Jossey-Bass.

Latham, G. P. (2006). *Work motivation: History, theory, research, and practice.* Thousand Oaks, CA: Sage Publications.

Marks, M. A., DeChurch, L. A., Mathieu, J. E., Panzer, F. J., & Alonso, A. (2005). Teamwork in multi-team systems. *Journal of Applied Psychology, 90,* 964–971.

Nash, E. B., Edwards, G. W., Thompson, J. A., & Barfield, W. (2000). A review of presence and performance in virtual environments. *International Journal of Human-Computer Interaction, 12*(1), 1–41.

Naylor, J. C., Pritchard, R. D., & Ilgen, D. R. (1980). *A theory of behavior in organizations.* New York: Academic Press.

Pritchard, R. D. (1990). *Measuring and improving organizational productivity: A practical guide.* New York: Praeger.

Pritchard, R. D. (1992). Organizational productivity. In M. D. Dunnette & L. M. Hough (Eds.), *Handbook of industrial and organizational psychology* (2nd ed., Vol. 2, pp. 443–471). Palo Alto, CA: Consulting Psychologists Press.

Pritchard, R. D., & Ashwood, A. (2008). *A manager's guide to diagnosing and improving motivation.* New York: Psychology Press.

Pritchard, R. D., Harrell, M. M., DiazGranados, D., & Guzman, M. J. (2008). The productivity measurement and enhancement system: A meta-analysis. *Journal of Applied Psychology.*

Pritchard, R. D., Jones, S. D., Roth, P. L., Stuebing, K. K., & Ekeberg, S. E. (1989). The evaluation of an integrated approach to measuring organizational productivity. *Personnel Psychology, 42,* 69–115.

Pritchard, R. D., Paquin, A. R., DeCuir, A. D., McCormick, M. J., & Bly, P. R. (2002). Measuring and improving organizational productivity: An overview of ProMES, The Productivity Measurement and Enhancement System. In R. D. Pritchard, H. Holling, F. Lammers, & B. D. Clark (Eds.), *Improving organizational performance with the Productivity Measurement and Enhancement System: An international collaboration* (pp. 3–50). Huntington, NY: Nova Science.

Pritchard, R. D., van Tuijl, H., Bedwell, W., Weaver, S., Fullick, J., & Wright, N. (2008). *Maximizing controllability in performance measures.* Manuscript submitted for publication.

Rose, F. D., Attree, E. A., Brooks, B. M., Parslow, D. M., & Penn, P. R. (2000). Training in virtual environments: Transfer to real world tasks and equivalence to real task training. *Ergonomics, 43*(4), 494–511.

Salas, E., Rosen, M. A., Burke, S. C., Nicholson, D., & Howse, W. R. (2007). Markers for enhancing team cognition in complex environments: The power of team performance diagnosis. *Aviation, Space, and Environmental Medicine, 78*(5), B77–B85.

Smith-Jentsch, K., Zeisig, R. L., Acton, B., & McPherson, J. A. (1998). Team dimensional training: A strategy for guided team self-correction. In J. A. Cannon-Bowers & E. Salas (Eds.), *Making decisions under stress: Implications for individual and team training* (pp. 271–297). Washington, DC: American Psychological Association.

Tubre, T. C., & Collins, J. M. (2000). Jackson & Schuler (1985) revisited: A meta-analysis of the relationships between role ambiguity, role conflict, and job performance. *Journal of Management, 26,* 155–169.

Part VIII: Methods in Performance Assessment

Chapter 16

ASSESSMENT MODELS AND TOOLS FOR VIRTUAL ENVIRONMENT TRAINING

William L. Bewley, Gregory K. W. K. Chung, Girlie C. Delacruz, and Eva L. Baker

Although many consider the effectiveness of virtual environment (VE) training to be self-evident, the sad truth is that some training systems, including VE training systems, do not work, and some trainees do not learn, even from VE training systems. Assessments of learner performance can provide evidence for the effectiveness of training systems and for trainee learning, as well as information supporting training system improvement, guidance of instruction, trainee placement decisions, and certification of skill. It is also sadly true, however, that in some cases where assessments are used, they do not work. The problem is poor design. Assessments of learner performance must be designed to measure the entire range of knowledge and skills addressed by the training, and they must be validated as sources of evidence to support the interpretations and uses of assessment results. This chapter describes a model based approach to the design and validation of assessments of performance, with a focus on assessments in VE training. It begins with a discussion of validity, the essential requirement for any assessment, and then describes Baker's (1997) model based approach to design and validation of performance assessments. The chapter concludes with a discussion of future directions in assessment for VE training and a summary and discussion.

INTRODUCTION

Virtual environment systems enable interaction with a simulated, often three-dimensional computer-generated environment. A typical VE includes representations of objects, people, paths, tools, and information sources and provides facilities for interaction with the environment through gestures and other body movements, speech, and such input devices as a glove, a mouse or joystick, or a keyboard. In VE training, the VE represents the real environment at some level of fidelity. It supports training on complex tasks that traditionally have required hours, days, weeks, or even months of training and practice in the real environment to learn, with assistance and feedback from a human instructor or mentor. Examples of such tasks are firefighting, architectural design, surgery, equipment maintenance, ship handling, flying an airplane, and battle planning. This is not to say that VE training can replace training in the real environment supervised by a knowledgeable instructor—nobody would want a surgeon who had trained only in a VE—but a useful level of knowledge and skill can be developed cost-effectively and safely with VE training in preparation for training in the real environment.

What these examples have in common is the complexity of the task in terms of environmental cues, trainee responses, and the interaction of trainee responses and the behavior of the environment over an extended period of time. They have all the characteristics of complex tasks as defined by Williamson, Bejar, and Mislevy (2006): multiple, nontrivial, domain-relevant steps and/or cognitive processes, high potential variability in task performance, and interdependent task features. In addition, performance involves recognition and use of complex situational cues and affordances represented in the environment and feedback providing new situational cues. In ship handling, for example, maneuvers such as underway replenishment and docking require the use of perceptual cues for ship location and speed relative to other objects in the environment, and environmental characteristics such as ship dynamics in response to trainee actions, weather and ocean conditions, and the predicted effects of commands to other humans during different task phases.

Why Use VE for Training?

VE training is usually faster, less costly, and less dangerous than training in the real environment—compare training in a space shuttle simulator to on-the-job training en route to the space station. The benefits of cost and risk avoidance are convincing, but there are also benefits to learning due to the ability to unobtrusively collect detailed data on the process used by the learner in performing the task, data providing assessment information that can be used to automatically score performance and diagnose learning problems. The VE can also be used to provide an experience not possible in the real environment that benefits learning. One example is practice with parts of a task that cannot be isolated in the real world, for example, repeated takeoffs and landings under controlled conditions in a flight simulator (Carretta & Dunlap, 1998). Another is the ability to see the environment from a viewpoint not possible without a VE, for example, an external view of the airplane to observe the effects of actions during a landing (Wickens & May, 1994).

Why Assess Performance in VE Training?

Although some enthusiasts consider the effectiveness of VE training to be self-evident, the sad truth is that some training systems, including VE training systems, do not work, and some trainees do not learn, even from VE training systems (Clark & Estes, 2002). Performance assessments produce results that can be used to draw inferences about ability or competence, inferences that may be used for multiple purposes. They may be used to determine ability or competence at the beginning of instruction in support of placement decisions, to diagnose knowledge and skill gaps during training in order to guide instructional decisions, to predict future performance in other settings, to certify competence after training, to evaluate the training program as a whole, or to measure the impact of specific attributes of the training program (see Lampton, Bliss, & Morris, 2002).

Overview of the Chapter

This chapter describes a model based approach to the design and validation of assessments of performance in VE training. It begins with a discussion of validity and then describes a model based approach to the design and validation of assessments. The chapter concludes with an overview of future directions in assessment for VE training and a summary and discussion.

VALIDITY

Assessments of learner performance in VE training environments must be designed to measure the entire range of knowledge and skills at the same level of complexity addressed by the training, and they must be validated for the purposes and situations to which they are applied. Validation is the fundamental requirement in assessment development. Its importance cannot be overemphasized. Assessments can be developed with little difficulty. Developing *valid* assessments—assessments that have been demonstrated to provide evidence appropriate to the uses of their results—requires a rigorous methodology involving significant analysis and testing.

Just as assessments are used for different purposes, the evidence to support the validation of assessments will differ depending on its intended use. The validity of an assessment is the key indicator of its technical quality and provides evidence for the appropriateness of the interpretations and uses of the assessment results. Validity is not a general quality of the assessment that applies to all uses for all time, nor is it based on a single procedure, such as always correlating an existing set of scores with those of another measure. Rather, the validity of an assessment depends on the context and inferences to be drawn. Validation should be thought of more as the job of an attorney making a legal case than as the calculation of a statistic. A *validity argument* must be developed that marshals a wide range of evidence to make the case (American Educational Research Association, American Psychological Association, and National Council for Measurement in Education, 1999).

This is very different from early models of validity, in which specific validity types are considered. Some traditional validity types and the question each answers are listed below (see Messick, 1993, pp. 16–19, for a discussion of early conceptions of validity):

- Face validity: Does the test performance look like what is supposed to be measured?
- Content validity: Is the performance measured related to content goals or domains?
- Predictive validity: Do people with higher scores do better on a distal criterion measure?
- Criterion validity: Does performance on the new measure relate in predictable ways with an existing measure of known quality?

While all these questions may be considered in making a validity argument, one no longer looks at a list of validity types and chooses one or two as most appropriate or, more likely, easiest to implement.

The validity of assessment is often seen as separate from the instruction and instructional goals, something to do after the instructional materials are developed. In the interest of saving time and money, developers may decide to use an existing, readily available assessment without attempting to validate it for a specific purpose. This approach is indeed fast and easy, but it brings the significant risk that the measure selected will not be valid for the tasks, learners, situations, or purposes addressed by the training. One likely consequence is that the measure is not aligned with the goals of the instruction, which makes the assessment irrelevant to its planned use. Another is that the assessment is insensitive to the training experiences of interest—that is, people who receive instruction perform as well as people who do not.

Because assessments are used for placement, guidance of instruction, certification, and training program improvement, their quality should be the last element of the training program to be compromised. The measures, their characteristics, and how they are to be used must be considered an integral part of the vision that motivates the training program design. The design of assessments and training should be an integrated activity rather than be approached separately.

ASSESSMENT DESIGN

Baker's (1997) model based assessment design methodology helps ensure that assessments are valid. As shown in Figure 16.1, it begins with the specification of the type of learning outcomes to be measured—the cognitive demand—and then describes how cognitive demand influences the domain representation, the task representation, and the scoring model.

Cognitive Demand

The cognitive demands are the domain-independent and domain-dependent knowledge and cognitive skills that an assessment should target (American Educational Research Association et al., 1999; Baker, 1997; Baker & Mayer, 1999). It must be clearly defined at the beginning of the assessment design process. Only after designers have addressed what the assessment is intended to measure should decisions be made about details that too often are the only concern in assessment design: format (for example, multiple-choice or true-false questions), number of items, and the amount of testing time.

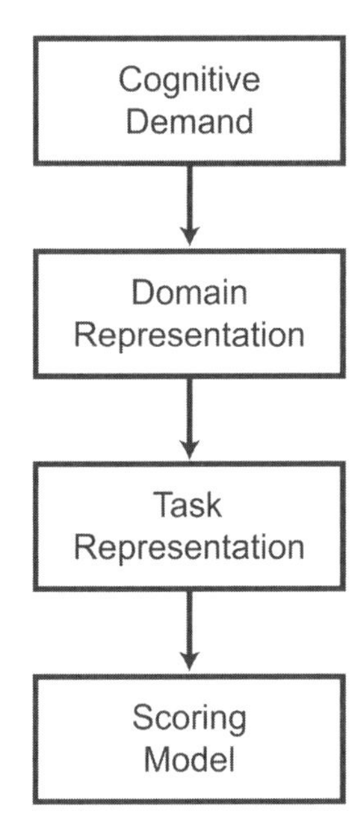

Figure 16.1. Baker's model based assessment design methodology begins with specification of the cognitive demand, which influences the domain representation, the task representation, and the scoring model.

Specification of cognitive demand helps determine how the assessment task can elicit performance demonstrating the desired knowledge and skills. For example, if we are measuring factual knowledge, then an appropriate test format might be multiple-choice or true-false questions, but if we are measuring conceptual understanding, an essay or knowledge map may be appropriate (Baker, Freeman, & Clayton, 1991). But for tasks requiring VE training, the cognitive demand is typically complex, requiring high level cognitive processing involving activity that would be placed in the create and evaluate categories of Anderson and Krathwohl (2001), in Mayer's (1992) strategic and schematic categories, or in van Merriënboer's (1997) strategic knowledge supported by cognitive schemata; it may include the use of perceptual cues embedded in the environment with complex interaction of trainee responses and the behavior of the environment, and the use of precise motor skills.

Domain Representation

A domain representation combines cognitive demand and a representation of domain content. It is an explicit description of the content, knowledge, skills, abilities, interests, and attitudes contained in the construct to be assessed (American Educational Research Association et al., 1999; Baker, 1997; Baker & Mayer, 1999; Baker & O'Neil, 1987). It should be explicit, precise, and externalized, and it should capture the essential elements of what is to be tested with respect to the target environment. As pointed out by Williamson et al. (2006), the emphasis is on "essential elements." Detailed models are not required if the purpose of the assessment is to support summative decisions. The granularity of the model depends on the targets of inference. A simple estimate of ability may be sufficient for a summative decision. A complex model would be required for cognitive diagnosis and prescription of remediation such as might be used by an intelligent tutoring system.

The domain representation is the basis for sampling assessment tasks, the referent against which to evaluate the relevance and representativeness of the task, and it reflects the universe that assessment performance represents. It can also support identification of student deficiencies, which could be used to guide remediation efforts, with domain-referenced comparisons providing diagnostic information on what individuals can and cannot do. For example, if a domain is composed of different subdomains (for example, threat-assessment and threat-sector definition skills belonging to the larger domain of air defense planning), then tasks can be sampled from each subdomain and scales developed for each. Student performance on the scales can be used to infer the degree to which students have mastered those skills. Finally, with an explicit domain model, the model itself can be tested and validated (Gitomer & Yamamoto, 1991).

See Baker and O'Neil (1987) for an in-depth discussion of domain-referenced testing issues, and see Baker, Chung, and Delacruz (2008) for a discussion of approaches to developing domain representations, including ontologies, the rule-space method, and Bayes nets.

Task Representation

The task representation specifies what the assessment task asks the examinee to do. For cognitive demand at the level of facts and concepts, the task could be based on selected-response test items (for example, multiple-choice questions) or constructed-response items (for example, short answer, essay question, or knowledge maps). For high level cognitive processing, it might be a simulation or VE based task. The assessment task is designed to elicit performance that will provide evidence of the examinee's knowledge and skills. It is the testbed used to observe and gather evidence about performance and is the basis for drawing inferences about the examinee's competence (American Educational Research Association et al., 1999; Baker, 2002; Baker et al., 1991; Baker & Herman, 1983; Messick, 1995).

Task design is driven by the domain representation. If the tasks are not aligned with the domain representation, or the domain representation is not representative of the cognitive demand and domain content, then inferences based on students' performance on the tasks will not be valid. And, of course, when examinees are tested on content they have not been exposed to, or not tested on content they have been exposed to, the assessment results will not accurately reflect achievement (Baker et al., 1991).

To maximize fidelity to the target task, assessments in VE training should be based on tasks identical to the training tasks, which should be designed to be as similar as possible to the task as performed in the real environment. One of the great advantages of an assessment embedded in a VE or a simulation is the ability to unobtrusively measure performance of the task as it is delivered in the training. This can provide valuable information on the process used by the examinee, a potentially valuable addition to measures focused on the outcome of the process such as a rating of overall success, the number of errors, and time to complete. In tasks performed by manipulating objects on a computer screen, this may be done by recording the clickstream—the responses of the examinee in performing the task, usually mouse clicks, with the associated location, time, and task context of the clicks as appropriate. In tasks performed by manipulating simulated controls in a VE, for example, turning a wheel, moving a lever, or pulling a trigger, the manipulations are recorded, again with the associated location, time, and task context. Additional measures—mostly unobtrusive—can also be collected to correlate with process measures, such as video of the trainee's performance, audio of think-alouds, sensor based measures of gaze (using eye trackers), motor performance (using pressure and motion sensors attached to the device operated by the trainee), and psychophysiological measures of stress and attention, such as electroencephalography, electromyography, an electrocardiogram, electrodermal activity, blood pressure, respiration, or heart rate.

Bewley, Lee, Munro, and Chung (2007) describe the use of clickstream data in assessments of air defense planning knowledge and skill. Plans are generated by selecting and moving plan elements, for example, ships, aircraft, and threat sectors. The time and location of each selection and placement are recorded, and a scoring algorithm reduces the data for interpretation by comparison to an expert's judgment of appropriate planning behavior.

Another approach to unobtrusively measuring task performance is the use of sensors. Sensor based measures are being used by Greg Chung and colleagues at CRESST (National Center for Research on Evaluation, Standards, and Student Testing) in assessments of rifle marksmanship performance on a virtual shooting range. The trainee fires a rifle instrumented with pressure sensors on the trigger to measure trigger squeeze, a motion sensor on the muzzle to measure steadiness, and an eye tracker to determine focus at the time of the shot. All measures are correlated with the location of the strike on a laser target. Measures being added in current research include an electroencephalograph, galvanic skin response, heart rate, and respiration.

Scoring Model

A disadvantage of the ability to easily measure process performance in assessments for VE training is the ease of collecting too much data. The model based approach to assessment helps mitigate this risk by requiring that tasks and measures map to the domain representation, which maps back to the cognitive demand and content representation—the purpose and goals of the assessment—

which helps ensure that data collected will be relevant to the interpretations and uses of the assessment. But a scoring model must also be developed that translates observations of examinee performance into scores that can be used to draw inferences about knowledge or skills. A scoring model includes an information measurement scale, scoring criteria, performance descriptions of each criterion at each point on the scale, and sample responses that illustrate the various levels of performance (American Educational Research Association et al., 1999).

Scoring issues for assessments in VE training are complex, as they can generate a rich set of observations that are fine grained, interrelated, and process oriented (Baker & Mayer, 1999; Bennett, 1999; Chung & Baker, 2003b; Clauser, 2000; National Research Council, 2001). It is important to define how the observations are combined and how they are scored and scaled. Evidence needs to be collected on how the measures relate to other measures of the construct and how the measures discriminate between high and low performers.

Three major approaches to automated scoring have been used: expert based methods, data-driven methods, and domain-modeling methods.

Expert Based Methods

There are two expert based methods: using expert performance and modeling expert judgment.

In the first approach, actual expert performance is considered the gold standard against which to compare student performance (Baker, 1997; Baker et al., 1991), not what experts say should be competent performance or how experts rate student performance. This approach has been used to develop tasks for content understanding using essays (Baker et al., 1991) and knowledge maps (Herl, O'Neil, Chung, & Schacter, 1999).

A related approach is to model experts' rating of examinees' performance on various task variables. Expert judgment is considered the gold standard against which to compare student performance, not actual expert performance. This scoring approach has been used successfully to model expert and rater judgments in a variety of applications, including essays (Burstein, 2003), patient management skills (Margolis & Clauser, 2006), and air defense planning (Bewley et al., 2007).

One of the major issues with expert based scoring is the selection of the expert (Bennett, 2006; Bennett & Bejar, 1998). Problems include experts' biases, the influences of the experts' content and world knowledge, linguistic competency, expectations of student competency, and instructional beliefs (Baker & O'Neil, 1996).

Data-Driven Techniques

In data-driven techniques, performance data are subjected to statistical or machine learning analyses (for example, an artificial neural network with hidden Markov models). Using artificial neural network and hidden Markov model technologies, Ron Stevens and colleagues have developed a method for identifying learner problem solving strategies and modeling learning trajectories, or sequences of performance states (Stevens, Soller, Cooper, & Sprang, 2004). Applying the method to chemistry, they were able to identify trajectories revealing learning problems, for example, not thoroughly exploring the problem space early, reaching a performance state that makes it unlikely to reach a more desirable end state, and reaching a state from which the learner could transition to a better or worse state with equal likelihood. With this information, it may be possible to perform a fine-grained diagnosis of what learners do not know and to use learning trajectories to guide the sequence of instruction and the type and form of remediation, and do it on the fly.

Validation of data-driven methods is complicated because there is no a priori expectation of what scores mean and no inherent meaning of the classification scheme. Interpretation is post hoc, which creates the potential for the introduction of bias in assignments to groups after the groups have been defined. A second problem is that machine learning techniques can be highly sample dependent, and the scoring process is driven by statistical rather than theoretical issues (Bennett, 2006). Because of these issues, validity evidence is particularly important when using data-driven techniques to score student responses.

Domain Modeling

This approach attempts to model the cognitive demands of the domain itself. The model specifies how knowledge and skills influence each other and the task variables on which observations are being made. The approach relies on a priori linking of student performance variables to hypothesized knowledge and skill states. Student knowledge and skills are then interpreted in light of the observed student performance. This approach has been used successfully in a variety of domains and modeling types, from canonical items (for example, Hively, Patterson, & Page, 1968), to Tatsuoka's rule-space methodology (for example, Birenbaum, Kelly, & Tatsuoka, 1993), to the use of Bayes nets to model student understanding in such domains as dental hygiene skills, hydraulic troubleshooting, network troubleshooting, Web searching, circuit analyses, and rifle marksmanship (for example, Bennett, Jenkins, Persky, & Weiss, 2003; Chung, Delacruz, Dionne, & Bewley, 2003; Mislevy & Gitomer, 1995; Mislevy, Steinberg, Breyer, Almond, & Johnson, 2002; Williamson, Almond, Mislevy, & Levy, 2006).

The most important issue in domain modeling is identifying the essential concepts and their interrelationships. This can be mitigated through cognitive task analyses and direct observation of performance, but it is critical to gather validity evidence to validate the structure of and inferences drawn by the Bayes net. For examples of empirical validation techniques, see Chung, Delacruz, et al. (2003) and Williamson, Almond, and Mislevy (2000).

FUTURE DIRECTIONS

The need for efficient and cost-effective development of quality assessments has motivated several efforts to create automated or partially automated supports for assessment design (Baker, 2002; Chung et al., 2008). Some have been focused on very specific topics, such as algebra (Koedinger & Nathan, 2004), some involve systems of training (Mislevy & Riconscente, 2005), and others have focused on the development of cognition and content templates and objects. CRESST's Assessment Design and Delivery System (ADDS) provides the capability to create assessments using assessment components, for example, new or preexisting prompts, and information sources (Vendlinski, Niemi, & Wang, 2005). ADDS users have been found to focus more on measuring conceptual knowledge and to create more appropriate rubrics and coherent prompts that address critical ideas. Additional work should be done to develop assessment design tools.

A second important trend is the use of formative assessments embedded in training to diagnose knowledge gaps and guide instructional decisions. Complex modeling can be used during instruction, post-instruction, retention trials, and generalization and transfer measurement to understand and locate specific performance problems and diagnose the causes as a combination of lack of knowledge, attention, motivation, or integration of content and skill. Using clickstream methods (Chung & Baker, 2003a; Stevens & Casillas, 2006), one can now pinpoint some of these areas. Because of the growing sophistication of computationally supported data collection, and the importance of formative information about the trainee's process during learning, the future of outcome assessment will merge with process information to create learner profiles rather than scores or classifications. We anticipate that these will have domain-independent components that may predict a learner's likely success in a range of other tasks. We also expect greater use of ontologies for domain representation and see the study of expertise continuing to add to our knowledge of performance measurement and its validity.

Finally, we predict an increased use of artificial intelligence and advanced decision analysis techniques to support assessment. These include ontologies, Bayes nets, artificial neural networks, hidden Markov models, lag sequential analysis, and constraint networks.

SUMMARY AND DISCUSSION

This chapter has described a model based approach to assessment design and validation, with a focus on assessments for VE training environments. We have argued for the need to assess performance in VE training, and to do so using valid assessments. We have discussed the concept of validity and described Baker's (1997) model based methodology for assessment design and validation. The methodology begins with defining cognitive demand, which combines with the content representation to influence definition of the domain representation. The domain representation influences the task representation. Finally, the scoring model specifies how observations of task performance are translated to scores that can be used to draw inferences about the knowledge, skills, abilities, attitudes, and other properties of the construct to be assessed—the cognitive demands.

The chapter's central take-away message is the importance of validity as the fundamental requirement for any assessment and the key indicator of the assessment's technical quality. To ensure training effectiveness, learner performance must be assessed, and the assessments must be valid.

The validity message is linked to three supporting ideas:

1. *Validity is not a general quality of an assessment.* An assessment does not possess a general quality called validity. An assessment's validity depends on the context of its use and the inferences to be drawn based on the results. A validity argument must be made using a wide range of evidence for the appropriateness of the inferences for the particular context.
2. *Begin with a definition of cognitive demand.* The first step in assessment design (and instructional design) is a definition of the cognitive demands of the task—the set of processes and performances required for success. This leads to designing methods of measuring these processes and learning outcomes, including designing tasks that will elicit the desired performance, defining performance measures, and operationalizing the scoring algorithm for measuring constructs such as "understanding" or "problem solving" and then validating the approach with empirical evidence.
3. *Designing a valid assessment cannot be separated from the design of instruction.* Assessment design is not something to do after the training is developed. The assessment, including tasks, measures, and scoring, and how the results are to be used, must be included in the training program design. The design of assessments and training should be an integrated activity rather than approached separately.

Virtual environment training has great promise for training on complex high value tasks that have in the past required extended periods of training using expensive equipment and manpower, sometimes in environments placing trainees and instructors in harm's way. Great promise and impressive technical capability are not sufficient to conclude effectiveness, however. To realize the promise, practitioners must assess the systems and the learning they help produce, and the assessments must be valid. The model based assessment methodology can help make this happen.

REFERENCES

American Educational Research Association, American Psychological Association, and National Council for Measurement in Education. (1999). *Standards for educational and psychological testing.* Washington, DC: American Educational Research Association.

Anderson, L. W., & Krathwohl, D. R. (2001). *A taxonomy for learning, teaching, and assessing: A revision of Bloom's taxonomy of educational objectives.* New York: Addison Wesley Longman, Inc.

Baker, E. L. (1997). Model-based performance assessment. *Theory Into Practice, 36*(4), 247–254.

Baker, E. L. (2002). Design of automated authoring systems for tests. In National Research Council, Board on Testing and Assessment, Center for Education, Division of Behavioral and Social Sciences and Education (Eds.), *Technology and assessment: Thinking ahead—Proceedings from a workshop* (pp. 79–89). Washington, DC: National Academy Press.

Baker, E. L., Chung, G. K. W. K., & Delacruz, G. C. (2008). Design and validation of technology-based performance assessments. In J. M. Spector, M. D. Merrill, J. J. G. van Merriënboer, & M. P. Driscoll (Eds.), *Handbook of research on educational communications and technology* (pp. 595–604). New York: Lawrence Erlbaum.

Baker, E. L., Freeman, M., & Clayton, S. (1991). Cognitive assessment of history for large-scale testing. In M. C. Wittrock & E. L. Baker (Eds.), *Testing and cognition* (pp. 131–153). Englewood Cliffs, NJ: Prentice-Hall.

Baker, E. L., & Herman, J. L. (1983). Task structure design: Beyond linkage. *Journal of Educational Measurement, 20,* 149–164.

Baker, E. L., & Mayer, R. E. (1999). Computer-based assessment of problem solving. *Computers in Human Behavior, 15,* 269–282.

Baker, E. L., & O'Neil, H. F., Jr. (1987). Assessing instructional outcomes. In R. M. Gagné (Ed.), *Instructional technology* (pp. 343–377). Hillsdale, NJ: Erlbaum.

Baker, E. L., & O'Neil, H. F., Jr. (1996). Performance assessment and equity. In M. B. Kane & R. Mitchell (Eds.), *Implementing performance assessment: Promises, problems, and challenges* (pp. 183–199). Mahwah, NJ: Erlbaum.

Bennett, R. E. (1999). Using new technology to improve assessment. *Educational Measurement: Issues and Practice, 18*(3), 5–12.

Bennett, R. E. (2006). Moving the field forward: Some thoughts on validity and automated scoring. In D. M. Williamson, I. I. Behar, & R. J. Mislevy (Eds.), *Automated scoring of complex tasks in computer-based testing* (pp. 403–412). Mahwah, NJ: Erlbaum.

Bennett, R. E., & Bejar, I. I. (1998). Validity and automated scoring: It's not only the scoring. *Educational Measurement, 17*(4), 9–17.

Bennett, R. E., Jenkins, F., Persky, H., & Weiss, A. (2003). Assessing complex problem solving performances. *Assessment in Education: Principles, Policy & Practice, 10,* 347–359.

Bewley, W. L., Lee, J. J., Munro, A., & Chung, G. K. W. K. (2007, April). *The use of formative assessments to guide instruction in a military training system.* Paper presented at the annual meeting of the American Educational Research Association, Chicago, IL.

Birenbaum, M., Kelly, A. E., & Tatsuoka, K. K. (1993). Diagnosing knowledge states in algebra using the rule-space model. *Journal of Educational Measurement, 20,* 221–230.

Burstein, J. (2003). The e-rater scoring engine: Automated essay scoring with natural language processing. In M. D. Shermis & J. Burstein (Eds.), *Automated essay scoring: A cross-disciplinary perspective* (pp. 113–122). Mahwah, NJ: Erlbaum.

Carretta, T. R., & Dunlap, R. D. (1998). *Transfer of effectiveness in flight training: 1986 to 1997* (Rep. No. AFRL-HE-AZ-TR-1998-0078). Mesa, AZ: U.S. Air Force Research Laboratory.

Chung, G. K. W. K., & Baker, E. L. (2003a). An exploratory study to examine the feasibility of measuring problem-solving processes using a click-through interface. *Journal of Technology, Learning, and Assessment, 2*(2). Available from http://jtla.org

Chung, G. K. W. K., & Baker, E. L. (2003b). Issues in the reliability and validity of automated scoring of constructed responses. In M. D. Shermis & J. E. Burstein (Eds.), *Automated essay grading: A cross-disciplinary approach* (pp. 23–40). Mahwah, NJ: Erlbaum.

Chung, G. K. W. K., Baker, E. L., Delacruz, G. C., Bewley, W. L., Elmore, J., & Seely, B. (2008). A computational approach to authoring problem-solving assessments. In E. L. Baker, J. Dickieson, W. Wulfeck, & H. F. O'Neil (Eds.), *Assessment of problem solving using simulations* (pp. 289–307). Mahwah, NJ: Erlbaum.

Chung, G. K. W. K., Delacruz, G. C., Dionne, G. B., & Bewley, W. L. (2003). Linking assessment and instruction using ontologies. *Proceedings of the I/ITSEC, 25,* 1811–1822.

Clark, R. E., & Estes, F. (2002). *Turning research into results: A guide to selecting the right performance solutions.* Atlanta, GA: CEP Press.

Clauser, B. E. (2000). Recurrent issues and recent advances in scoring performance assessments. *Applied Psychological Measurement, 24,* 310–324.

Gitomer, D. H., & Yamamoto, K. (1991). Performance modeling that integrates latent trait and class theory. *Journal of Educational Measurement, 28,* 173–189.

Herl, H. E., O'Neil, H. F., Jr., Chung, G. K. W. K., & Schacter, J. (1999). Reliability and validity of a computer-based knowledge mapping system to measure content understanding. *Computers in Human Behavior, 15,* 315–334.

Hively, W., Patterson, H. L., & Page, S. H. (1968). A "universe defined" system of arithmetic achievement tests. *Journal of Educational Measurement, 5,* 275–290.

Koedinger, K. R., & Nathan, M. J. (2004). The real story behind story problems: Effects of representations on quantitative reasoning. *Journal of the Learning Sciences, 13,* 129–164.

Lampton, D. R., Bliss, J. P., & Morris, C. S. (2002). Human performance measurement in virtual environments. In K. M. Stanney (Ed.), *Handbook of virtual environments: Design, implementation, and applications* (pp. 701–720). Mahwah, NJ: Erlbaum.

Margolis, M. J., & Clauser, B. E. (2006). A regression-based procedure for automated scoring of a complex medical performance assessment. In D. M. Williamson, I. I. Behar, & R. J. Mislevy (Eds.), *Automated scoring of complex tasks in computer-based testing* (pp. 123–167). Mahwah, NJ: Erlbaum.

Mayer, R. E. (1992). *Thinking, problem solving, cognition* (2nd ed.). New York: W. H. Freeman and Company.

Messick, S. (1993). Validity. In R. Linn (Ed.), *Educational measurement* (3rd ed., pp. 13–103). Phoenix, AZ: The Oryx Press.

Messick, S. (1995). Standards of validity and the validity of standards in performance assessment. *Educational Measurement: Issues and Practice, 14*(4), 5–8.

Mislevy, R., & Gitomer, D. H. (1995). The role of probability-based inference in an intelligent tutoring system. *User Modeling and User-Adapted Interaction, 5,* 253–282.

Mislevy, R., & Riconscente, M. (2005). *Evidence-centered assessment design: Layers, structures, and terminology* (PADI Tech. Rep. No. 9). Menlo Park, CA: SRI International.

Mislevy, R. J., Steinberg, L. S., Breyer, F. J., Almond, R. G., & Johnson, L. (2002). Making sense of data from complex assessments. *Applied Measurement in Education, 15,* 363–389.

National Research Council. (2001). *Knowing what students know: The science and design of educational assessment.* Washington, DC: National Academy Press.

Stevens, R., Soller, A., Cooper, M., & Sprang, M. (2004). Modeling the development of problem solving skills in chemistry with a web-based tutor. *Proceedings of the 7th International Conference on Intelligent Tutoring Systems* (pp. 580–591). Berlin: Springer-Verlag.

Stevens, R. H., & Casillas, A. (2006). Artificial neural networks. In D. M. Williamson, I. I. Behar, & R. J. Mislevy (Eds.), *Automated scoring of complex tasks in computer-based testing* (pp. 259–312). Mahwah, NJ: Erlbaum.

van Merriënboer, J. J. G. (1997). *Training complex cognitive skills: A four-component instructional design model for technical training.* Englewood Cliffs, NJ: Educational Technology Publications.

Vendlinski, T., Niemi, D., & Wang, J. (2005). Learning assessment by designing assessments: An on-line formative assessment design tool. In C. Crawford, R. Carlsen, I. Gibson, K. McFerrin, J. Price, & R. Weber (Eds.), *Proceedings of Society for Information Technology and Teacher Education International Conference 2005* (pp. 228–240). Norfolk, VA: AACE.

Wickens, C. D., & May, P. (1994). *Terrain representation for air traffic control: A comparison of perspective with plan view displays* (Tech. Rep. No. ARL-94-10/FAA-94-2). Savoy: University of Illinois, Aviation Research Laboratory.

Williamson, D. M., Almond, R. G., & Mislevy, R. J. (2000). Model criticism of Bayesian networks with latent variables. In C. Boutilier & M. Goldzmidt (Eds.), *Uncertainty in artificial intelligence; Proceedings of the 16th conference* (pp. 634–643). San Francisco: Morgan Kaufmann.

Williamson, D. M., Almond, R. G., Mislevy, R. J., & Levy, R. (2006). An application of Bayesian networks in automated scoring of computerized simulation tasks. In D. M. Williamson, I. I. Behar, & R. J. Mislevy (Eds.), *Automated scoring of complex tasks in computer-based testing* (pp. 201–257). Mahwah, NJ: Erlbaum.

Williamson, D. M., Bejar, I. I., & Mislevy, R. J. (2006). Automated scoring of complex tasks in computer-based training: An introduction. In D. M. Williamson, R. J. Mislevy, & I. I. Behar (Eds.), *Automated scoring of complex tasks in computer-based testing* (pp. 1–13). Mahwah, NJ: Lawrence Erlbaum.

Chapter 17

AUTOMATED PERFORMANCE ASSESSMENT OF TEAMS IN VIRTUAL ENVIRONMENTS

Peter Foltz, Noelle LaVoie, Rob Oberbreckling, and Mark Rosenstein

Multiplayer virtual environments provide an excellent venue for distributed team training. They provide realistic, immersive, engaging situations that can elicit the complex behaviors that encompass teamwork skills. These environments permit trainers the opportunity to target particular skills in order to assess and improve a team's performance in situations that are difficult to create in live environments. In addition, because virtual environments provide fine-tuned control of the training situation and automate the collection of data, training teams in virtual environments can save effort and money when compared to live training. As the military and other large collaborative organizations incorporate greater network centric methods, operations, tactics, and technologies, virtual environments become an essential means to monitor, train, and assess teams.

However, there are numerous challenges to effectively identify, track, analyze, and report on teams in complex virtual environments. For example, many current methods of assessing team and group performance rely on both global outcome metrics and handcrafted assessment techniques. These metrics often lack information rich enough to diagnose failures, detect critical incidents, or suggest improvements for the teams for use in their collaborative aids. It is also problematic for these techniques to produce assessments in the near real time frame that is necessary for effective training feedback because of the reliance on time consuming hand coding. Thus, while there has been an explosive increase in the availability of team information that can be obtained from a virtual environment, there needs to be a concomitant development in tools that can leverage the data to monitor, support, and enhance team performance. In this chapter, we discuss the issues of evaluating teams in virtual environments, describe an automated communications based analysis approach that we have found fruitful in tackling these issues, and finally detail the application and evaluation of this approach in predicting team performance in the context of three task domains.

TEAM PERFORMANCE MEASUREMENT IN VIRTUAL ENVIRONMENTS

Complex team virtual environments provide an ideal venue for team training. Orsanu and Salas (1993) identify a number of critical characteristics for training teams, including having interdependent members with defined roles, using multiple information sources, and sharing common goals. Because of the inherent automation in virtual environments, they afford better ability to measure performance of teams, both in recording what is being done by the team as well as what is communicated by team members. Nevertheless, while a virtual environment can produce a record of what team members have done and said, there are challenges in converting that information into measures of performance and difficulties in determining how those measures can be used to give feedback.

Team performance can be seen as a combination of taskwork and teamwork. Taskwork, which is the work a team does to accomplish its mission, is often more amenable to automated analysis from a virtual environment event log. For example, a system can provide information on whether a person moved from *x* to *y* at time *t* and whether an objective was completed. Teamwork, on the other hand, encompasses how the team members coordinate with each other. In order to measure teamwork within virtual environments, the critical aspects of teamwork must be identified along with how they can be measured, assessed, and trained (for example, Salas & Cannon-Bowers, 2001). These skills include leadership, monitoring, backup behavior, coordination, and communication (for example, Cannon-Bowers, Tannenbaum, Salas, & Volpe, 1995; Curtis, Harper-Sciarini, DiazGranados, Salas, & Jentsch, 2008; Freeman, Diedrich, Haimson, Diller, & Roberts, 2003; Hussain et al., 2008). Curtis et al. (2008) identify three teamwork processes that have major impacts on teamwork and that appear to be strongly predictive of team performance: communication, coordination, and team leadership. These processes are typically assessed by subject matter experts (SMEs) watching and checking off behaviors associated with the processes. This protocol can be quite time consuming and is often performed after the exercise is completed rather than in real time, limiting the ability to incorporate teamwork performance measurement into virtual environments or provide timely feedback to teams. Thus, methods are required to automatically measure teamwork in an accurate and responsive manner. This chapter focuses on the aspects of communication that can be used to predict performance and how analyses of communications can be automated to provide rapid measurement of teamwork.

COMMUNICATION AS AN INDICATOR OF PERFORMANCE

Networked teams in virtual environments provide a rich source of information about their performance through their verbal communication. The communication data contain information both about the actual structure of the network and the flow of meaning through the network over time. The structure and communication patterns of the network can provide indications of team member roles, paths of information flow, and levels of connectedness within and across teams.

The content of the information communicated provides detailed indications of the information team members know, what they tell others, whom they tell, and their current situation. Thus, communication data provide information about team cognitive states, knowledge, errors, information sharing, coordination, leadership, stress, workload, intent, and situational status. Indeed, within the distributed training community, trainers and subject matter experts typically rely on listening to a team's communication in order to assess that team's performance. Nevertheless, to effectively exploit the communication data, technologies need to be available that can assess both the content and patterns of the verbal information flowing in the network and convert the analyses into results that are usable by teams, instructors, and commanders.

In this chapter, we provide an overview of ongoing research and development of a set of tools for the automatic analysis of team verbal communication and discuss their application in measuring team performance in virtual environment training systems. The tools exploit team communication data and use language technologies to analyze the content of communication, thereby permitting characterization of the topics and quality of information being transmitted. To explore these ideas further, we describe how these tools were incorporated into three application environments and the results of their use.

VERBAL COMMUNICATION ANALYSIS

The overall goal of automated verbal communication analysis is to apply a set of computational modeling approaches to verbal communication in order to convert the networked communication into useful characterizations of performance. These characterizations include metrics of team performance, feedback to commanders, or alerts about critical incidents related to performance. This type of analysis has several prerequisites. The first is the availability of sources of verbal communication. Second, there must be performance measures that can be used to associate the communication to standards of actual team performance. These prerequisites can then be combined with computational approaches to perform the analysis. These computational approaches include computational linguistics methods to analyze communication, machine learning techniques to associate communication to performance measures, and finally cognitive and task modeling techniques.

By applying the computational approaches to the communication, we have a complete communication analysis pipeline as represented in Figure 17.1. Proceeding through the tools in the pipeline, spoken and written communication are converted directly into performance metrics that can then be incorporated into visualization tools to provide commanders and soldiers with applications, such as automatically augmented after action reviews (AARs) and briefings, near real time alerts of critical incidents, timely feedback to commanders of poorly performing teams, and graphic representations of the type and the quality of information flowing within a team. We outline the approach to this communication analysis below.

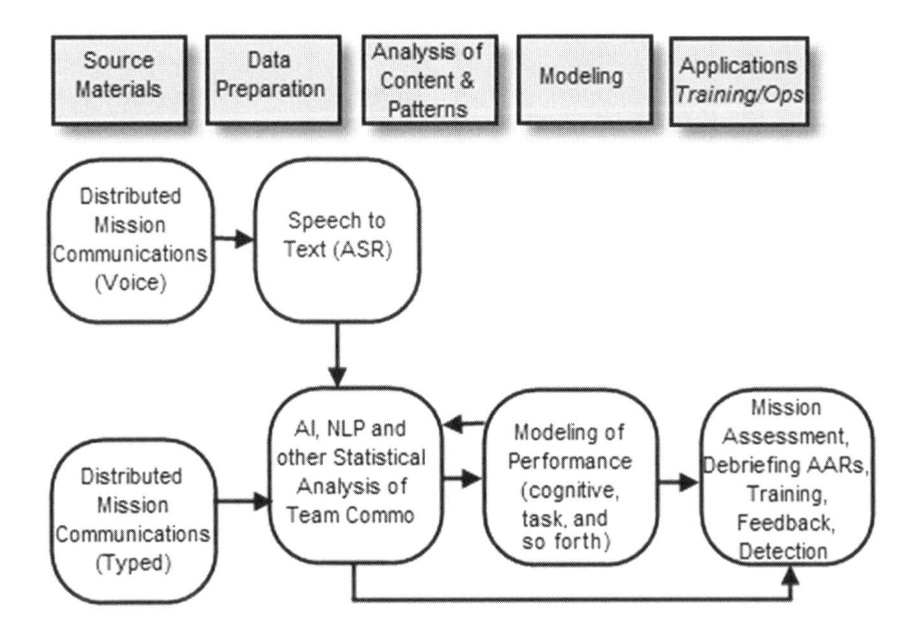

Figure 17.1. The Communication Analysis Pipeline

Communication Data

For analysis purposes, communication data include most kinds of verbal communication among team members. Typed communication (for example, chat, e-mail, or instant messages) can be automatically formatted for input into the analysis tools. Audio communication includes the capture of many kinds of spoken data, including use of voice over Internet protocol systems, radios, and phones.

Because a majority of communication in virtual environments is typically spoken, two classes of information can be gleaned from the audio stream: content and audio features. Automatic speech recognition (ASR) systems convert speech to text for analysis of content, while audio analysis extracts such characteristics as stress or excitement levels from the audio. ASR systems often also provide measures such as rate of speech and ASR uncertainty. All this processed information can be input into the communication analysis system.

Performance Metrics

In order to provide feedback on team performance, the toolset learns to associate team performance metrics with the communication streams from those teams. Thus, the system typically requires one or more metrics of team performance. There is a wide range of issues in determining appropriate metrics for measuring team performance (for example, Brannick, Salas, & Prince, 1997). For example, metrics need to be associated with key outcomes or processes related to the team's tasks; they should indicate and provide feedback on deficiencies for individuals and/or teams, and they need to be sufficiently reliable so that experts can agree on both the value of the metric and on how it should be scored for different teams (Paris, Salas, & Cannon-Bowers, 2001).

Objective measures of performance can be used as metrics to indicate specific aspects of team performance. These measures can include threat eliminations, deviations from optimal solution paths, number of objectives completed, and measures derived from task-specific artifacts, such as Size, Acuity, Location, Unit, Time, and Equipment Report and Anterior Cingulate Cortex Report. One advantage of computer based environments is that they are able to automatically track and log events and then generate such objective measures.

Subjective measures of performance can also be used as metrics. These can include subject matter experts' ratings of such aspects as command and control, management of engagement, following doctrine, communication quality, and situation understanding. Additionally, SME evaluations of AARs and identification of specific critical incidents, failures, or errors can be used to measure performance. Care must be taken, as all metrics will have varying levels of reliability as well as validity. For new metrics, it is often advisable to obtain ratings from more than one SME in order to determine reliability.

Computational Modeling Tools

Communication data are converted into a computational representation that includes measures of the content (what team members are talking about), quality (how well team members seem to know what they are talking about), and fluency (how well team members are talking about it). This process uses a combination of computational linguistics and machine learning techniques that analyze semantic, syntactic, relational, and statistical features of the communication streams.

While we will discuss a number of tools, the primary underlying technology used in this analysis is a method for mimicking human understanding of the meaning of natural language called Latent Semantic Analysis (LSA) (see Landauer, Foltz, & Laham, 1998, for an overview of the technology). LSA is automatically trained on a body of text containing knowledge of a domain, for example, a set of training manuals, and/or domain relevant verbal communication. After such training, LSA is able to measure the degree of similarity of meaning between two communication utterances in a way that closely mimics human judgments. This capability can be used to understand the verbal interactions in much the same way a subject matter expert compares the performance of one team or individual to others. The technique has been widely used in other machine understanding applications, including commercial search engines, automated scoring of essay exams, and methods for modeling human language acquisition.

The results from the LSA analysis are combined with other computational language technologies, including techniques to measure syntactic complexity, patterns of interaction and coherence among team members, audio features, and statistical features of individual and team language (see Jurafsky & Martin, 2000). The computational representation of the team language is then combined with machine learning technology to predict the team performance metrics. In a sense, the overall method learns which features of team communication are associated with different metrics of team performance and then predicts scores for new sets of communication data employing those features.

Performance Prediction with the Communication Analysis Toolkit

Tests of the toolkit's use for communication analysis have shown great promise. Tests are performed by training the system on one set of communication data and then testing its prediction performance on a new dataset. This procedure tests that the models generalize to new communication. Over a range of communication types, the toolkit is able to provide accurate predictions of the overall team performance and individual team metrics. It makes reliable judgments of the type of statements each team member is making, and it can predict team performance problems based on the patterns of communication among team members (Foltz, 2005; Gorman, Foltz, Kiekel, Martin, & Cooke, 2003). In addition to the approaches described above, there have been other approaches used to analyze communication in teams that have shown great promise. These have included modeling communication flow patterns to predict team performance and cognitive states (see Gorman, Weil, Cooke, & Duran, 2007; Kiekel, Gorman, & Cooke, 2004).

The communication analysis toolkit has been tested in many environments, including an unmanned aerial vehicle synthetic task environment (see Gorman et al., 2003; Foltz, Martin, Abdelali, Rosenstein, & Oberbreckling, 2006), in air force simulators of F-16 missions (Foltz, Laham, & Derr, 2003; Foltz et al., 2006), and in Navy Tactical Decision Making Under Stress (TADMUS) exercises (Foltz et al., 2006). The tools predicted both objective team performance scores and SME ratings of performance at very high levels of reliability (correlations ranged from $r = 0.5$ to $r = 0.9$ over 20 tasks). It should be noted that the agreement between the toolkit's predictions and SMEs is typically within the range of one SME to another. In addition, the tools are able to characterize the type of communication for individual utterances (for example, planning, stating facts, and acknowledging; Foltz et al., 2006).

Issues and Current Limitations of This Approach

While the next section describes successful applications of this approach, there are a number of issues and current limitations. For verbal communication, this approach requires automatic speech recognition, and that technology currently has a number of limitations. The state of the art requires building acoustic models and speaker-independent but task-specific models currently requiring about 20 hours of speech to train the ASR system, which increases the startup time for a new task domain.

The second prerequisite of the approach is performance measures. If objective measures are available, then as soon as the ASR is available, teams can begin to execute the task, communication data and performance data can be collected, and then a performance model can be built. If expert ratings are preferred, then protocols for scoring communications need to be developed and then SMEs must score a set of missions to be used as a training set, though this limitation will impact any approach that uses experts.

Besides these startup costs, there is also the issue of the accuracy of the ASR. The communication analysis technologies have been tested with ASR input from a number of datasets of spoken communication (see Foltz et al., 2003). The results indicate that even with typical ASR systems degrading word recognition by 40 percent, the model prediction performance degraded less than 10 percent. Thus, the approach appears to be quite robust to typical ASR errors.

APPLICATIONS OF THE COMMUNICATION ANALYSIS TOOLKIT IN VIRTUAL ENVIRONMENTS

A number of applications have been developed to test the performance and validate the use of the toolkit in virtual and live training situations. Below we describe three applications, one monitoring and assessing learning in online discussion environments, another providing real time analyses and visualizations of multinational stability and support operation simulation exercises, and the third providing automated team performance metrics and detection of critical incidents in both convoy operations in simulators and in live training environments. These applications cover the range of immersion in virtual environments. At one end are collaborative discussion environments, which permit use and evaluation of the planning, communication, and coordination aspects of teams, but do not provide the full immersive qualitics of a simulator environment. At the other end are virtual convoy environments and similar live training environments where the approach is tested as teams move from virtual to real world training and operations.

Knowledge Post

In large networked organizations, it is difficult to track performance in distributed exercises. Knowledge Post is designed for monitoring, moderating, and assessing collaborative learning and planning. The tools within Knowledge Post have been tested in a series of studies at the U.S. Army War College and the U.S. Air Force Academy (LaVoie, Psotka, Lochbaum, & Krupnick, 2004; LaVoie et al., in press; Lochbaum, Streeter, & Psotka, 2002). The application consists of an off-the-shelf threaded discussion group that has been substantially augmented with latent semantic analysis based functionality to evaluate and support individual and team contributions in simulated planning operations.

Knowledge Post supports the following abilities:

- To automatically notify the instructor when the discussion goes off track,
- To enhance the overall quality of the discussion and consequent learning level by having expert comments or library articles automatically interjected into the discussion at appropriate places,
- To locate material in the discussion or electronic library similar in meaning to a given posting,
- To automatically summarize contributions, and
- To assess the quality of contributions made by individuals and groups.

The utility of each of the aforementioned functions was empirically evaluated with senior officers, either in research sessions or participating in distributed learning activities at the U.S. Army War College, or with cadets at the U.S. Air Force Academy. Among the findings of the studies was the superiority of learning in a Knowledge Post environment over a face-to-face discussion with significantly improved quality of discussion and the usefulness to the participants of the Knowledge Post searching and summarizing features (Lochbaum et al., 2002). The research conducted with the Army War College established the usefulness and accuracy of a software agent that automatically alerts moderators when groups and individuals are floundering by identifying on- and off-topic comments in a discussion (LaVoie et al., 2004). A human rater coded over 1,000 comments as either on topic or off. A second rater coded a random 10 percent of these comments. The correlation between the two raters for this task was r (162) = .85, $p < .001$, while the correlation between the LSA based model and one human rater was r (1,605) = 0.72, $p < .001$, showing that the model was able to accurately determine when a group's discussion was off topic. The work with the Air Force Academy demonstrated improved solution quality of a group of cadets as a result of exposure to automatically interjected relevant expert comments (LaVoie et al., in press). Cadets participated in a discussion of a challenging leadership scenario. The discussion was conducted in one of three ways: (1) face-to-face in a classroom with a live human moderator, (2) in Knowledge Post with an automated moderator that added relevant comments from experts, or (3) in Knowledge Post without the automated moderator. The quality of the discussions was evaluated by using LSA to determine the similarity of the cadets' discussion to that of senior military officers, and the highest quality discussions were found for groups that used Knowledge Post with the automated moderator to conduct their discussions (see Figure 17.2).

Although customized for distributed learning activities, the tools developed within Knowledge Post can be incorporated into other virtual environments for automated analysis and monitoring of teams performing planning based discussions.

TeamViz

The TeamViz application provides teams and evaluators ways of monitoring performance in large collaborative environments using a set of visualization tools and enhancements built on the Knowledge Post toolset. TeamViz ran live during

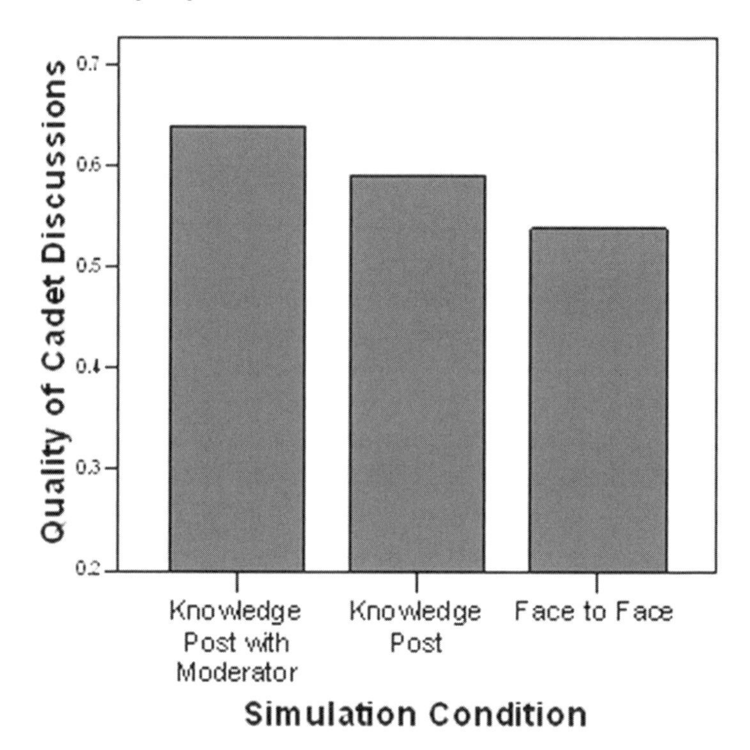

Figure 17.2. Quality of Discussion Comments in the Three Discussion Conditions

a U.S.–Singapore simulation exercise designed to evaluate collaboration among joint, interagency, and multinational forces conducting combat and stability operations (Pierce et al., 2006). The system automatically analyzed the content and patterns of information flow of the networked communication. It also provided automated summarizations of the ongoing communications as well as network visualization tools to improve situation understanding of team members. Analyses showed that the technology could track the flow of commander's intent among the team members by comparing the commander's briefing to the content of communication of different parts of the team. For example, the commander stressed the importance of naval facility defense in his briefing to three groups: two brigades under his command and the coalition task force (CTF) command staff. Comparing the content of the communications in each group following this briefing shows that Brigade 1 followed the commander's intent more closely than did Brigade 2 (see Figure 17.3).

It was also possible to detect the effects of scenario information injects on performance within the coalition task force and brigades by comparing the communication within each group to the content of the scenario inject. Figure 17.4 shows the response to an inject about a chemical weapons attack. It is clear that the coalition task force responded more quickly to the inject, and with a greater degree of discussion, than did either brigade.

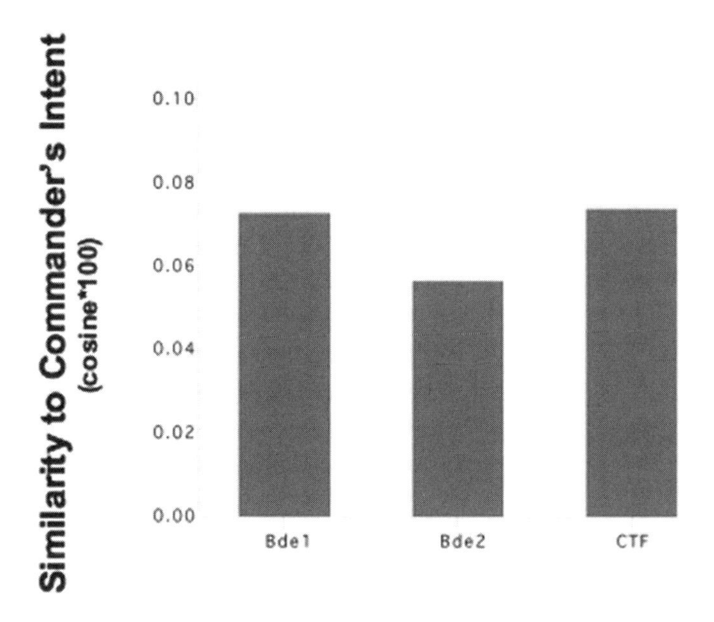

Figure 17.3. Communication analysis shows that Brigade 1 (Bde1) followed the commander's intent more closely than Brigade 2 (Bde2).

Singapore staff officers used TeamViz in real time to monitor the communication streams and to inform their commanders of important information flowing in the network, as well as to indicate perceived information bottlenecks. Overall, the TeamViz technologies permit knowledge management of large amounts of communication, as well as improved cognitive interoperability in distributed operations where communication among ad hoc teams is critical.

Competence Assessment and Alarms for Teams

Convoy operations require effective coordination among a number of vehicles and other elements, while maintaining security and accomplishing specific goals. However, in training for convoy operations it is difficult to monitor and provide feedback to team members in this complex environment. The Defense Advanced Research Projects Agency (DARPA) Automated Competence Assessment and Alarms for Teams (DARCAAT) program was designed to automate performance assessment and provide alarms for live and virtual convoy operations training. As part of the program, communication data and SME based performance measurements were collected, and then specialized tools to assess and visualize performance in convoy operations were developed.

The DARCAAT program collected voice communication data during convoy training operations and then collected SME based performance measurements on that data. From these, the DARCAAT program developed specialized tools to assess and visualize convoy operation performance. Two sources of data were

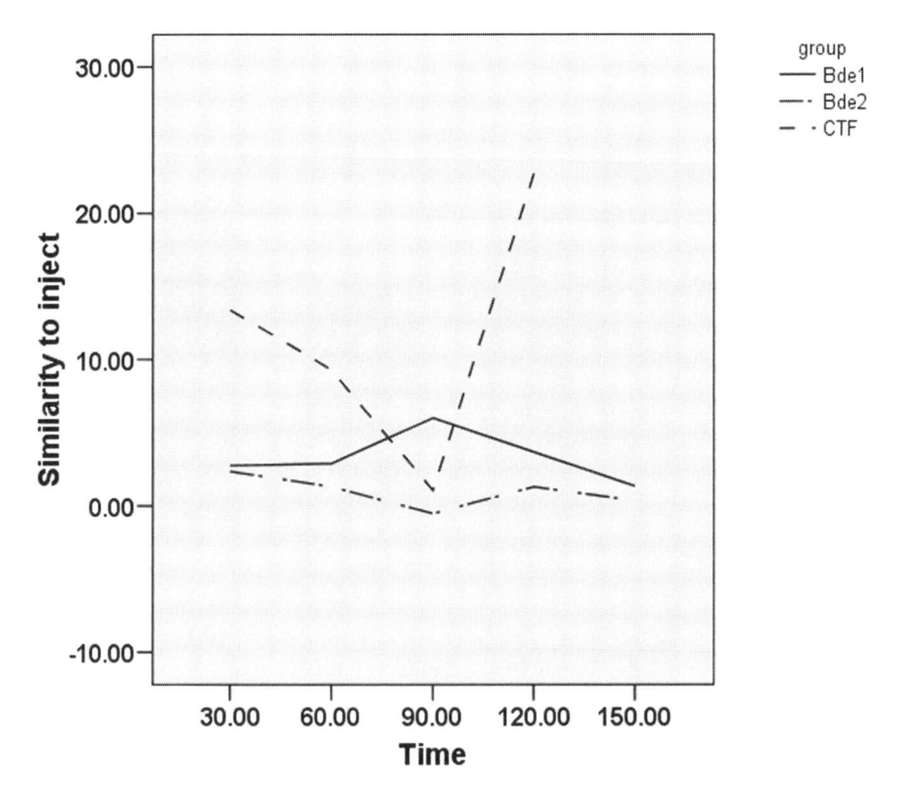

Figure 17.4. Communication analysis shows that the coalition task force responded more quickly to a scenario inject than either brigade.

used: one from teams in a virtual environment and one from teams in live training environments. The goal was to evaluate how well performance assessment tools could be applied to a single domain across both virtual and live training. For the virtual environment, communication data were collected from the Fort Lewis Mission Support Training Facility, which uses the DARWARS Ambush! virtual environment for convoy training. DARWARS Ambush! is a widely used game based training system and has been integrated into training for many brigades prior to deployment in Iraq (Diller, Roberts, Blankenship, & Nielson, 2004; Diller, Roberts, & Wilmuth, 2005). DARWARS Ambush! provides an excellent environment for team training and performance analysis because it provides reasonably controlled scenarios and environment and has the ability to instrument teams for voice communications, video, and environmental event data collection. In this environment, up to 60 soldiers can jointly practice battle drill training and leader/team development during convoy operations. Figure 17.5 shows the training environment for DARWARS Ambush! and Figure 17.6 shows a typical user's view during training.

In addition to the virtual environment DARWARS Ambush! data, the DARCAAT program collected live convoy Situational Training Exercise lane training data from the National Training Center (NTC) at Fort Irwin. The data included digital audio recordings of FM radio communication among the convoy team members, as well as videos of the convoy operations. Using the virtual and live convoy communications data, subject matter experts rated team performance on a number of metrics (battle drills, adherence to standard operating procedures [SOPs], situation understanding, command and control, and overall team performance) as well as indicated places in the scenario in which a critical event occurred (that is, "an event that significantly alters the battleground"). Prediction models were then built by analyzing the communication data using the full team communication analysis pipeline shown in Figure 17.1.

The results indicate that the DARCAAT toolset is able to accurately match SME ratings of team performance as well as detect critical events (for example, performance alarms). Using the DARWARS Ambush! data, the system could automatically detect 87 percent of the SME-rated critical events with a false positive rate of 19 percent. Thresholds for detecting critical events can be adjusted to allow them to be used as performance alarms enabling a commander to set lower thresholds that provide alerts for any case in which a team might be having performance problems at the cost of only a slightly higher false alarm rate. The DARCAAT model also predicted the SME ratings of team performance on each of the performance metrics. Table 17.1 shows the correlations between the SME ratings of overall team performance and the predictions generated by the DARCAAT toolset from analyzing the teams' communications based on

Figure 17.5. DARWARS Ambush! at the Fort Lewis Mission Support Training Facility

Figure 17.6. Screen from DARWARS Ambush! Training Scenario

Table 17.1. Correlation between SME Ratings and DARCAAT Predictions for Overall Team Performance

Metric	NTC & Ambush! ($n = 51$)	Ambush! ($n = 45$)
Battle drills	0.74	0.73
Command and control	0.71	0.70
Situation understanding	0.83	0.81
SOPs	0.73	0.79
TEAM	0.78	0.72

45 Ambush! missions and 6 NTC missions (all significant $p < .01$). It should be noted that the correlations between the SMEs and the toolset were equivalent to those found between multiple SMEs rating the same missions.

As a demonstration of the application of the DARCAAT toolset, an after action review application was developed that could be integrated into a training program to allow observer/controllers (OCs) and commanders to monitor teams and receive feedback on the team's performance. The application provides efficient automatic augmentation of AARs assisting the OCs in choosing the most appropriate segments of missions to illustrate training points. Figure 17.7 shows one screen from the AAR tool.

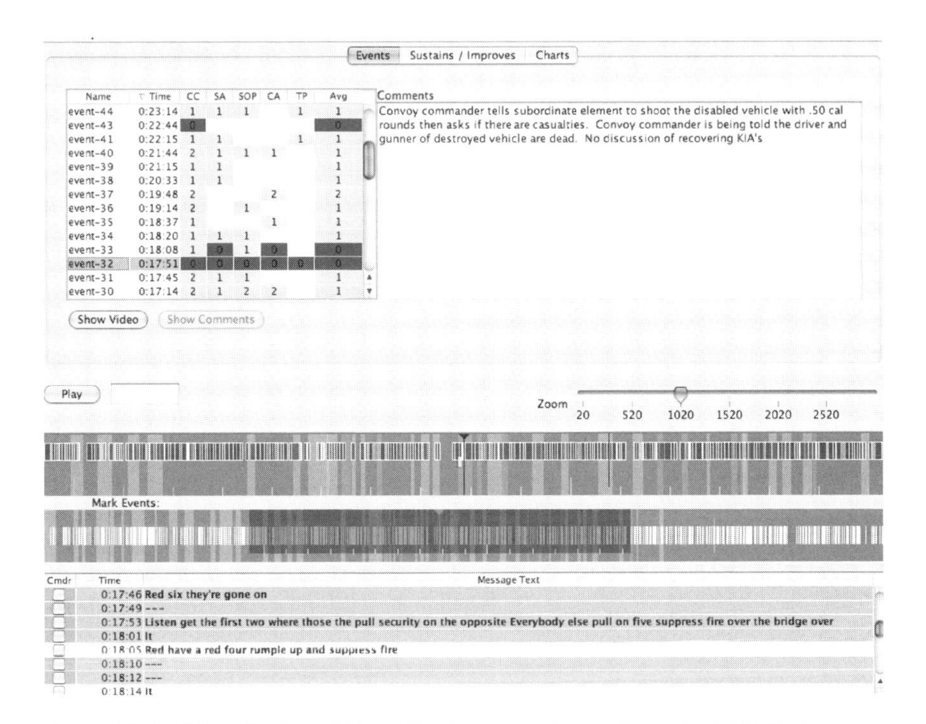

Figure 17.7. Visualization of Team Performance Scores from the AAR Tool

The tool processes the incoming communication data from a team and then allows an OC or commander to load any mission and provides immediate access to several critical pieces of information:

- The top left portion of the AAR tool displays the mission divided into a list of sequenced events with highlighted critical events.
- Each event in the list is scored on a series of metrics: CC (command and control), SA (situation awareness), SOP (adherence to standard operating procedures), CA (combat action/battle drills), and TP (overall team performance).
- The event list can be sorted by score, allowing rapid identification of the most serious issues.
- The lower portion of the AAR tool shows a mission timeline linked to the event list, with facility to play audio files and view an ASR transcript of each event.

Overall, the results from the DARCAAT project illustrate that performance measures can be automatically and accurately generated from communication in teams performing in multiuser virtual and live environments. These performance measures can then be incorporated into visualization and training tools that permit trainers to monitor and assess team status in real time.

CONCLUSIONS

Communication is the glue that holds teams together in networked virtual environments. It is also one of the richest sources of information about the performance of the team. The content and patterns of a team's communication provide a window into performance and cognitive states of the individuals and the team as a whole. Analysis of the complex cascades of communication requires tools that can assess both the content and patterns of information flowing in the network. The approach described in this chapter can automatically convert the communication into specific metrics of performance thereby permitting a better picture of the state of teams in virtual environments at any point in time. The tools use language technologies to analyze the content of communication thereby permitting characterization of the topics and quality of information being transmitted.

The toolkit allows the analysis and modeling of both objective and subjective performance metrics, and it is able to work with large amounts of communication data. Indeed, because of its machine learning foundation, it works better with more data. The toolkit can automatically extract measures of performance by modeling how subject matter experts have rated similar communication in similar situations, as well as modeling objective performance measures. Further, because the methods used are automatic and do not rely on any hand-coded models, they allow performance models to be developed without the extensive effort typically involved in standard task analysis or cognitive modeling approaches. Notably, the approach can be integrated with traditional assessment methods to develop objective and descriptive models of distributed team performance. Overall, the toolset has the ability to provide near real time (within seconds) assessment of team performance, including measures of situation understanding, knowledge gaps, workload, and detection of critical incidents. It can be used for tracking teams' behaviors and cognitive states, for determining appropriate feedback, and for automatically augmenting after action reviews.

New Directions

There remain a number of challenges to incorporating automated analysis of the content of communication into full-scale virtual environments for training venues. First, virtual environments must provide technology to allow easy collection of communication data to allow analysis by toolsets. In addition, virtual environments need to make log files of participant actions, locations, and movements easily accessible so that tools can derive and analyze additional performance measures. Second, while the results described in this chapter use teams ranging in size from 3 to 70 soldiers, it is important to understand the challenges of scaling up to even larger operations. Finally, a number of other technologies can be included to improve and help generalize modeling performance. These include better modeling of network structures, incorporation of additional modalities of information (for example, event and action information), improved computational modeling tools, and leveraging of other advances in measuring performance in complex virtual environments.

The automated analysis of communication can be applied in a wide range of virtual environment applications beyond those described here. This approach can be integrated into and make possible adaptable training systems that automatically adjust the level of difficulty of the training based on performance of the team. Finally, the overall approach helps in understanding the role of communication in complex human networks. Results from analyses of teams in realistic situations can help clarify both how communication affects team performance and how performance is reflected through communication.

ACKNOWLEDGMENTS

This work was supported in part by grants and contracts from DARPA, the U.S. Army Research Institute, the U.S. Army Research Laboratory, the Office of Naval Research, and the Air Force Research Laboratory. The authors are grateful for the contributions of Terry Drissell, Marita Franzke, Brent Halsey, Kyle Habermehl, Tim McCandless, Chuck Panaccione, Manju Putcha, and David Wroblewski for development and data analyses.

REFERENCES

Brannick, M. T., Salas, E., & Prince, C. (1997). *Team performance assessment and measurement: Theory, methods, and applications.* Mahwah, NJ: LEA.

Cannon-Bowers, J. A., Tannenbaum, S. I., Salas, E., & Volpe, C. E. (1995). Defining team competencies and establishing team training requirements. In R. Guzzo & E. Salas

(Eds.), *Team effectiveness and decision making in organizations* (pp. 330–380). San Francisco: Jossey-Bass.

Curtis, M. T., Harper-Sciarini, M., DiazGranados, D., Salas, E., & Jentsch, F. (2008). Utilizing multiplayer games for team training: Some guidelines. In H. F. Oneil & R. S. Perez (Eds.), *Computer games and team and individual learning* (pp. 145–165). Oxford, United Kingdom: Elsevier.

Diller, D. E., Roberts, B., Blankenship, S., & Nielsen, D. (2004). DARWARS Ambush!—Authoring lessons learned in a training game. *Proceedings of the Interservice/Industry Training, Simulation and Education Conference.* Arlington, VA: National Training Systems Association.

Diller, D. E., Roberts, B., & Willmuth, T. (2005, September). *DARWARS Ambush! A case study in the adoption and evolution of a game-based convoy trainer with the U.S. Army.* Paper presented at the Simulation Interoperability Standards Organization, Orlando, FL.

Foltz, P. W. (2005). Tools for enhancing team performance through automated modeling of the content of team discourse. *Proceedings of the HCI International Conference.* Saint Louis, MO: Mira Digital Publishing.

Foltz, P. W., Laham, R. D., & Derr, M. (2003). Automated speech recognition for modeling team performance. *Proceedings of the 47th Annual Human Factors and Ergonomic Society Meeting.* Santa Monica, CA: Human Factors and Ergonomics Society.

Foltz, P. W., Martin, M. A., Abdelali, A., Rosenstein, M. B., & Oberbreckling, R. J. (2006). Automated team discourse modeling: Test of performance and generalization. *Proceedings of the 28th Annual Cognitive Science Conference.* Bloomington, IN: Cognitive Science Society.

Freeman, J., Diedrich, F. J., Haimson, C., Diller, D. E., & Roberts, B. (2003). Behavioral representations for training tactical communication skills. *Proceedings of the 12th Conference on Behavior Representation in Modeling and Simulation.* Scottsdale, AZ.

Gorman, J., Weil, S. A., Cooke, N., & Duran, J. (2007). Automatic assessment of situation awareness from electronic mail communication: Analysis of the Enron dataset, *Proceedings of the Human Factors and Ergonomics Society 51st Annual Meeting* (pp. 405–409). Santa Monica, CA: Human Factors and Ergonomics Society.

Gorman, J. C., Foltz, P. W., Kiekel, P. A., Martin, M. A., & Cooke, N. J. (2003). Evaluation of Latent Semantic Analysis-based measures of communications content. *Proceedings of the 47th Annual Human Factors and Ergonomic Society Meeting.* Santa Monica, CA: Human Factors and Ergonomics Society.

Hussain, T. S., Weil, S. A., Brunyé, T. T., Sidman, J., & Alexander, A. L., & Ferguson, W. (2008). Eliciting and evaluating teamwork within a multi-player game-based training environment. In H. F. Oneil & R. S. Perez (Eds.), *Computer games and team and individual learning* (pp. 77–104). Oxford, United Kingdom: Elsevier.

Jurafsky, D., & Martin, J. (2000). *Speech and language processing: An introduction to natural language processing, computational linguistics, and speech recognition.* New York: Prentice Hall.

Kiekel, P.A., Gorman, J. C., & Cooke, N. J. (2004). Measuring speech flow of co-located and distributed command and control teams during a communication channel glitch. *Proceedings of the Human Factors and Ergonomics Society 48th Annual Meeting.*

Landauer, T. K., Foltz, P. W., & Laham, D. (1998). An introduction to Latent Semantic Analysis. *Discourse Processes, 25*(2&3), 259–284.

LaVoie, N., Psotka, J., Lochbaum, K. E., & Krupnick, C. (2004, February). *Automated tools for distance learning.* Paper presented at the New Learning Technologies Conference, Orlando, FL.

LaVoie, N., Streeter, L., Lochbaum, K., Wroblewski, D., Boyce, L., Krupnick, C., & Psotka, J. (in press). Automating expertise in collaborative learning environments. *Journal of Asynchronous Learning Networks.*

Lochbaum, K., Streeter, L., & Psotka, J. (2002, December). *Exploiting technology to harness the power of peers.* Paper presented at the Interservice/Industry Training, Simulation and Education Conference, Orlando, FL.

Orsanu, J., & Salas, E. (1993). Team decision making in complex environments. In G. A. Klein, J. Orsanu, R. Calderwood, & C. E. Zambok (Eds), *Decision making in action: Models and methods* (pp. 327–345). Norwood, NJ: Ablex Publishing.

Paris, C. R., Salas, E., & Cannon-Bowers, J. A. (2001). Teamwork in multi-person systems: A review and analysis. *Ergonomics, 43*(8), 1052–1075.

Pierce, L., Sutton, J., Foltz, P. W., LaVoie, N., Scott-Nash, S., & Lauper, U. (2006, July). *Technologies for augmented collaboration.* Paper presented at the CCRTS, San Diego, CA.

Salas, E., & Cannon-Bowers, J. A. (2001). The science of training: A decade of progress. *Annual Review of Psychology, 52,* 471–499.

Chapter 18

A PRIMER ON VERBAL PROTOCOL ANALYSIS

Susan Trickett and J. Gregory Trafton

We have been using verbal protocol analysis for over a decade in our own research, and this chapter is the guide we wish had been available to us when we first started using this methodology. The purpose of this chapter is simply to provide a very practical, "how-to" primer for the reader in the science—and art—of verbal protocol analysis.

It is beyond the scope of this chapter to discuss the theoretical grounding of verbal protocols; however, there are several resources that do so, as well as providing some additional practical guidelines for implementing the methodology: Ericsson and Simon (1993); van Someren, Barnard, and Sandberg (1994); Chi (1997); Austin and Delaney (1998); and Ericsson (2006). In addition, numerous studies—too many to list here—have effectively used verbal protocols, and these may provide additional useful information about the application of this technique.

Briefly put, verbal protocol analysis involves having participants perform a task or set of tasks and verbalizing their thoughts ("talking aloud") while doing so. The basic assumption of verbal protocol analysis, to which we subscribe, is that when people talk aloud while performing a task, the verbal stream functions effectively as a "dump" of the contents of working memory (Ericsson & Simon, 1993). According to this view, the verbal stream can thus be taken as a reflection of the cognitive processes in use and, after analysis, provides the researcher with valuable information about not only those processes, but also the representations on which they operate. In addition, verbal protocols can reveal information about misconceptions and conceptual change, strategy acquisition, use, and mastery, task performance, affective response, and the like.

Verbal protocol analysis can be applied to virtual environments in several ways. For example, verbal protocols collected during performance by both experts and novices can inform the design of virtual environments to be used for training. By comparing expert and novice performance, a designer could identify specific areas where learners would benefit from support, remediation, or instruction. Similarly, having experts talk aloud while performing a task could help a curriculum designer identify and develop the content to be delivered in the virtual environment. Verbal protocols are likely to be especially useful in evaluation—first, in determining how effective the virtual environment is as a training tool, and second, in assessing the student's learning and performance, where verbal protocols can provide valuable information about the learner's cognitive processes, beyond simple measures of accuracy and time on task.

Before explaining the "nuts and bolts" of verbal protocol analysis, we should note that there are several caveats to using this method. First, it is important to distinguish between *concurrent* and *retrospective* verbal protocols. Concurrent protocols are delivered at the same time as the participant performs the task and are ideally unprompted by the experimenter. Retrospective protocols are provided *after* the task has been completed in response to specific questions posed by the experimenter, such as "How did you solve this problem, and why did you choose that particular strategy?" The chief problem with retrospective protocols is that people do not necessarily have access either to what they did or why they did it (see Nisbett and Wilson, 1977, for a full discussion of this issue; Ericsson and Simon, 1993, for an overview of research on retrospective versus concurrent protocols; and Austin and Delaney, 1998, for a discussion of when retrospective protocols may be reliably used).

Second, it is important to differentiate between having a participant think aloud while problem solving and having him or her describe, explain, or rationalize what he or she is doing. The main problem with participants providing explanations is that this removes them from the *process* of problem solving, causing them to think about what they are doing rather than simply doing it, and as a result, such explanations may change their performances. Ericsson and Simon (1993) provide a complete discussion of the effects of such "socially motivated verbalizations" on problem solving behavior, and later in this chapter we suggest ways to reduce the likelihood that they will happen.

Third, it is important to be aware of the potential for subjective interpretation at all stages of collecting and analyzing verbal protocol data, from interpreting incomplete or muttered utterances to assigning those utterances a code. The danger is that a researcher may think he or she "knows" what a participant intended to say or meant by some utterance and thus inadvertently misrepresent the verbal protocol. We will describe ways to handle the data that reduce the likelihood of researcher bias affecting interpretation of the protocols, and we will suggest some methodological safeguards that can also help minimize this risk.

Another important consideration for the researcher is the level of time commitment involved in verbal protocol studies. Simply put, this type of research involves a serious investment of time and resources, not only in collecting data (resources in terms of experimenters, participants, and equipment), but also in transcribing, coding, and analyzing data. It is simply not possible for one researcher to manage all the tasks associated with verbal protocol analysis of a complex task/domain with many participants. Furthermore, both collecting and coding the data require a high level of training, and thus it is important that members of the research team are willing to make a commitment to complete the project. A coder who quits midstream causes a major setback to the research, because a new recruit must be found and trained before the work can continue.

Verbal protocol data are "expensive data." Depending on the task to be performed, experimental sessions generally last one or two hours, and participants must usually be "run" singly. In some studies, participants in an undergraduate psychology pool are not suitable candidates for the research; thus, time must be spent identifying and recruiting appropriate participants. In the case of expert studies or "real world" performance, data collection often involves travel to the participant's work site and may require coordination with supervisors and coworkers, or even special permission to enter a site. Those who agree to participate in the research usually must give up valuable work time, and schedule changes or other unexpected occurrences may prevent their participation at the last minute. Thus, perhaps even more than in laboratory research, it is crucial to plan as much as possible and to take into account many contingencies that can arise. Such contingencies range from the mundane (for example, triple-checking that the equipment is functioning properly, that backup batteries are available, and that all the relevant forms and materials are in place) to the unpredictable (for example, having backup plans for cancellation or illness).

Despite these caveats, we believe that collecting verbal protocol data is a flexible methodology that is appropriate for research in any research endeavor for which the goals are to understand the *processes* that underlie performance. We think it will be especially fruitful in the domain of virtual reality (VR)/virtual environments (VEs). Not much protocol analysis has been done in this area; however, the technique offers many insights into cognitive processing and is likely to be very useful in comparing VR/VEs to real world performance, for example. Researchers involved in education and training, including VE training, may have many goals—for example, understanding how effective a given training program is, what are its strengths and weaknesses, how it can be improved, how effectively students use the system, whether students actually learn what the instruction is designed to teach, what difficulties they have in mastering the material to be learned, what difficulties they have using the delivery system, and so on. Outcome measures, such as time on task or accuracy on a knowledge assessment test, provide only partial answers and only to some of these questions. Process measures, on the other hand, provide a much richer, more detailed picture of what occurs during a learning or problem solving session and allow a much finer-grained analysis of both learning and performance. Verbal protocols are the raw data from which extremely useful process data can be extracted. Although our emphasis in this chapter is only on verbal protocol data, combining different types of process data (such as verbal protocols, eye-track data, and mouse-click data) can create an extremely powerful method to understand not only *what* is learned, but also *how* learning occurs.

DATA COLLECTION

A number of issues related to data collection should be addressed before the first participant even arrives! Several confidentiality issues surround the act of collecting data, which stem from the potentially sensitive nature of video and/or audio recording participants as they perform tasks. First, in obtaining informed consent, it is crucial that participants fully understand the recording process and have a specific option to agree or disagree to have their words, actions, and bodies recorded. Thus, consent forms need to contain two lines, as follows:

____ I agree to have my voice, actions, and upper body videotaped.

____ I do not agree to have my voice, actions, and upper body videotaped.

Of course, the actual wording will depend on the scope of recording planned—it might additionally include gestures or walking patterns, for example.

Second, agreement to have one's words and actions recorded should not be taken as agreement to have one's words and actions shared with others. Although participants might be quite willing to participate in a study and to be recorded while doing so, they might be uncomfortable at the thought of their data being shared, for example, at conferences and other research presentations. If the researcher is planning to use video clips as illustrations during research talks, it is important to have an additional section of the consent form that specifically asks participants whether or not they agree to have their data shared. They have the right to refuse, and this right must be honored. As mentioned earlier, this does not necessarily disqualify participants from participating in the study; it just means that a participant's video cannot be shared with others. The third issue relating to confidentiality is that of storing the data. It is especially important to reassure participants that the data will be stored confidentially—that data will be identified by a number rather than a participant's name, that data will be locked away or otherwise secured, and that access to the data will be strictly limited to members of the research team.

One of the advantages of verbal protocol data is its richness; the downside of this richness is that the data quickly become voluminous. Even a relatively short session can result in pages of transcribed protocol that must be coded and analyzed. In addition, participants, such as experts, may be scarce and difficult to recruit. For these reasons, verbal protocol studies generally have fewer subjects than other kinds of psychological studies. How many participants are needed? The answer depends on the nature of the study. We (and others) have had as few as one person participate in a particular study. Such single-subject case studies are extremely useful for generating hypotheses that can then be tested experimentally on a larger sample. In general, however, we have found that the more expert the participants, the less variance there is in relevant aspects of performance. With less expert participants, we generally find slightly higher numbers of participants work well. Ideally, we try to include 5 to 10 novices, although this is not always possible. Studies with larger numbers of participants generally involve less exploratory research, for which a coding scheme is already well established and for which only a subset of the data needs to be coded and analyzed. As with all research, the precise number of participants will rest on the goals of the study (exploratory, confirmatory, case study, and so forth), the nature of the participants (experts, novices, or naive subject pool participants), and practical considerations concerning recruitment and accessibility of participants.

COLLECTING DATA

Adequate recording equipment is a must. Although we have occasionally used audio recording alone, we have found that the additional information contained in video recording can be very valuable when it comes to analyzing the data. Furthermore, we have found that having more than one camera is often worthwhile, for several reasons. First, an additional camera (or cameras) provides an automatic backup in the case of battery or other equipment failure. In addition, a second camera can be aimed at a different area of the experimental setup, thus reducing the need for the experimenter to attempt to follow the participant around while he or she performs the task. In some cases, we have even used three cameras, as additional cameras recording from different angles allow the experimenter to reconstruct more precisely what was happening. We are also somewhat extravagant in our use of videotapes. Even when a session is quite short and leaves a lot of unused space on a tape, we use a new videotape for each participant or new session. Especially with naturalistic tasks, it is difficult to anticipate how long a participant will take. Some people work quite slowly, and underestimating the time they will take can lead to having to change a tape mid-session. Our view is that videotapes and batteries are relatively cheap, whereas missing or lost data are extremely expensive. In short, we have found that thinking ahead to what data we *might* need has helped us to have fewer regrets once data collection is completed.

The most critical aspect of verbal protocol data is the sound, and a high quality lapel microphone is well worth the cost. Many participants are inveterate mutterers, and most participants become inaudible during parts of the task. These periods of incoherence often occur at crucial moments during the problem solving process, and so it is important to capture as much of what may be transcribed as possible. The built-in microphone on most cameras is simply not powerful enough to do so. Using a lapel microphone also allows the researcher to place the camera at a convenient distance behind the participant, capturing both the sound and the participant's interactions with experimental materials but not his or her face. Attaching an external microphone (such as a zoom or omnidirectional microphone) to any additional cameras that are used will allow the second, or backup, camera to capture sound sufficiently well in most cases, although muttering or very softly spoken utterances may be lost. Having such a reliable backup sound recording system is always a good idea.

It is crucial to begin the session with good batteries and to check, before the participant begins the task, that sound is actually being recorded. The experimenter should also listen in during the session to make sure that nothing has gone amiss. Although this point may seem obvious, most researchers who collect verbal protocols have had the unhappy experience of reviewing a videotape only to find there is no sound. We hope to spare you that pain!

As with any psychological research, participants should be oriented to the task before beginning the actual study. In the case of verbal protocols, this orientation includes training in actually giving verbal protocols. We have a standard training procedure that we use with all participants.[1] During the task, if a participant is silent for more than three or four seconds, it is important to jump in with a prompt. The researcher should simply say, "Please keep talking" or even "Keep talking." More "polite" prompts, such as "What are you thinking?" are actually detrimental, because they invite a more social response from the participant (such as, "Well, I was just wondering whether . . .") and thus remove the participant from the process of problem solving into the arena of social interaction.

Occasionally, a participant will be unable to provide a verbal protocol, and this difficulty generally manifests itself in one of two ways: either the participant remains essentially silent throughout the task, talking aloud only when prompted and then lapsing back into silence, or the participant persists in explaining what he or she is doing or about to do. In the first case, there is very little the experimenter can do, and depending on the situation, it may be better to bail out rather than subject the participant to this uncomfortable situation, since the data will be unusable in any case. One way to help prevent this from happening is to ensure beforehand that participants are fully at ease in whatever language the protocol is to be given. In the second case, the experimenter must make a judgment call as to whether to pause the session and provide some retraining. Retraining will be disruptive and may make the participant even more nervous; however, it is unlikely that a participant will change strategy midstream. In either case, protocols that consist mostly of explanations cannot be used. These problems should not be confused with protocols that contain a great deal of muttering, broken grammatical structures, or incomplete or incoherent thoughts. Although a challenge to the transcriber and the coder, such protocols are quite acceptable and, in fact, represent "good" protocols, since people rarely think in complete sentences.

The objective during verbal protocol collection is to keep the participant focused on the task, without distraction. One obvious way to do this is to make sure that the room in which the session is taking place is not in a public place and that a sign on the door alerts others not to interrupt. Less obviously, it means that the experimenter should resist the urge to ask questions during the session. If domain related questions do arise, the experimenter should make a note of them and ask after the session is completed. Querying participants' explicit knowledge after task completion does not jeopardize the integrity of the verbal protocols, because it taps into stable domain knowledge that is not altered by the retrospective process. Asking questions during problem solving, however, is disruptive and may alter the participants' cognitive processes.

[1]A copy of this procedure is available at http://www.nrl.navy.mil/aic/iss/aas/cog.complex.vis.php

PROCESSING THE DATA

Once data collection is completed, verbal protocols must be transcribed and coded before data analysis can begin. Unfortunately, there are no consistently reliable automated tools to perform transcription. Fortunately, transcription can be carried out by less-skilled research assistants, with a couple of caveats. First, transcribers should be provided with a glossary of domain relevant terms; otherwise, when they encounter unfamiliar words, they are likely to misinterpret what is said. Second, transcribers should be trained in segmenting protocols. Both these steps will reduce the need for further processing of the transcriptions before coding can begin.

How protocols are segmented depends on the grain size of the analyses to be performed and will therefore vary from project to project. Careful thought should be given to the appropriate segmentation, because this process will affect the results of the analyses. It is beyond the scope of this chapter to illustrate in detail the many options for segmenting protocols; however, Chi (1997) provides a thorough discussion of different methods and levels of segmentation and their relationships with the type of analyses to be performed.

In general, we have found that segmenting according to complete thought provides us with the flexibility to code data on a number of different dimensions. By "complete thought" we basically mean a clause (a subject and a verb), although given the incomplete nature of human speech, utterances do not always fall neatly into a clausal structure. However, we have had excellent agreement among transcribers by using the complete thought rubric. Table 18.1 shows an example

Table 18.1. Example of Segmenting by Complete Thought

Utterance
OK, 0Z surface map
shows a low pressure in the Midwest moving up
there's a little bubble high ahead of it
that's pretty much stacked well with 500 millibars with another low
so that's looking at severe weather in the Midwest, easily
it's at 700 millibars
just shows increasing precipitation
clouds are going to be increasing for the next 24 hours in western Ohio
Oh, what level are you?
850
they're kind of in a little bit of a bubble high
so 0Z looking at low cloud cover
no precipitation at all

of this segmentation scheme from the meteorology domain. As the illustration shows, the length of each utterance can vary from a single word(s) (for example, "850") to fairly lengthy (for example, "clouds are going to be increasing for the next 24 hours in western Ohio"). In this domain, we are generally interested in the types of mental operation participants perform on the visualizations, such as reading off information, transforming information (spatially or otherwise), making comparisons, and the like, and these map well to our segmentation scheme. The utterances "it's at 700 millibars" and "clouds are going to be increasing for the next 24 hours in western Ohio" would each be coded as one "read-off information" event, even though they differ quite a bit in length. Of course, it would be possible to subdivide longer utterances further (for example, "clouds are going to be increasing // for the next 24 hours // in western Ohio"); however, according to our coding scheme this is still one read-off information event, which would now span three utterances. Having a single event span more than one utterance makes data analysis harder. The two most important guidelines in dividing protocols are (1) consistency and (2) making sure that the segments map readily to the codes to be applied.

Although transcription cannot be automated, a number of video analysis software programs are available to aid in data analysis, and some of these programs allow protocols to be transcribed directly into the software. We have not found the perfect protocol analysis program yet, though we have used MacShapa, Transana, and Noldus Observer. Looking ahead and deciding what kind of analysis software will be used will also save a great deal of time by reducing the need for further processing of transcripts in order to successfully import them to the video analysis software. Another option that we have successfully used is to transcribe protocols in a spreadsheet program, such as Excel, with each segment on a different row. We then set up columns for our different coding categories and can easily perform frequency counts, create pivot tables, and import the data into a statistical analysis program. This method works very well if the video portion of the protocol is irrelevant; however, in most cases we are interested not only in what people are saying, but what they are doing (and looking at) while they are saying it. In this case, using the spreadsheet method is disadvantageous, because the transcription cannot easily be aligned with the video. Video analysis software has the advantage that video timestamps are automatically entered at each line of transcription, thus facilitating synchronizing text and video.

Once data have been transcribed and segmented, they are ready for coding. Unless the research question can be answered by some kind of linguistic coding (for example, counting the number of times a certain word or family of words is used), a coding scheme must be developed that maps to the cognitive processes of interest. Establishing and implementing an effective and reliable coding scheme lies at the heart of verbal protocol analysis, and this is usually the most difficult and time consuming part of the whole process. In some cases, researchers will have strong a priori notions about what to look for. However, verbal protocol analysis is a method that is particularly useful in exploratory research, and in these cases, researchers may approach the protocol with only a general idea.

Regardless of the stage of research, coding schemes must frequently be elaborated or refined to match the circumstances of a given study.

To illustrate this point, consider a study we performed to determine how meteorologists handle uncertainty in the forecasting task and in the visualizations they use. Our hypothesis was that when uncertain, they would use more spatial transformations than when certain. Our coding scheme for spatial transformations was already in place. However, we had several possible ways that we could code uncertainty. One option was to code it linguistically. We developed a range of linguistic pointers to uncertainty, such as markers of disfluency (um, er, and so forth), hedging words (for example, "sort of," "maybe," and "somewhere around"), explicit statements of uncertainty (for example, "I have no idea" and "what's that?"), and the like. Although this scheme proved highly reliable, with very little disagreement between coders, it turned out to be less useful for the research question we were trying to answer. A forecaster could be highly *uncertain* overall, but because he or she had no doubt about that uncertainty, the utterances contained few, if any, linguistic markers of uncertainty. Table 18.2 shows an example of this mismatch. The participant was trying to forecast precipitation for a particular period, and the two models (ETA and Global Forecast System [GFS]) she was using disagreed. Model disagreement is a major source of uncertainty for forecasters, so we were confident that at a more "global" level, the participant was highly uncertain at this juncture. Her uncertainty was also supported by her final utterance (that the forecast was going to be hard). However, the

Table 18.2. Mismatch between Linguistic and Global Coding Schemes: Uncertain Episode but Certain Utterances

Utterance	Linguistic Coding of Uncertainty	Global Coding of Uncertainty
OK, now let's compare to the ETA . . . and they differ Interesting Well, they don't differ too much they differ on when the precipitation's coming Let's go back to 36 hours OK, 42, 48, 54 Yeah, OK, so they have precip coming in 48 hours from now Let me try to go back to GFS and go to 48 hours and see what they have Well, OK, well they don't differ Well, yeah, cause they're calling for more so maybe this is gonna be a hard precip forecast	No linguistic uncertainty expressed in any utterance	Model disagreement indicates overall uncertainty (explicitly articulated in last utterance at left)

individual utterances that comprise this segment express no uncertainty on a linguistic level. Our hypothesis about the use of spatial transformations mapped to a more global concept of uncertainty than could be captured by this linguistic coding scheme. By using contextual clues (the model disagreement, the participant's dithering between whether the models disagreed or not, and her frustration at the difficulty of the forecast), we were able to correctly code this sequence as an episode of uncertainty.

Conversely, a forecaster could be overall highly *certain* about something, but use a lot of uncertain linguistic markers. Table 18.3 illustrates this aspect of the problem.

We solved our coding problem by dividing the protocol into segments, or episodes, based on participants' working on a particular subtask within the general forecasting task, such as figuring out the expected rainfall or temperature for a certain period, or resolving a discrepancy between two models. We then coded each episode as *overall* certain, uncertain, or mixed. Tables 18.2 and 18.3 illustrate uncertain and certain episodes, respectively. Using this more global coding of uncertainty, we could evaluate the use of spatial transformations within certain and uncertain episodes. This level of coding uncertainty was a much better match for our purposes, because spatial transformations occur within a broader context of *information* uncertainty than can be captured by mere expressions of *linguistic* uncertainty.

Unfortunately, it is often only when the scheme is applied that its weaknesses emerge; consequently, developing a coding scheme is a highly iterative process. Generally, our procedure is to begin with the smallest possible number of codes, which are then described in rather broad terms. We then attempt to apply the coding scheme to a small portion of the data. At this stage, we usually find problems

Table 18.3. Certain Episode but Uncertain Utterances (Linguistic Markers of Uncertainty in Italics)

Utterance	Linguistic Marker of Uncertainty	Global Coding of Uncertainty
Look at the AVN MOS . . . and they're in agreement say it gets down to 43 degrees tonight *um,* scattered clouds, *it may be* partly cloudy *uh,* tomorrow 72 . . . the other model had upper 60s, lower 70s that's *about right* *uh,* Wednesday night . . . clouds *I think* it said getting *around,* down, *around* 52 Thursday, 64 *maybe*	um, it may be uh about right uh I think, around around maybe	Models agree, indicating forecaster is certain about the temperature, explicitly articulated in utterance "That's about right"

coding specific utterances. It is important to keep very careful, detailed notes during this phase as to the nature of the coding problem. We can then look for patterns in the problematic segments, which often leads to subdividing or otherwise revising a particular code.

There are two equally important goals to keep in mind in developing a coding scheme. First, the scheme must be *reliable;* that is, it should be sufficiently precisely delineated that different coders will agree on the codes applied to any given utterance. Second, the scheme must be *useful;* that is, it must be at the right "level" to answer the question of interest. The linguistic scheme for coding uncertainty described above was highly reliable, but did not answer our need. The episodic coding scheme captured the relevant level of uncertainty, but required several refinements before we were confident that it was reliable. First, we had to obtain agreement on our division of the protocols into episodes, and then we had to obtain agreement on our coding of each episode as certain, uncertain, or mixed.

Obtaining agreement, or establishing *inter-rater reliability,* is essential in order to establish the validity of a coding scheme (see Cohen, 1960, for a discussion of establishing agreement). Once a coding scheme appears to be viable, the next step is to have two independent coders code the data and then compare their results. Obviously, coders must be well trained prior to this exercise. We establish a set of training materials, based on either a subset of the data, or, preferably, from a different dataset to which the same coding scheme can be applied. Using a different dataset reduces bias when it comes to the final coding of the data, because the coders are not prejudiced by previously having seen and worked with—and having discussed—the transcripts. The training materials consist of a set of written instructions describing each of the codes and a set of examples and nonexamples of the code. We find examples that are as clear and obvious as possible, but we also use "rogue" examples that meet some aspect of the coding criteria, but are nonetheless *not* illustrations of the code. These "near-misses" help define the boundaries of what does and does not count as a particular example of a code. Beside each example or nonexample, we provide an explanation of the coding decision. Once the coders believe they understand the coding scheme, they code a small subset of the data, and we meet to review the results. Once again, it is important for the coders to keep detailed notes about issues they encounter during this process. The more specific the coder can be, the more likely the fundamental cause of the coding problem can be addressed and resolved.

Initially, we look simply at points of agreement and disagreement in order to get a general sense of how consistently the codes are being applied. Although it is tempting to ignore points of agreement, we find it useful to look at individual coders' reasonings in these instances in order to make sure that agreement is as watertight as possible. Points of disagreement, of course, provide a wealth of useful information in refining and redefining the coding scheme. Often, resolution is simply a matter of providing a crisper definition of the code. Sometimes, however, resolution involves subdividing one coding category into two or collapsing two separate codes into one when the distinction proves blurry or otherwise not useful. The purpose of this stage is to identify and resolve the coders' uncertainties about the coding scheme in order that inter-rater reliability (IRR) might be established.

In order to obtain IRR, one coder codes the entire dataset, and the second coder codes a subset of the data. How much double-coding must be done depends in some measure on the nature of the transcripts and the coding scheme. The more frequently the codes occur, the less double-coding is necessary—usually 20–25 percent of the data is sufficient. For codes that occur rarely, a larger portion of the data must be double-coded in order to have sufficient instances of the code. If a code is infrequent, it is less likely to occur in any given subset of the data, and coders are more likely to be in agreement. However, they are agreeing that the code did *not* occur, which does not speak to the consistency with which they identify the code. After both coders have completed this exercise, we investigate their level of agreement.

There are two approaches to establishing agreement. The first is simply to count the number of instances in which coders agreed on a given code and calculate the percent agreement. This approach, however, generally results in an inflated measure of agreement, because it does not take into account the likelihood of agreement by chance. A better measure of agreement is Cohen's kappa (Cohen, 1960). This involves constructing a contingency table and marking the number of times the coders agreed on the code's occurrence or nonoccurrence[2] and the number of instances in which each coder said the code occurred but the other coder did not. Even if codes are scarce and the majority of instances fall into the "no-occurrence" cell, Cohen's kappa makes the appropriate adjustment for agreement by chance. Cohen's kappa is easily calculated as follows:

$$\kappa = (Po - Pc)/(1 - Pc)$$

Po is the proportion of observed agreement, and *Pc* is the proportion of agreement predicted by chance. Some researchers define poor reliability as a kappa of less than 0.4, fair reliability as 0.4 to 0.6, good reliability as 0.6 to 0.8, and excellent as greater than 0.8. We believe that kappa values lower than 0.6 indicate unacceptable agreement and that in such cases the coding scheme must undergo further revision.

A low value for kappa need not be considered a deathblow to the coding scheme. Most coding schemes go through a cycle of construction, application, evaluation, and revision several times before they can be considered valid. Although the process can be tedious, it is a vital part of using verbal protocols, because of the dangers of subjectivity and researcher bias raised earlier in this chapter. It should also be noted that coding schemes will most likely never achieve perfect agreement, nor is perfect agreement necessary. When coders disagree, there are two options. First, all disagreements can be excluded from analysis. The advantage of this approach is that the researcher can be confident that the

[2]Nonoccurrence is coded in addition to occurrence in order for coding to be symmetrical. Overlooking agreement about nonoccurrence can lead to a lower estimate of agreement than is warranted. This is especially an issue if instances of a code are relatively rare.

included data are reliably coded. The disadvantage is obviously that some valuable data are lost. A second option is to attempt to resolve agreements by discussion. Often, for example, one coder may simply have misunderstood some aspect of the transcript; in such cases, the coder may decide to revise his or her original coding. If, after discussion, coders still disagree, these instances will have to be excluded from analysis. However, the overall loss of data using this method will be less than if disagreements are automatically discarded. Either method is acceptable, although it is important to specify how disagreements were handled when presenting the results of one's research. After IRR is established and all the data are coded, the analysis proceeds as with any other research, using quantitative statistical analyses or qualitative analytical methods, depending on the research questions and goals.

PRESENTING RESEARCH RESULTS

There are some specific issues in presenting research based on verbal protocol analysis. Although the use of verbal protocols has been established as a sound method and has therefore gained increased acceptance in the last few years, such research may nevertheless be difficult to publish. Some reviewers balk at the generally small sample sizes. One approach to the sample size issue is to conduct a generalizability analysis (Brennan, 1992), which can lay to rest concerns that variability in the data is due to idiosyncratic individual differences rather than to other, more systematic factors.

Another concern is the possibility of subjectivity in coding the data and the consequential danger that the results will be biased by the researcher's predilection for drawing particular conclusions. There are two ways to address this concern. The first is to take extreme care not only in establishing IRR, but also in describing how the process was conducted, making sure, for example, that a sufficient proportion of the data was double-coded and that this is reported. Specifying how coders were trained, how coders worked independently, without knowledge of each other's codings, and how disagreements were resolved can also allay this concern. The second action is to provide very precise descriptions of the coding scheme, with examples that are crystal clear, so that the reader feels that he or she fully understands the codes and would be able to apply them. Although this level of detail consumes a great deal of space, many journals now have Web sites where supplementary material can be posted. When describing a coding scheme, good examples may be worth many thousands of words!

Even with these safeguards, research using verbal protocols may still be difficult to publish, especially if it is combined with an in vivo methodology. In vivo research is naturalistic in that it involves going into the environment in which participants normally work and observing them as they perform their work. The stumbling block to publication is that the research lacks experimental control. However, the strength of the method is precisely that it is less likely to change people's behavior, as imposing experimental controls may do. We have found that an excellent solution to this vicious cycle is to follow our in vivo studies with controlled laboratory experiments that test the conclusions we draw from the in vivo work. In one project, we observed several expert scientists in a variety of domains as they analyzed their scientific data. Our observations suggested that they used a type of mental model we called "conceptual simulation" and that they did so especially when the hypothesis involved a great deal of uncertainty. However, our conclusions were correlational; in order to test whether the uncertainty was causally related to conceptual simulation, we conducted an experimental laboratory study in which we manipulated the level of participants' uncertainty. In this type of follow-up study, it is especially important to design a task and materials that accurately capture the nature of the original task in order to elicit behavior that is as close as possible to that observed in the natural setting.

CONCLUSION

We have found verbal protocol analysis an invaluable tool in our studies of cognitive processes in several domains, such as scientific reasoning, analogical reasoning, graph comprehension, and uncertainty. Our participants have ranged from undergraduates in a university subject pool, to journeymen meteorology students, to experts with advanced degrees and decades of experience. Although we have primarily used verbal protocol analysis as a research tool, we believe that it can be used very effectively in applied settings, such as the design, development, and evaluation of virtual environments for training, where it is important to discern participants' cognitive processes.

There are other methods of data collection that also require participants' verbal input, such as structured and unstructured interviews, knowledge elicitation, and retrospective protocols. These methods assume that people have accurate access to their own cognitive processes; however, this is not necessarily the case. Consider, for example, the difficulty of explaining how to drive a stick-shift vehicle, compared with actually driving it. People frequently have faulty recall about their actions and motives or may describe what they *ought*—or were taught—to do rather than what they *actually* do. In addition, responses may be at a much higher level of detail than meets the researcher's interest, or they may skip important steps in the telling that they would automatically perform in the doing. Although in some interview settings, such as semistructured interviews, the interviewer has the opportunity to probe the participant's answers in more depth, unless the researcher already has some rather sophisticated knowledge of the task, the questions posed may not elicit the desired information.

The strength of concurrent verbal protocols is that they can be considered a reflection of the actual processes people use *as they perform a task*. Properly done, verbal protocol analysis can provide insights into aspects of performance that might otherwise remain inside a "black box" accessible only by speculation. Although they are time consuming to collect, process, and analyze, we believe that the richness of the data provided by verbal protocols far outweighs the costs. Ultimately, as with all research, the purpose of the research and the resources available to the researcher will determine the most appropriate method; however, the addition of verbal protocol analysis to a researcher's repertoire will open the door to a potentially very productive source of data that invariably yield interesting and often surprising results.

REFERENCES

Austin, J., & Delaney, P. F. (1998). Protocol analysis as a tool for behavior analysis. *Analysis of Verbal Behavior, 15,* 41–56.

Brennan, R. L. (1992). *Elements of generalizability theory* (2nd ed.). Iowa City, IA: ACT Publications.

Chi, M. T. H. (1997). Quantifying qualitative analyses of verbal data: A practical guide. *The Journal of the Learning Sciences, 6*(3), 271–315.

Cohen, J. (1960). A coefficient of agreement for nominal scales. *Educational and Psychological Measurement, 20,* 37–46.

Ericsson, K. A. (2006). Protocol analysis and expert thought: Concurrent verbalizations of thinking during experts' performance on representative tasks. In K. A. Ericsson, N. Charness, P. J. Feltovich, & R. R. Hoffman (Eds.), *The Cambridge handbook of expertise and expert performance* (pp. 223–241). New York: Cambridge University Press.

Ericsson, K. A., & Simon, H. A. (1993). *Protocol analysis: Verbal reports as data* (2nd ed.). Cambridge, MA: MIT Press.

Nisbett, R. E., & Wilson, T. D. (1977). Telling more than we can know: Verbal reports on mental processes. *Psychological Review, 84,* 231–259.

van Someren, M. W., Barnard, Y. F., & Sandberg, J. A. C. (1994). *The think aloud method: A practical guide to modeling cognitive processes.* London: Academic Press.

Part IX: Capturing Expertise in Complex Environments

Chapter 19

DEVELOPMENT OF SIMULATED TEAM ENVIRONMENTS FOR MEASURING TEAM COGNITION AND PERFORMANCE

Jamie Gorman, Nancy Cooke, and Jasmine Duran

ASSESSING TEAMS IN SIMULATED ENVIRONMENTS

From network centric warfare and intelligence analysis to emergency response and modern medical care, today's sociotechnical systems are rife with tasks comprising teams of human and machine players interacting on many different levels while working interdependently toward a common goal. Catastrophic failures in today's high technology sociotechnical systems are often due to factors in the social components of these systems (Weir, 2004). This highlights the need for a deeper scientific understanding of the social and team related causes of such failures. However, given the level of complexity of interactions among human and machine players in these environments, application of scientific principles is challenging from both traditional laboratory and naturalistic study perspectives. Through the parallel development of embedded measurement and scenarios highlighting interaction behaviors of interest, simulations provide a middle ground between the real world and the lab for advancing the scientific study of teams of human and machine players and, ultimately, an understanding of how to train or intervene in order to correct deficient team interaction behaviors.

Teams have been defined specifically as "a distinguishable set of two or more people who interact dynamically, interdependently, and adaptively toward a common and valued goal/object/mission, who have each been assigned specific roles or functions to perform, and who have a limited life span of membership" (Salas, Dickinson, Converse, & Tannenbaum, 1992, p. 4). Team cognition is the process by which the heterogeneously skilled team members interact to reason, decide, think, plan, and act as a unit. The necessity of teams for accomplishing complex tasks and an appreciation for the uniqueness of the team *as a unit* have turned attention toward team research and training. Team members are generally exposed to some individual training followed by team training either using classroom approaches, such as Crew Resource Management (Salas, Wilson, Burke, &

Wightman, 2006), and/or via exposure to the team task through multiplayer simulations or games (Cannon-Bowers, Salas, Duncan, & Halley, 1994; Salas, Bowers, & Rhodenizer, 1998). Simulations and virtual environments not only provide training solutions for teams, but also serve as testbeds for research on teams. Research conducted in these testbeds is typically more ecologically valid than the highly controlled tasks of traditional laboratory research. However, simulations and games need to be designed to exercise the targeted behavior. Further, there is little value in this technology without some means to measure and assess the constructs of interest, in this case, team performance and team cognition.

Our definition of team cognition as the process by which heterogeneously skilled team members interact to reason, decide, think, plan, and act as a unit equates interaction processes to team cognition. Because process variance is unique to teams and directly related to team performance, we will emphasize process measurement (holistic) rather than the aggregate cognition of the team members (collective). In this chapter we show how simulations can be designed to optimally exercise team cognition and how measurement can be integrated within these team simulations.

Team Cognition

Theoretically, team cognition has been viewed either as collective cognition, which is the sum of the cognitive resources and abilities each individual brings to the task (for example, Langan-Fox, Code, & Langfield-Smith, 2000), or holistically as an emergent property of the interaction processes of team members (for example, Cooke, Gorman, & Rowe, in press). Most views do not entirely exclude the influence of collective or holistic cognition, but vary more as a matter of emphasis of one over the other. Whether a collective or holistic perspective on team cognition is taken has particular implications for how team cognition is measured and the types of interventions chosen to augment team cognition.

Collective cognition may work well for homogeneous groups of individuals (for example, juries). However this perspective can be inadequate for explaining cognition in teams with highly heterogeneous team members. Alternatively, empirical evidence in the domain of heterogeneous team command and control suggests that interactions, which are the focus of the holistic perspective, are predictive of team performance, whereas collective knowledge metrics are not (Cooke et al., 2004; Cooke, Gorman, Duran, & Taylor, 2007; Cooke, Gorman, Pedersen, & Bell, 2007; Cooke, Kiekel, & Helm, 2001). Based on these findings, in this chapter we emphasize measures based on the holistic perspective of team cognition. Likewise it is important to develop simulation testbeds that are capable of exploiting team interactions for the purpose of measurement.

Scope of Team Simulation

Team simulation encompasses a variety of possible technologies. Table 19.1 lays out some of the possibilities in a matrix in which the simulation can be focused on the team members, the team task, or both.

Table 19.1. Space of Possible Training Technologies Involving Teams

	Real Team Members	Simulated Team Members
Real Team Task	Operational environment and on-the-job training	Synthetic agents and training by computer
Simulated Team Task	Games, training environments, and synthetic tasks	Synthetic tasks with one or more synthetic agents or opposition forces

As exemplified in Table 19.1, simulation technology can be used to reproduce the task (or aspects of the task) or team members and in some cases entire teams (for example, simulated opposition forces). The upper left quadrant reflects the case in which team training occurs in the actual context (on the job) with the actual team members. Due to demands of the domain for personnel and resource shortages, training often occurs in this "seat-of-the-pants" context. If we accept very high fidelity simulation as close to a "real" task, then this quadrant may also include live training exercises, such as red versus blue force exercises in the military and emergency response exercises for emergency management. In these exercises measurement is critical, and more is needed; however, it is difficult to stage such exercises, and thus it is difficult to conduct controlled research in real team task environments. The use of virtual environments, including simulations, therefore becomes perhaps the most viable option for meeting the challenge of balancing experimental control with ecological validity. In this chapter we do not address the upper left quadrant, but instead advocate the use of simulation in cases when adequate resources (for example, team members) exist.

Most of the research and development on team simulation would fall into the lower left quadrant of Table 19.1. This chapter deals primarily with the lower left quadrant of Table 19.1. This is the case in which the task or part of the task is simulated and the actual team members interact in this context for training or research purposes. Some examples include the Air Force Research Laboratory's F-16 four-ship simulator (Schreiber & Bennett, 2006), gaming (for example, multiplayer Internet video games such as *Counter-Strike*), and STEs (synthetic task environments) for research (Schiflett, Elliott, Salas, & Coovert, 2004).

It is important to point out, however, that simulation of team members or entire teams is becoming increasingly prevalent. A simulated team member may interact with an actual team member in an operational setting (top right quadrant of Table 19.1). For example, a simulated team member could serve as a synthetic agent that meets the specific needs and requests of an individual (for example, searching for relevant information at the right time). The simulated team member or agent could also reproduce a source of expertise that is missing from the team. GPS (global positioning system) navigation systems in cars provide an example of this kind of simulated team member that interacts in the context of a real situation.

Finally, as represented by the bottom right quadrant in Table 19.1, another increasingly common trend in simulation technology for teams is to simulate the task as well as one or more team members (Gluck et al., 2006). Although this chapter focuses on synthetic tasks and real team members, some of the measurement methods considered are also applicable to this last quadrant (that is, synthetic tasks with synthetic team members).

Purpose of Assessing Team Cognition through Simulation

Even if synthetic environments and their simulated scenarios are designed to elicit team level thinking or to exercise team level cognitive skills, this technology is of little use without comparable measurement technology. It is critical to measure team cognition in these team simulations for several reasons: (1) The results of measures of team cognition will not only provide feedback on whether we are achieving training objectives (that is, team performance is improving), but should also provide diagnostic information that can help in understanding the basis for successes or failures in this regard. (2) The assessment of team cognition can also be used to evaluate the success or failure of design or training interventions directed at improving team cognition or to compare two or more such interventions. (3) The assessment of team cognition if embedded in the task and processed in real or near real time can facilitate online monitoring of team cognition and performance and in some cases lead to real time intervention (for example, monitoring team communications may suggest that a team is losing team situation awareness and actions can be taken to mitigate this possibility).

ISSUES AND CHALLENGES OF MEASUREMENT IN TEAM SIMULATION

Simulated environments provide a rich context for a variety of measurement opportunities. Issues concerning measurement and data collection, such as deciding what to measure, embedding measures, and balancing scenario flexibility with experimental control, are particularly relevant when conducting research in simulated environments. These issues pose their own unique sets of challenges related to conducting research in simulated environments.

What to Measure

Of primary relevance to assessment are criterion measures of team performance. Performance measurement is primary because a valid and reliable performance measure provides a benchmark for a researcher to disambiguate experimental treatments, as well as to make inferences about relationships with other team cognitive variables, such as team process. Some recommendations for developing performance measures in simulations include unobtrusiveness (for example, not interrupting the task) and the collection of objective measures at the team level (holistic) versus generating a composite (collective) team member score. Further, embedding performance measures into the task prevents having to use subjective measures while providing instant performance feedback to participants. Later in the chapter, we describe a team performance measure that is based on several parameters embedded in a command and control team simulation.

From the holistic perspective, assessment of team process provides perhaps the most direct insight into sources of variance that are most specific to team cognition: team member interactions. A number of interaction-oriented process behaviors have been classed as team process behaviors. Using the multitrait-multimethod approach (Campbell & Fiske, 1959), Brannick, Prince, Prince, and Salas (1995) demonstrated construct validity for team process behaviors related to assertiveness, decision making, adaptability, situation awareness, leadership, and communication. Thus a variety of behaviors can be measured independently as team process, although high correlations between process dimensions may necessitate collapsing across dimensions (for example, Cooke, Gorman, Duran, et al., 2007; compare Smith-Jenstch, Johnston, & Payne, 1998). Emphasizing the holistic perspective, the most direct route to measuring any aspect of team process is through interaction/communication data. Later in the chapter we present team process measures of coordination, team situation awareness (team SA), and communication process measurement in detail.

Embedding Measures

Wherever possible, embedded measures are desirable. There are several benefits to embedding measures in the simulation. Embedded measures mitigate subjective bias in post-processing. Since embedded measures occur during task performance, they do not require retrospective judgments to be made. Further, the use of embedded measurement allows for ease of data collection and processing. For example, if a researcher is interested in measuring team communication, then he or she can record team communication in such a way that timestamps are collected in real time. This saves the time and cost of having someone go through communication data to insert timestamps during post-processing.

Flexibility versus Control

Often, research conducted in simulated environments is less controlled compared to laboratory experiments due to the very nature of the phenomena being studied. One may view simulated environments as the midpoint of a continuum anchored by naturalistic observation at one end and the laboratory at the other. Conventional psychological laboratories are highly controlled in order to isolate causality under very specific conditions. Although these characteristics are important for experimental validity, it often comes at the expense of ecological validity; specifically, it is difficult to study realistic team phenomena under such tightly controlled conditions. With respect to validity, simulated environments may therefore entail a need for higher ecological validity at the expense of some experimental validity. As a result, simulated team environments should strike a balance between control and ecological validity in order to elicit behaviors that emerge only under comparatively "real world" conditions.

STEPS TOWARD ADDRESSING ISSUES AND CHALLENGES: A CASE STUDY

Uninhabited Aerial Vehicle Synthetic Task Environment (UAV-STE)

A synthetic task environment for teams in the context of UAV ground control was developed for the purpose of studying team performance and cognition (Cooke & Shope, 2004). This work has been greatly influenced by the assumption that synthetic tasks provide ideal environments for cognitive engineering research on complex tasks in that they serve as a middle ground between the difficult to control naturalistic study and the highly controlled tasks typically found in the lab. The UAV-STE was designed to facilitate experimentation and to this end special attention was given to the exercising of team cognition in the context of UAV-STE scenarios and its measurement. Therefore, this simulation is a good example of the principles and methods discussed in this chapter.

The UAV-STE development was based on a cognitive task analysis (Gugerty, DeBoom, Walker, & Burns, 1999) of ground control operations for the Predator at Indian Springs, Nevada (Cooke, Rivera, Shope, & Caukwell, 1999; Cooke & Shope, 2005; Cooke & Shope, 2002a, Cooke & Shope, 2002b; Cooke & Shope, 1998; Cooke, Shope, & Rivera, 2000). The UAV-STE emphasizes team aspects of the task, such as planning, replanning, decision making, and coordination. In fact, the scenarios emphasize or exercise these types of team level cognitive activities at the expense of individually oriented activities, such as takeoff and landing. In addition, it was important that team level cognition had the opportunity to occur multiple times throughout the scenario. This was accomplished by introducing multiple target waypoints as landmarks associated with team cognition during the course of any scenario run. That is, most of the team level cognitive activity takes place in the context of the target landmarks with the activity being initiated some time prior to arrival at the landmark and ending shortly after leaving the landmark. Multiple landmarks thus provide multiple opportunities to observe team cognition.

The UAV-STE is a three team-member (pilot, navigator, and photographer) task in which each team member is provided with distinct, though overlapping, training; has unique, yet interdependent roles; and is presented with different and overlapping information during the mission. The overall goal is to fly the UAV to designated target areas (the landmarks) and to take acceptable photos at these areas. The pilot controls airspeed, heading, and altitude and monitors UAV systems. The photographer adjusts camera settings, takes photos, and monitors the camera equipment. The navigator oversees the mission and determines flight paths under various constraints. To successfully complete a UAV-STE mission, which is comprised of multiple targets, the team members need to share information with one another in a coordinated fashion multiple times.

Most communication is done via microphones and headsets, although some involves computer messaging. Measures taken include audio records, video records, digital information flow data, embedded performance measures, team process behavior measures, situation awareness measures, and a variety of individual and team knowledge measures. Hardware and software features of the UAV-STE relevant to measurement considerations include the following experimenter console features:

- Video and audio recording equipment (including digital audio) for communication analysis,
- Intercom and software for logging communications flow,
- Embedded performance measures,
- Ability to disable or insert noise in channels of communication intercom,
- Easy to change start-up parameters and landmark/waypoint library that define a scenario,
- Software to facilitate measurement of team process behaviors,
- Software to facilitate team SA measurement,
- Coordination logging software, and
- Numerous possibilities for inserting team SA measurement opportunities (roadblocks) into a scenario.

Also relevant to measurement are the following participant console features:

- Participant computer event logging capabilities,
- Training software modules with tests,
- Software modules for offline knowledge measurement (participant ratings), and
- Software for administering questionnaires (participant debriefing, NASA TLX [National Aeronautics and Space Administration Task Load Index], and so forth).

The UAV-STE design features that illustrate some of the concepts described in this chapter include the following:

1. *Training needs.* STEs allow the trainer or researcher to take liberties in order to best exercise or study the behavior of interest. In the case of the UAV-STE, we were interested in eliciting team cognition. For that reason, it was important to minimize individual training and skill acquisition so that team members could be in a position to quickly acquire the skills needed to contribute to team cognition. In the UAV-STE, individuals train to criterion in 1.5 hours after which they are ready to begin interacting as a team. By comparison, the actual Predator interface is complex, requiring many months of individual training. To accomplish the goal of rapid individual skill acquisition, we modified the interface to simplify individual skill acquisition while preserving the functionality required for the team level cognitive tasks. Thus, part of designing the STE required that the team behaviors of interest be exercised, often at the expense of individual level fidelity.
2. *Experimental control.* As we have noted, the design of synthetic tasks and their scenarios entails a balancing act between ecological and experimental validity.

However, trade-offs also need to be managed between measurement and realism, or cognitive fidelity. In the case of the UAV-STE the interface was modified as mentioned previously in order to focus the task on team, rather than individual, cognition. As a result the individual operator fidelity is low; however, the focus on team aspects of the task allow for the measurement of team member interaction and team cognition. In this case a balance between measurement and fidelity involved maximizing the opportunity for measuring team level cognition at the expense of individual cognitive fidelity.

3. *Performance measure.* Although we were primarily interested in measuring team cognition, these measures should be validated with respect to team performance. Therefore we sought a performance measure that would allow us to evaluate team cognition-related interventions tied to variance in team performance. The team performance measure being fundamental, it was therefore important to identify a robust, reliable measure of team performance in the context of the UAV-STE. In the following sections we describe team performance and team cognition measurement in the UAV-STE in detail.

Team Performance

Team performance is measured using a composite score based on the result of critical mission variables, including time each individual spent in an alarm state, time each individual spent in a warning state, the rate with which critical waypoints were acquired, and the rate with which targets were successfully photographed. Penalty points for each of these components are weighted a priori in accord with importance to the task and subtracted from a maximum score of 1,000 (see Cooke et al., 2004, for the scoring algorithm).

Each individual role within a team (pilot, photographer, and navigator) also has a score based on various mission variables, including time spent in alarm or warning state, as well as variables that are unique to that role. As with team performance, penalty points for each of the components are weighted in accord with importance to the task and subtracted from a maximum score of 1,000. For example, the most important components for the pilot are time spent in alarm state and course deviations; for the navigator, they are critical waypoints missed and route planning errors; and for the photographer, duplicate good photos, time spent in an alarm state, and number of bad photos are the most important components. Like team performance, individual performance data for a role are collected within the context of each UAV-STE scenario run (that is, mission).

The team performance measure has been used in several UAV-STE studies and has been modified in order to take into account workload differences in scenarios (Cooke et al., 2004). The team performance measure serves as the criterion for assessing experimental interventions, as well as validating relationships with team cognition as indexed by coordination, team SA, and communication measures.

Coordination Measure

We define coordination as the timely pushing and pulling of information among team members. For measurement, coordination consists of the temporal relationships between team members performing different aspects of the team task, relative to a procedural model. In order to develop the measure, first a procedural model of team coordination was developed based on the standard operating procedure for taking pictures of the ground target landmarks.

Essentially the standard operating procedure is a function of ordering, timing, and mode of task elements for each target. Ordering corresponds to sequential ordering of task elements for each target. Timing corresponds to the onset time (measured in seconds) of each element for each target within a mission. Mode corresponds to the nature of the element, that is, *information* mode versus *negotiation* mode versus *feedback* mode (Figure 19.1). The procedural model provides a blueprint for team coordination for the repetitive task of taking pictures of ground targets, and the coordination score provides a measure of variation in the application of the procedural blueprint.

In the procedural model (Figure 19.1) the optimal coordination procedure for taking a picture of a ground target begins with the navigator informing the pilot of information concerning the upcoming target restrictions (task elements *a* through *c* in Table 19.2). The pilot and the photographer then negotiate the appropriate altitude and airspeed for taking the photograph through back-and-forth negotiation (task elements *d* through *g* in Table 19.2). Finally, the photographer tells the navigator and the pilot that the target has been photographed (task element *h* in Table 19.2) and, thus, that the UAV may continue to the next target, which starts the coordination cycle over again.

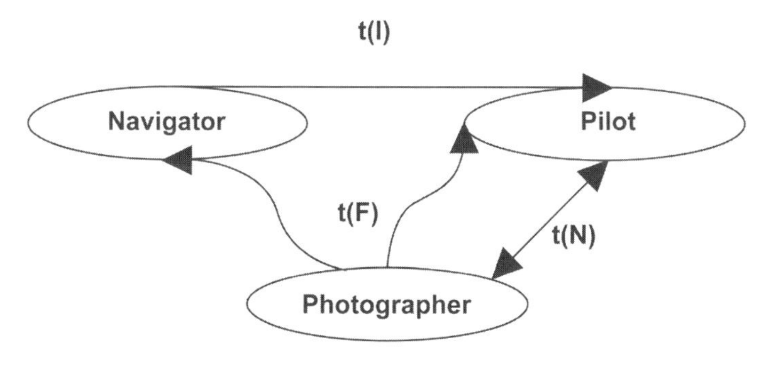

Figure 19.1. Model for Standard Operating Procedure for Photographing Uninhabited Aerial Vehicle–Synthetic Task Environment Ground Targets

Table 19.2. Uninhabited Aerial Vehicle–Synthetic Task Environment Target Procedure Task Elements

	a	Navigator tells pilot target restrictions
Information (I)	*b*	Navigator tells pilot target radius
	c	Navigator tells pilot target name
	d	Photographer coordinates altitude with pilot
Negotiation (N)	*e*	Photographer coordinates airspeed with pilot
	f	Pilot coordinates altitude with photographer
	g	Pilot coordinates airspeed with photographer
Feedback (F)	*h*	Photographer acknowledges good photo

Coordination scores (κ's; Gorman, Amazeen, Cooke, under review) were obtained by evaluating the relationship between the onset times of each task element in the procedural model at each target waypoint: $\kappa = F - I/F - N$. The timestamps for κ are collected by an experimenter monitoring team communication in real time using a coordination logger. The logger consists of one panel for each target landmark, and the timestamps for each button on the target panel correspond to one of the three procedural model task elements, *information, negotiation,* or *feedback* (Figure 19.1; Table 19.2). The dynamics of κ have provided evidence that teams that are very rigid in their coordination do not respond well to unexpected team SA roadblocks (Gorman et al., under review).

Coordinated Awareness of Situation by Teams (CAST)

The coordination based measure of team SA (CAST) is taken in the context of a UAV-STE mission. During a mission, a UAV-STE experimenter introduces a "roadblock" to team coordination at a prespecified event or time. The team is not told about the roadblock. Team communications are monitored for communication relevant to overcoming the roadblock. Interactions between specific team members, in response to the roadblock, are checked off by an experimenter on a CAST scoring sheet. Three types of interactions are checked off based on first-hand perception—recognition of some aspect of the roadblock without being told by another team member, coordinated perception—being told about some aspect of the roadblock by another team member, or coordinated action—sequence of interaction that mitigates the roadblock. Additionally, whether or not the roadblock was overcome is checked off if the team coordinated *around* the roadblock (Figure 19.2; Gorman, Cooke, Pedersen, Connor, & DeJoode 2005).

Figure 19.2 shows two CAST score sheets for a five minute communication channel glitch from navigator to pilot (but not pilot to navigator), which was introduced when teams reached a designated point during a UAV-STE mission (Kiekel, Gorman, & Cooke, 2004). The CAST scoring procedure consisted of listening to team communications around the five minute glitch and then checking appropriate boxes on the CAST score sheet. In Figure 19.2, the score sheet in

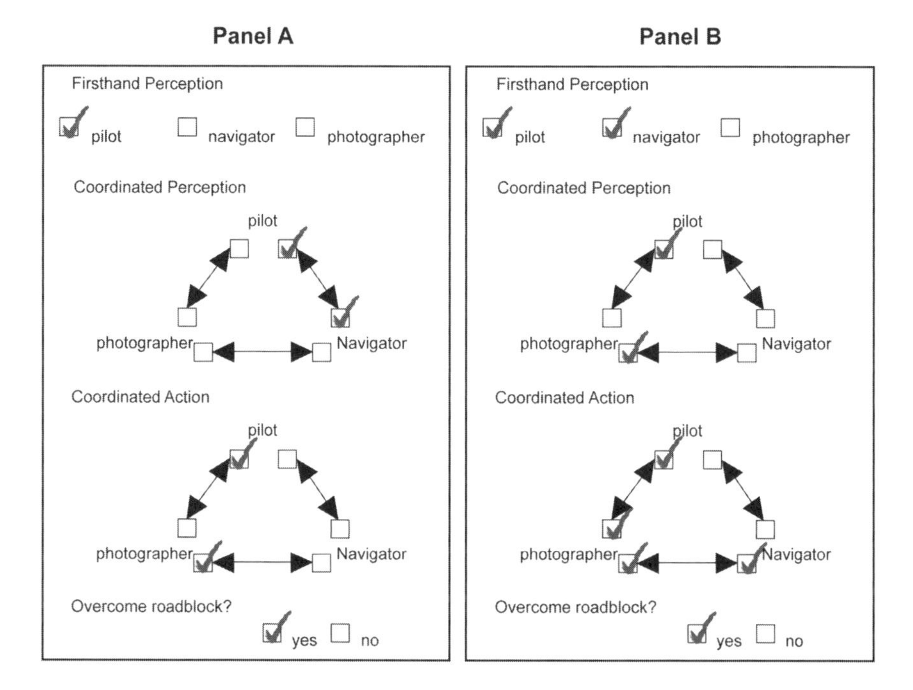

Figure 19.2. CAST Team SA Scoring Sheet; Panel A: Optimal Solution to Communication Glitch; Panel B: Suboptimal Solution to Communication Glitch

panel A shows a CAST result where the pilot perceived the glitch and coordinated his perceptions with the navigator: namely, that the pilot could not hear the navigator, but the navigator could hear the pilot. This led to the successful coordinated action of the navigator channeling communications to the pilot through the photographer.

In Figure 19.2 the score sheet in panel B shows an instance where both the navigator and the pilot perceived the glitch and established a coordinated perception via the photographer. Subsequently the photographer was involved in the coordinated action as a bidirectional conduit of pilot-navigator communications. While both the left and right solutions overcome the glitch, one is more efficient (panel A) than the other (panel B) corresponding to better team SA overall. The solution on the left is more efficient because it accurately reflects the true state of the world (that is, a one-way, not a two-way, glitch) and requires less effort on the part of the team. In fact, the team that reached the solution shown in panel B missed a target because the photographer had dedicated himself to relaying messages back and forth and missed an opportunity for a good target picture.

Communication Analysis and Real Time Assessment

Interaction based measures, such as team communication measurement, are akin to a team "thinking out loud." This is akin to a team verbal protocol analysis.

In addition to operationalizing the behavior of interest in a holistic manner, communication based measures are amenable to embedded measurement, because team communication occurs naturally during the course of task performance. The challenge that remains is to capture and assess these naturally occurring team member interactions.

The design of the UAV-STE provides facilities for not only capturing communication flow data, but capturing it in real time because team members must communicate over headsets. Communication flow (ComLog) data are generated unobtrusively in the background during UAV task performance by team members opening communication channels using push-to-talk buttons. In the UAV-STE there are nine possible communication channels across the team members, where each team member has access to three channels (for example, pilot à photographer, pilot à navigator, and pilot à all). The state of communication channels, characterized as a three-by-three matrix, is sampled eight times per second in order to capture all significant team member interactions, including very brief ones (for example, acknowledgements). Current efforts are focused on reproducing the coordination metric depicted in Figure 19.1 from the ComLog data. Success in this challenging endeavor would present a significant step forward in the assessment of team cognition from relatively low level communication flow data. Specifically, the goal tied to real time communication analysis is to achieve real time automated assessment of team cognition, such that online interventions can be used to augment deficient team cognition (for example, Figure 19.2, panel B).

CONCLUSION

Teams play an integral role in the operation of many of today's high technology, sociotechnical environments. Teams are brought together by the need for multiple, heterogeneous, yet interdependent operators to control these complex systems. In addition, the underlying mechanisms of team cognition (for example, communication) are ubiquitous across a variety of these task settings. The need for a scientific understanding, and support for team cognition, is a pressing challenge for both basic and applied research. Simulated environments may help us address team cognition on both of these research fronts concurrently.

Nevertheless, teams remain a challenging subject of study. With respect to possible trade-offs between ecological validity and control, team simulations that are too complex may lead to such poor control that it is difficult to replicate results, while simulations that are too highly controlled may not elicit the interaction behaviors of interest. A balance must be struck between ecological and experimental validity such that the simulated environment is as real as possible while providing a replicable research context.

The issues and challenges involved in developing a simulated team task were addressed in the context of the UAV-STE case study. First, the UAV-STE is flexible enough to provide ecological validity in terms of certain aspects of UAV operations (for example, reconnaissance and monitoring flight systems) while providing experimental control (for example, training interventions) that has produced replicable findings (Cooke et al., 2004). Second, the UAV-STE supports the elicitation of team cognition organized around critical landmarks (targets) while providing facilities for monitoring and measuring team interaction and team performance. Overall this case study emphasizes the need for embedded measurement tied to the simulation. Therefore, in the UAV-STE case study we outlined the scope of, and some measurement priorities for, the assessment of team cognition and performance in simulated environments. These include several issues and challenges that need to be addressed when developing a simulated team task, including the design of valid and reliable performance measures and embedding measurement in the context of the simulated environment wherever possible.

Valid and reliable performance measures serve several purposes in simulated environments. First, performance measures facilitate hypothesis testing. That is, they serve as dependent variables when manipulating various independent variables, such as training interventions. Second, valid and reliable performance measures make validation of knowledge, process, and other measures possible. With respect to embedded measurement, we emphasize the need for integration of scenario design and assessment *during* the development of the simulated environment. That is, a scenario designed to elicit specific behaviors at specific times (for example, team coordination organized around landmarks and team SA organized around roadblocks) suggests specific events around which measurement, and therefore assessment, can be organized. Having insight into how the scenario is designed to elicit specific types of behavior at specific times gives researchers the ability to identify what to measure and when. These features of simulated environments make them amenable to the development of valid, reliable embedded real time measures of team cognition, which we see as the next most important step toward mitigating the effects of deficient team interaction on complex sociotechnical systems, before catastrophic failures can occur.

ACKNOWLEDGMENTS

Support for this work was provided by AFOSR Grant No. FA9550-04-1-0234, AFRL Grant No. FA8650-04-6442, and ONR Grant No. N00014-05-1-0625.

REFERENCES

Brannick, M. T., Prince, A., Prince, C., & Salas, E. (1995). The measurement of team process. *Human Factors, 37,* 641–651.

Campbell, D. T. & Fiske, D. W. (1959). Convergent and discriminant validation by the multitrait-multimethod matrix. *Psychological Bulletin, 56,* 81–105.

Cannon-Bowers, J. A., Salas, E., Duncan, P., & Halley, E. J. (1994). Application of multimedia technology to training for knowledge-rich systems. *Proceedings of the 16th Annual Interservice/Industrial Training Systems Conference* (pp. 6–11). Washington, DC: National Training and Simulation Association.

Cooke, N. J., DeJoode, J. A., Pedersen, H. K., Gorman, J. C., Connor, O. O., & Kiekel, P. A. (2004). *The role of individual and team cognition in uninhabited air vehicle command-and-control* (Tech. Rep. for AFOSR Grant Nos. F49620-01-1-0261 and F49620-03-1-0024). Mesa: Arizona State University East.

Cooke, N. J., & Gorman, J. C. (2005). Assessment of team cognition. *International encyclopedia of ergonomics and human factors* (2nd ed., pp. 271–275). Boca Raton, FL: CRC Press.

Cooke, N. J., Gorman, J. C., Duran, J. L., & Taylor, A. R. (2007). Team cognition in experienced command-and-control teams. *Journal of Experimental Psychology: Applied, 13,* 146–157.

Cooke, N. J., Gorman, J., Pedersen, H., & Bell, B. (2007). Distributed mission environments: Effects of geographic distribution on team cognition, process, and performance. In S. Fiore and E. Salas (Eds.), *Toward a science of distributed learning* (pp. 147–167). Washington, DC: American Psychological Association.

Cooke, N. J., Gorman, J. C., & Rowe, L. J. (in press). An ecological perspective on team cognition. In E. Salas, J. Goodwin, & C. S. Burke (Eds.), *Team effectiveness in complex organizations: Cross-disciplinary perspectives and approaches.* Mahwah, NJ: Erlbaum.

Cooke, N. J., Kiekel, P. A., & Helm E. (2001). Measuring team knowledge during skill acquisition of a complex task. *International Journal of Cognitive Ergonomics: Special Section on Knowledge Acquisition, 5,* 297–315.

Cooke, N. J., Rivera, K., Shope, S. M., & Caukwell, S. (1999). A synthetic task environment for team cognition research. *Proceedings of the Human Factors and Ergonomics Society 43rd Annual Meeting* (pp. 303–307). Santa Monica, CA: Human Factors and Ergonomics Society.

Cooke, N. J., & Shope, S. M. (1998). *Facility for cognitive engineering research on team tasks* (Report for Grant No. F49620-97-1-0149). Washington DC: Bolling AFB.

Cooke, N. J., & Shope, S. M. (2002a). Behind the scenes. *UAV Magazine, 7,* 6–8.

Cooke, N. J., & Shope, S. M. (2002b). The CERTT-UAV task: A synthetic task environment to facilitate team research. *Proceedings of the Advanced Simulation Technologies Conference: Military, Government, and Aerospace Simulation Symposium* (pp. 25–30). San Diego, CA: The Society for Modeling and Simulation International.

Cooke, N. J., & Shope, S. M. (2004). Designing a synthetic task environment. In S. G. Schiflett, L. R. Elliott, E. Salas, & M. D. Coovert (Eds.), *Scaled worlds: Development, validation, and application* (pp. 263–278). Surrey, England: Ashgate.

Cooke, N. J., & Shope, S. M. (2005). Synthetic task environments for teams: CERTTS's UAV-STE handbook on human factors and ergonomics methods (pp. 46-1–46-6). Boca Raton, FL: CRC Press, LLC.

Cooke, N. J., & Shope, S. M., & Rivera, K. (2000). Control of an uninhabited air vehicle: A synthetic task environment for teams. *Proceedings of the Human Factors and Ergonomics Society 44th Annual Meeting* (p. 389). Santa Monica, CA: Human Factors and Ergonomics Society.

Gluck, K. A., Ball, J. T., Gunzelmann, G., Krusmark, M. A., Lyon, D. R., & Cooke, N. J. (2006, September). *A prospective look at synthetic teammate for UAV applications.* Invited talk for AIAA "Infotech@Aerospace" Conference on Cognitive Modeling, Arlington, VA.

Gorman, J. C., Amazeen, P. G., & Cooke, N. J. (under review). Dynamics of team coordination. Manuscript submitted to *Physics Letters A.*

Gorman, J. C., Cooke, N. J., Pedersen, H. K., Connor, O. O., & DeJoode, J. A. (2005). Coordinated awareness of situation by teams (CAST): Measuring team situation awareness of a communication glitch. *Proceedings of the Human Factors and Ergonomics Society 49th Annual Meeting* (pp. 274–277). Santa Monica, CA: Human Factors and Ergonomics Society.

Gugerty, L. DeBoom, D., Walker, R., & Burns, J. (1999). Developing a simulated uninhabited aerial vehicle (UAV) task based on cognitive task analysis: Task analysis results and preliminary simulator data. *Proceedings of the Human Factors and Ergonomics Society 43rd Annual Meeting* (pp. 86–90). Santa Monica, CA: Human Factors and Ergonomics Society.

Kiekel, P. A., Gorman, J. C., & Cooke, N. J. (2004). Measuring speech flow of co-located and distributed command and control teams during a communication channel glitch. *Proceedings of the Human Factors and Ergonomics Society's 48th Annual Meeting* (pp. 683–687). Santa Monica, CA: Human Factors and Ergonomics Society.

Langan-Fox, J., Code, S., & Langfield-Smith, K. (2000). Team mental models: Techniques, methods, and analytic approaches. *Human Factors, 42,* 242–271.

Salas, E., Bowers, C. A., & Rhodenizer, L. (1998). It is not how much you have but how you use it: Toward a rational use of simulation to support aviation training. *The International Journal of Aviation Psychology, 8,* 197–208.

Salas, E., Dickinson, T. L., Converse, S. A., & Tannenbaum, S. I. (1992). Toward an understanding of team performance and training. In R. W. Swezey & E. Salas (Eds.), *Teams: Their training and performance* (pp. 3–29). Norwood, NJ: Ablex.

Salas, E, Wilson, K. A., Burke, C. S., & Wightman, D. C. (2006). Does CRM training work? An update, extension, and some critical needs. *Human Factors, 48,* 392–412.

Schiflett, S. G., Elliott, L. R., Salas, E., & Coovert, M. D. (Eds.). (2004). *Scaled worlds: Development, validation, and applications.* Hants, England: Ashgate.

Schreiber, B. T., & Bennett, W. Jr. (2006). Distributed *mission operations within-simulator training effectiveness baseline study: Summary report* (Rep. No. AFRL-HE-AZ-TR-2006-0015-Vol I, 1123AS03). Air Force Research Laboratory, AZ: Warfighter Readiness Research Division.

Smith-Jentsch, K. Johnston, J. H., & Payne, S. C. (1998). Measuring team-related expertise in complex environments. In J. A. Cannon-Bowers & E. Salas (Eds.), *Decision making under stress: Implications for individual and team training* (pp. 61–87). Washington, DC: American Psychological Association.

Weir, D. (2004). Catastrophic failure in complex socio-technical systems. *International Journal of Nuclear Knowledge Management, 1,* 120–130.

Chapter 20

AFFECTIVE MEASUREMENT OF PERFORMANCE

James Driskell and Eduardo Salas

The goal of this chapter is to examine the use of affective measures of performance in simulation and training. Affect has emerged as a central topic in psychology relatively recently, and some have termed this resurgence of interest an "affective revolution" in psychology. Therefore, it is informative to briefly note some historical antecedents of this resurgence of interest in affect, especially as it relates to simulation and training. Over 50 years ago, Bloom and associates attempted to develop a taxonomy of educational objectives, culminating in the publication of separate handbooks addressing the cognitive domain (Bloom, Engelhart, Furst, Hill, & Krathwohl, 1956) and the affective domain (Krathwohl, Bloom, & Masia, 1964). Gagne (1984) also proposed multiple categories of learning outcomes, including attitudinal outcomes. Drawing on these perspectives, Kraiger, Ford, and Salas (1993) presented a comprehensive scheme for classification of learning outcomes, emphasizing cognitive, skill based, and affective learning outcomes. Even early on, it was noted that the attempt to structure the affective domain was a difficult task. Krathwohl et al. (1964) defined the affective domain as comprising those learning outcomes that "emphasize a feeling tone, an emotion, or a degree of acceptance or rejection" (p. 7) and including a large number of objectives, such as interests, attitudes, beliefs, and values.

It is also useful to examine how the term "affective measures" is used in practice. In other words, when other researchers discuss affect or affective measures, how do they describe this domain? In various reports that have examined affective measures, the topics addressed include affect, emotions, and moods (Humrichouse, Chmielewski, McDade-Montez, & Watson, 2007); feelings or sentiments (Heise, 2002); beliefs (Robinson & Clore, 2002); temperament (Ilies & Judge, 2005); personality or disposition (Thoresen, Kaplan, Barsky, Warren, & de Chermont, 2003); interests and self-perceptions (Cassady, 2002); attitudes and motivational outcomes, such as self-efficacy and goal setting (Kraiger et al., 1993); and affective states, such as trust, collective orientation, and cohesiveness (Stagl, Salas, & Day, 2008). So, we can conclude that affective constructs refer to emotions, moods, beliefs, interests, dispositions, attitudes, motivational states, self-perceptions, and preferences—quite a daunting list. Although some have drawn a simple distinction between cognition (thinking) and affect (feeling), others have noted that in practice, the use of the term *affective* is largely intuitive (Diener, Smith, & Fujita, 1995). Those who study emotions per se offer more exacting definitions of affective state (see Yik, Russell, & Barrett, 1999). However, for our purposes, it may be appropriate to simply concur with Krathwohl et al. (1964) by noting that the affective domain reflects an important but broad domain of learning and includes a host of constructs, such as attitudes, motivation, self-efficacy, and other noncognitive constructs—all which are important to determine the efficacy of simulation based training. The renewed attention to the affective domain reflects recent attempts to identify characteristics of the individual (and teams) other than cognitive ability that determine skill acquisition and learning in simulation based training environments.

Traditionally, affective measures have been viewed with at least some degree of reservation regarding their value or usefulness. First, affective constructs, such as expectations, attitudes, feelings, or beliefs, are not observable. They are instead theoretical constructs, and although they are not directly observable, one may observe their indicators. However, these are critical for learning (for example, motivation to learn; Salas & Cannon-Bowers, 2001). A second area of concern relates to the individual's access to and accuracy in reporting introspective data. For example, Robinson and Clore (2002) note that when asked to report current feelings, individuals rely on accessible episodic memories, whereas when asked to report on feelings not currently experienced (via prospective or retrospective queries), people access their beliefs about their affect rather than the affect itself. Therefore, on one hand, questions have been raised regarding the validity or accuracy of various types of affective measures. However, on the other hand, researchers argue that these types of measures are especially meaningful because they provide a window into the individual's affective experience and their reactions to the training or simulation. Gaining this perspective on the individual's affective state is valuable for several reasons in simulation based training. The first is to determine the effect of training on the affective state of the trainee in instances in which learning outcomes include changes in attitude, motivation, self-efficacy, and so on. This common use of affective measures provides valuable information on whether the goals of training have been met or what needs to change in order to engage and/or motivate the trainee to learn. A second way in which attention to the affective domain in training is useful is the examination of affective constructs as determinants of training effectiveness. Such affective constructs as self-efficacy or anxiety can be examined as predictors of training effectiveness and as mediators of training effects (Salas & Cannon-Bowers, 2001).

THE AFFECTIVE DOMAIN: A TRAINING MODEL

Figure 20.1 presents a framework to organize the various ways in which affective measures can inform our understanding of training effectiveness. Elaborating

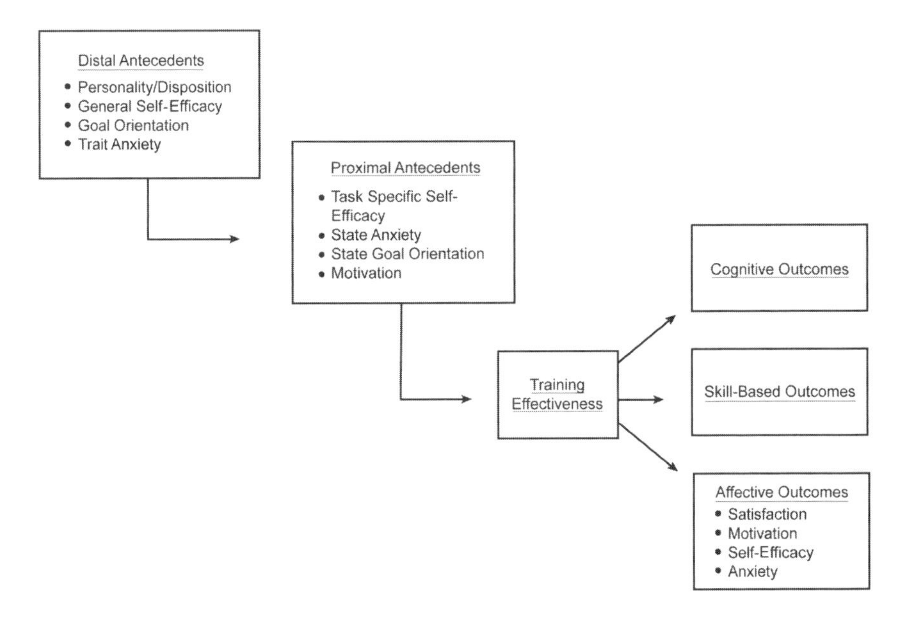

Figure 20.1. The Affective Domain: A Training Model

the discussion in the previous paragraph, we believe that affective measures are useful in answering several different types of questions. First, affective measures can be utilized to assess training outcomes in the affective domain. Kraiger et al. (1993) noted the distinction between the processes of *training effectiveness* and *training evaluation*. Training evaluation refers to the examination of whether a training event has achieved certain learning outcomes. Affective outcomes are one type of learning outcome. Thus, the use of affective measures allows us to address questions related to training outcome and specifically whether a training intervention has accomplished or not accomplished stated learning objectives. As shown in Figure 20.1, affective outcomes that are typically examined include measures of motivation, self-efficacy, and satisfaction.

Training effectiveness refers to the examination of why training does or does not accomplish its intended learning outcomes. For example, if we think in input-process-output terms, questions related to training effectiveness involve attention to input and process variables that may determine training outcome. In Figure 20.1, we draw attention to both distal and proximal determinants of training effectiveness. Affective measures can be used to assess distal antecedents of training effectiveness to measure such individual difference trait-like constructs as disposition, general self-efficacy, or goal orientation, that impact training as a more distal input variable. Furthermore, affective measures can be used to assess more proximal antecedents of training effectiveness to measure such state-like constructs as task-specific self-efficacy or state anxiety that are more proximal to performance and that may emerge during training.

Moreover, certain constructs, such as goal orientation, may be viewed as both a distal individual difference trait-like variable (at the input level) and as a proximal state-like variable that emerges during training (at the process level). In sum, we believe affective measures may be usefully examined as distal measures (measures of input variables, such as individual differences), proximal measures (measures of training process), and outcome measures (measures of training outcome). In the following, we discuss the use of affective constructs within each of these three categories.

AFFECTIVE CONSTRUCTS AS DISTAL ANTECEDENTS

In addressing training effectiveness—why training works—there is considerable practical value in understanding the role that individual differences play in predicting training outcomes. Most individual difference models focus on differences in cognitive ability, and there is good reason for this. Research indicates that predictions of training success from cognitive measures are consistent and positive; for the most part, higher ability individuals perform better in training—they learn more and faster. However, individuals differ not only in ability, but also in terms of achievement orientation, conscientiousness, and ambition—noncognitive factors that are likely to affect training outcome. Yet, Barrick and Mount (1991) have noted that "very little research has investigated the relation of individual measures of personality to measures of training readiness and training success" (p. 22).

There are at least two reasons for this. The first concerns the conceptual foundation of personality and its early uses. Historically, personality theory and research emphasized psychopathological and neuropsychic conceptualizations of personality structure. Personality was equated with psychopathology, and personality measurement was used to assess some underlying set of neurotic structures governing behavior. There was considerably less interest in understanding task performance in normal populations, and specifically in identifying desirable characteristics that define effective performance (Driskell, Hogan, & Salas, 1987).

A second reason that personality has made relatively little contribution to the examination of training performance is methodological in nature. Until recently, personality psychologists failed to reach any consensus regarding how to define personality and accordingly how it should be measured. Every theory of personality provided its own set of variables or constructs and its own measurement procedures. However, beginning with Fiske (1949), Tupes and Christal (1961), and Norman (1963), personality researchers have converged on five broad dimensions that constitute normal personality. The multitude of personality descriptors identified in previous literature can be expressed in terms of these five factors, often termed the "Big Five."

This development is significant for several reasons. First, research has accumulated that supports the robustness of this five-factor model across different populations and settings. Second, this model establishes a common vocabulary for both describing and measuring personality and can serve as a useful taxonomy for classifying personality. Third, research suggests that these factors are relatively independent of measures of cognitive ability (McCrae & Costa, 1987), and thus these factors may contribute a unique variance to the prediction of training performance.

Although there is some divergence on how personality traits should be labeled and organized, personality theorists are in general agreement on the nature of the structure of personality. Most theorists propose a hierarchical model of personality, with broad, higher order factors or traits that subsume and organize more specific lower level facets (compare Saucier & Ostendorf, 1999). For example, the Big Five factor model represents a broad set of traits that are themselves a collection of many facets that have something in common. Whereas the broad, higher level constructs offer an efficient and parsimonious way of describing personality, the more specific facets can offer higher fidelity of trait descriptions and greater predictive validity (Saucier & Ostendorf, 1999; Stewart, 1999). The higher level Big Five dimensions include the following:

Neuroticism. This trait is also termed emotional stability or adjustment and refers to a lack of anxiety and nervous tendencies. Those who are emotionally stable tend to be well-adjusted, calm, secure, and self-confident. Viewed from the negative pole of neuroticism, those who score low on this trait tend to be moody, anxious, paranoid, nervous, insecure, depressed, and high-strung (Barrick & Mount, 2001). Adjustment has been defined by Hogan (1986) as freedom from anxiety, depression, and somatic complaints. Watson, Clark, and Tellegen (1988) have viewed lack of adjustment as *negative affect,* a general dimension of subjective distress and unpleasurable engagement. Hogan and Hogan (1989) found that adjustment was a significant predictor of success in naval explosive ordance disposal training.

Extraversion. The higher level Big Five trait of extraversion has been viewed as a combination of assertiveness/dominance and sociability/affiliation. Some theorists view dominance as the primary marker of extraversion, and some view sociability as the primary component of extraversion (Hough, 1992; Saucier & Ostendorf, 1999). The dominance component has also been referred to as ascendance, assertiveness, or surgency (Costa & McCrae, 1992), and high scores reflect those who are active, outgoing, and gregarious. The sociability component describes those who are sociable, friendly, interested in social interaction, and interpersonally adept. Persons low on sociability are withdrawn, reserved, aloof, and prefer solitary tasks to social interactions in which they are less comfortable. In their meta-analysis, Barrick and Mount (1991) found that extraversion was positively related to training proficiency and noted that this relationship likely stemmed from the tendency for those who excelled in training to be both more active/outgoing and more sociable. Of course, this relationship would be expected to be stronger in training situations that require more collaboration and weaker in training situations that do not involve social interaction.

Openness. This dimension has been termed openness to experience, intellect, or intellectance and reflects intellectual, cultural, or creative interests. From the positive pole, openness refers to a preference for intellectual curiosity and interest in new ideas and experiences. McRae and Costa (1997) claimed that, from the negative pole, the trait of openness is related to rigidity in behavior and unwillingness to accept change. Barrick and Mount (1991) noted that those who score high on openness are likely to have more positive attitudes toward learning and are more willing to engage in training experiences. In fact, Barrick and Mount (1991) and Hough, Eaton, Dunnette, Kamp, and McCloy (1990) found that openness was positively related to training performance. Gully, Payne, Koles, and Whiteman (2002) reported a positive relationship between openness and training outcome, stating that those high on this characteristic are likely to excel in training environments because they are more curious and imaginative and willing to engage in new approaches to learning. Driskell, Hogan, Salas, and Hoskin (1994) found that the intellectance and ambition scales from the Hogan Personality Inventory predicted training performance in naval electronics training. They also reported that personality variables provided incremental prediction of training success above that provided by cognitive predictors alone, and further that personality predicted other training difficulties, such as nonacademic infractions, that were not predicted by cognitive measures.

Agreeableness. The trait of *agreeableness* is defined by such terms such kindness, trust, and warmth versus selfishness, distrust, and hostility. Persons high on agreeableness are considerate, honest, helpful, and supportive. Persons low on agreeableness are uncaring, intolerant, unsympathetic, and critical. Some researchers have claimed that agreeableness may be the best primary predictor of performance in interpersonal settings (Mount, Barrick, & Stewart, 1998). Thus, agreeableness seems to have high predictive validity for tasks that involve cooperation and that involve smooth relations with others. In a study of Australian Air Force trainees, Sutherland and Watkins (1997) found that agreeableness, conscientiousness, and neuroticism all made significant contributions to predicting training performance beyond that accounted for by cognitive ability.

Conscientiousness. The Big Five trait of conscientiousness is comprised of two primary components. Moon (2001) has noted that some researchers emphasize the achievement orientation component of conscientiousness (that conscientious persons persevere and are motivated to achieve), whereas others have viewed conscientiousness in terms of responsibility/dependability (that conscientious persons are dependable, reliable, responsible, and trustworthy). Dependability refers to a tendency toward planfulness and discipline in carrying out tasks to completion. Those high in dependability are responsible, organized, planful, careful, and trustworthy. Those low in dependability are irresponsible, disordered, and impulsive. The achievement component of the Big Five trait of conscientiousness refers to the desire to work hard to achieve goals and to master difficult tasks. Those who score high on this trait set challenging goals, work hard to achieve these goals, and persist in the face of hardships rather than give up or quit. Those who score low on this trait avoid difficult or challenging tasks, work only as hard as necessary, and give up when faced with difficult obstacles.

Although the Big Five approach is the most current and influential formulation of individual differences in personality, it is certainly not the only useful model, nor does it capture all there is to say about personality. Although Saucier and Goldberg (1998) argued that virtually all facets of personality fall within the Big Five factor space, Paunonen and Jackson (2000) have claimed that there are a number of traits not represented within the Big Five models. We concur with McAdams and Pals (2006), who note that although the Big Five model is arguably the most recognizable contribution personality psychology has to offer, understanding personality in a more finely grained sense requires going beyond the personality trait concept to include other motivational, social-cognitive, and developmental concerns.

AFFECTIVE CONSTRUCTS AS PROXIMAL ANTECEDENTS

We have noted that from a dispositional perspective, there are several traits that directly predict learning outcome. However, this brief overview masks the more complex relationships among trait-like individual differences, state-like individual differences, and training outcomes. For example, DeShon and Gillespie (2005) note that such a construct as goal orientation has been viewed as a relatively stable individual difference trait and as a more malleable quasi-trait, among other definitions. These authors feel that this conceptual confusion provides an unstable foundation for understanding this construct and note that "the literature on this construct is in disarray" (p. 1096). However, this may also simply reflect the fact that a number of such variables may be viewed as having a more distal impact on training outcome, as well as having a more proximal impact on training outcome, or as Payne, Youngcourt, and Beaubien (2007) note, may exist as both a trait and a state. Furthermore, Colquitt and Simmering (1998) argue that such variables as conscientiousness and goal orientation may serve as distal variables that influence training through more proximal mechanisms such as motivation to learn. Goal orientation may take two forms: (a) a learning orientation characterized by the desire to increase competence by developing new skills and (b) a performance orientation characterized by a desire to gain success and meet standards in a task setting. Further, Colquitt and Simmering note that this construct has both trait and state properties. That is, individuals may have dispositional goal orientations that we may view as input factors that they bring with them to the situation, as well as state goal orientations that are impacted by situational and training variables.

Chen, Gully, Whiteman, and Kilcullen (2000) also propose a more comprehensive training model in which distal individual differences serve as predictors of proximal motivational processes and performance. They present a model in which trait-like constructs that are distal from performance (such as general self-efficacy and goal orientation) impact state-like constructs that are proximal to performance (such as state self-efficacy and state anxiety) to determine training outcomes. They further note that the primary value of such trait-like constructs as general self-efficacy stems from its ability to predict state-like constructs rather than directly influencing training outcome.

Payne et al. (2007) present a similar overarching model that views goal orientation as a "compound" trait that is composed of various facets of the Big Five, including achievement, self-esteem, and general self-efficacy. Goal orientation, viewed as one's dispositional goal preferences, influences proximal outcomes, such as state goal orientation, state self-efficacy, and state anxiety, which in turn impact more distal consequences, such as learning or training outcomes. In their research, they identified three dimensions of goal orientation, including (a) learning goal orientation, or LGO, (b) prove performance goal orientation, or PPGO, defined as the desire to prove one's competence and gain favorable judgments, and (c) avoid performance goal orientation, or APGO, defined as the desire to avoid the disapproving of one's competence and avoid negative judgments. They found that LGO was positively related to learning outcomes, APGO was negatively related to learning outcomes, and PPGO was unrelated to learning outcomes. Moreover, they found that LGO was most strongly predicted by high openness and conscientiousness, whereas APGO and PPGO were both associated with low emotional stability.

AFFECTIVE CONSTRUCTS AS MEASURES OF LEARNING OUTCOME

As noted, learning is a dynamic, multidimensional and multilevel phenomenon. Learning is the desired outcome of any simulation based training, so careful consideration needs to be given to its assessment and measurement. The "triangulation" of different kinds of measures is needed. The simulation based training field has paid much attention to skills, behaviors, and cognitive actions (what trainees "do" and "think") taken by trainees during simulations, but little to affective measures (what trainees "feel"). We argue that these could be as diagnostic about learning as the more skill based or cognitively based measures. The training effectiveness field has moved into that direction—the deeper understanding of the role affective constructs have in learning and skill acquisition. The simulation based training community could benefit by incorporating the findings of emergent research in this domain. We illustrate some of that next.

Turning now to the question of *training evaluation* (that is, what types of learning outcomes are achieved by training), there have been several models of training evaluation that address affective measures. Affective measures are examined as a training outcome measure, at least in a basic sense, in Kirkpatricks's (1976) model as training *reaction* criteria. Reaction measures represent the trainee's subjective evaluation of satisfaction with the training experience. Although such reaction measures are often viewed as a narrow and somewhat superficial way to assess training outcome, Sitzmann, Brown, Casper, Ely, and Zimmerman (2008) reported results of a meta-analysis indicating that trainee reactions predicted cognitive learning outcomes of declarative and procedural knowledge, as well as predicted pre-to-post-training changes in motivation and self-efficacy.

Although these results provide evidence counter to the prevailing notion that trainee reaction measures are not useful, Kirkpatricks's model was limited in that it did not consider affect other than as satisfaction with training, and it conceptualized trainee reactions as distinct from learning outcomes.

Kraiger et al. (1993) have presented a more comprehensive model of training that views learning outcomes as multidimensional and includes (a) cognitive, (b) skill based, and (c) affective outcomes. This construct-oriented approach defines several key affective learning outcomes, including attitudinal outcomes and motivational outcomes, such as disposition, self-efficacy, and goal setting. According to this perspective, affective measures can be viewed as specific goals or objectives of training and simulation interventions. That is, according to this model, affective measures can be viewed as learning outcomes, in addition to being viewed as distal or proximal determinants of learning. Kraiger et al. (1993) described several types of affective learning outcomes, including the development of such attitudes as safety and changes in such motivational outcomes as goal orientation, self-efficacy, and goal setting.

We will briefly describe two areas in which the emphasis on affective outcome measures is particularly salient, stress training and team training. Stress exposure training (SET) is a simulation based approach to mitigating negative stress effects that has been developed for military training applications (see Driskell, Salas, Johnston, & Wollert, 2008; Driskell, Salas, & Johnston, 2006). Extensive laboratory research has documented the effectiveness of the SET training approach in reducing stress effects and enhancing performance (Inzana, Driskell, Salas, & Johnston, 1996; Saunders, Driskell, Johnston, & Salas, 1996).

SET incorporates three stages or phases of training: (a) *information provision,* an initial training stage in which information is provided to the trainee regarding stress, stress symptoms, and likely stress effects in the performance setting; (b) *skills acquisition,* in which specific skills required to maintain effective performance in a stress environment are taught and practiced; and (c) *application and practice,* the final stage of application and practice of these skills under simulated conditions that increasingly approximate the criterion environment. Johnston and Cannon-Bowers (1996) defined two specific types of affective training outcomes in the SET training model: a decrease in anxiety and an increase in performance confidence.

One primary objective of stress exposure training is a reduction in anxiety. The construct of anxiety has most often been operationally defined in terms of self-report responses. These self-reports have typically taken the form of Likert-type, or rating, scales or adjective checklists variously labeled as anxiety, tension, or arousal. Although these scales and checklists may be variously labeled, they all require self-report on highly similar sets of items (for example, uneasy, anxious, restless, tense, aroused, and nervous). These items tend to be of roughly equivalent emotionality ratings and to share relatively high free-association frequency (for example, John, 1988), which supports the notion that these various self-report indexes are tapping into a common underlying construct of arousal/anxiety. A second primary objective of stress exposure training is an increase in confidence or self-efficacy. Self-efficacy refers to the belief in one's capacity to perform successfully in a range of task situations (Chen et al., 2000). Self-efficacy is related to perceptions of confidence, capability, mastery, and control. Bennett, Alliger, Eddy, and Tannenbaum (2003) reported that the predictive power of measures of confidence was remarkably strong (r's of 0.68 and 0.86) in evaluating the effectiveness of two military training programs.

A substantial body of research has accumulated in recent years on team performance and team training (see Salas, Nichols, & Driskell, 2007; Stagl et al., 2008). Although team performance outcomes are multifaceted, considerable recent emphasis has been placed on affective measures of team functioning. Salas, Sims, and Burke (2005) have described the Big Five components of teamwork, focusing attention on the importance of measures of leadership, adaptability, mutual performance monitoring, backup behavior, and team orientation. Stagl et al. (2008) have described team learning outcomes related to trust, collective orientation, collective efficacy, and cohesion. For example, Driskell and Salas (1992) found that collective orientation, the extent to which team members attend to one another's task inputs, was a critical factor in effective team performance, noting that collectively oriented team members "benefit from the advantages of teamwork, such as the opportunity to pool resources and correct errors—factors that make teamwork effective" (p. 285). Driskell, Goodwin, Salas, and O'Shea (2006) posed the question "What makes a good team player?" and suggested that team training interventions focus on learning outcomes, such as cooperation, flexibility, responsibility, and cohesiveness. Moreover, attention to affective outcomes in teams may be particularly important in virtual environments given that the technological mediation of team interaction can impact such affective processes as cohesion and trust (Driskell, Radtke, & Salas, 2003).

CONCLUDING REMARKS

We conclude this brief overview of the use of affective measures of performance in simulation based training by emphasizing three points. First, we submit that affective constructs can be usefully examined as distal measures (measures of input variables, such as individual differences), proximal measures (measures of training process), and outcome measures (measures of training outcome). Second, although Humrichouse et al. (2007) noted that the basic issues in assessing affect are no different from those involved in assessing any psychological construct, there are critical concerns of validity and reliability that must be addressed by the researcher. Finally, we laud the new generation of training models that take a multidimensional view of learning as incorporating cognitive, behavioral, and affective outcomes.

The science of simulation based training is evolving and maturing at a rapid pace. And as we learn more about how, when, and what to measure individuals and teams during training, the greater the benefit of simulation based training to such complex settings as health care, aviation, and the military—where peoples' lives depend on effective skill and cognitive *and* affective performance. We hope this chapter motivates more research into the diagnostic value of affective measures.

REFERENCES

Barrick, M. R., & Mount, M. K. (1991). The Big Five personality dimensions and job performance: A meta-analysis. *Personnel Psychology, 44,* 1–26.

Barrick, M. R., & Mount, M. K. (2001). Select on conscientiousness and emotional stability. In E. A. Locke (Ed.), *Handbook of principles of organizational behavior* (pp. 15–28). Malden, MA: Blackwell.

Bennett, W., Alliger, G. M., Eddy, E. R., & Tannenbuam, S. I. (2003). Expanding the training evaluation criterion space: Cross aircraft convergence and lessons learned from evaluation of the Air Force Mission Ready Technician program. *Military Psychology, 15,* 59–76.

Bloom, B. S., Engelhart, M. D., Furst, E. J., Hill, W. H., & Krathwohl, D. R. (1956). *Taxonomy of educational objectives, Handbook I: Cognitive domain.* New York: David McKay.

Cassady, J. (2002). Learner outcomes in the affective domain. In J. Johnston & L. Baker (Eds.), *Assessing the impact of technology in teaching and learning* (pp. 35–65). Ann Arbor, MI: Institute for Social Research, University of Michigan.

Chen, G., Gully, S. M., Whiteman, J., & Kilcullen, R. N. (2000). Examination of relationships among trait-like individual differences, state-like individual differences, and learning performance. *Journal of Applied Psychology, 85,* 835–847.

Colquitt, J. A., & Simmering, M. J. (1998). Conscientiousness, goal orientation, and motivation to learn during the learning process: A longitudinal study. *Journal of Applied Psychology, 83,* 654–665.

Costa, P. T., & McCrae, R. R. (1992). *Revised NEO Personality Inventory (NEO-PI-R) and Five Factor Inventory (NEO-FFI) professional manual.* Odessa, FL: Psychological Assessment Resources.

DeShon, R. P., & Gillespie, J. Z. (2005). A motivated action theory account of goal orientation. *Journal of Applied Psychology, 90,* 1096–1127.

Diener, E., Smith, H., & Fujita, F. (1995). The personality structure of affect. *Journal of Personality and Social Psychology, 69,* 130–141.

Driskell, J. E., Goodwin, G. F., Salas, E., & O'Shea, P. G. (2006). What makes a good team player? Personality and team effectiveness. *Group Dynamics, 10,* 249–271.

Driskell, J. E., Hogan, R., & Salas, E. (1987). Personality and group performance. In C. Hendrick (Ed.), *Review of Personality and Social Psychology* (Vol. 9, pp. 91–112). Newbury Park, CA: Sage.

Driskell, J. E., Hogan, J., Salas, E., & Hoskin, B. (1994). Cognitive and personality predictors of training performance. *Military Psychology, 6,* 31–46.

Driskell, J. E., Radtke, P. H., & Salas, E. (2003). Virtual teams: Effects of technological mediation on team performance. *Group Dynamics, 7,* 297–323.

Driskell, J. E., & Salas, E. (1992). Collective behavior and team performance. *Human Factors, 34,* 277–288.

Driskell, J. E., Salas, E., & Johnston, J. (2006). Decision-making and performance under stress. In T. W. Britt, C. A. Castro, & A. B. Adler (Eds.), *Military Life: The psychology of serving in peace and combat: Vol. Military performance* (pp. 128–154). Westport, CT: Praeger.

Driskell, J. E., Salas, E., Johnston, J. H., & Wollert, T. N. (2008). Stress exposure training: An event-based approach. In P. A. Hancock & J. L. Szalma (Eds.), *Performance under stress* (pp. 271–286). London: Ashgate.

Fiske, D. W. (1949). Consistency of the factorial structures of personality ratings from different sources. *Journal of Abnormal and Social Psychology, 44,* 329–344.

Gagne, R. M. (1984). Learning outcomes and their effects: Useful categories of human performance. *American Psychologist, 39,* 377–385.

Gully, S. M., Payne, S. C., Koles, K., & Whiteman, J. (2002). The impact of error training and individual differences on training outcomes: An attribute-treatment interaction perspective. *Journal of Applied Psychology, 87,* 143–155.

Heise, D. R. (2002). Understanding social interaction with affect control theory. In J. Berger & M. Zelditch (Eds.), *New directions in sociological theory* (pp. 17–40). Boulder, CO: Rowman & Littlefield.

Hogan, J., & Hogan, R. (1989). Noncognitive predictors of performance during explosive ordinance disposal training. *Military Psychology, 1,* 117–133.

Hogan, R. (1986). *Hogan personality inventory*. Minneapolis, MN: National Computer Systems.

Hough, L. M. (1992). The "Big Five" personality variables: Construct confusion-Description versus prediction. *Human Performance, 5,* 139–155.

Hough, L. M., Eaton, N. K., Dunnette, M. D., Kamp, J. D., & McCloy, R. A. (1990). Criterion-related validities of personality constructs and the effect of response distortion on those validities. *Journal of Applied Psychology, 75,* 581–595.

Humrichouse, J., Chmeilewski, M., McDade-Montez, E., & Watson, D. (2007). Affect assessment through self-report methods. In. J. Rottenberg & S. Johnson (Eds.), *Emotion and psychopathology: Bridging affective and clinical science* (pp. 13–34). Washington, DC: American Psychological Association.

Ilies, R., & Judge, T. A. (2005). Goal regulation across time: The effects of feedback and affect. *Journal of Applied Psychology, 90,* 453–467.

Inzana, C. M., Driskell, J. E., Salas, E., & Johnston, J. (1996). Effects of preparatory information on enhancing performance under stress. *Journal of Applied Psychology, 81,* 429–435.

John, C. H. (1988). Emotionality ratings and free association norms of 240 emotional and nonemotional words. *Cognition and Emotion, 2,* 49–70.

Johnston, J. H., & Cannon-Bowers, J. A. (1996). Training for stress exposure. In J. E. Driskell & E. Salas (Eds.), *Stress and human performance* (pp. 223–256). Mahwah, NJ: Erlbaum.

Kirkpatrick, D. L. (1976). Evaluation of training. In. R. L. Craig (Ed.), *Training and development handbook: A guide to human resource development* (2nd. ed., pp. 18-1–18-27). New York: McGraw-Hill.

Kraiger, K., Ford, J. K., & Salas, E. (1993). Application of cognitive, skill-based, and affective theories of learning outcomes to new methods of training evaluation [Monograph]. *Journal of Applied Psychology, 78,* 311–328.

Krathwohl, D. R., Bloom, B. S., & Masia, B. B. (1964). *Taxonomy of educational objectives, Handbook II: Affective domain.* New York: David McKay.

McAdams, D. P., & Pals, J. L. (2006). A new Big Five: Fundamental principles for an integrative science of personality. *American Psychologist, 61,* 204–217.

McCrae, R. R., & Costa, P. T. (1987). Validation of the five-factor model of personality across instruments and observers. *Journal of Personality and Social Psychology, 52,* 81–90.

McCrae, R. R., & Costa, P. T. (1997). Conceptions and correlates of openness to experience. In R. Hogan, J. Johnson, & S. Briggs (Eds.), *Handbook of personality psychology* (pp. 825–847). San Diego, CA: Academic Press.

Moon, H. (2001). The two faces of conscientiousness: Duty and achievement-striving within escalation of commitment dilemmas. *Journal of Applied Psychology, 86,* 533–540.

Mount, M. K., Barrick, M. R., & Stewart, G. L. (1998). Five-factor model of personality and performance in jobs involving interpersonal interactions. *Human Performance, 11,* 145–165.

Norman, W. T. (1963). Toward an adequate taxonomy of personality attributes: Replicated factor structure in peer nomination personality ratings. *Journal of Abnormal and Social Psychology, 66,* 574–583.

Paunonen, S. V., & Jackson, D. N. (2000). What is beyond the Big Five? Plenty! *Journal of Personality, 68,* 821–835.

Payne, S. C., Youngcourt, S. S., & Beaubien, J. M. (2007). A meta-analytic examination of the goal orientation nomological net. *Journal of Applied Psychology, 92,* 128–150.

Robinson, M. D., & Clore, G. L. (2002). Belief and feeling: Evidence for an accessibility model of emotional self-report. *Psychological Bulletin, 128,* 934–960.

Salas, E., & Cannon-Bowers, J. A. (2001). The science of training: A decade of progress. *Annual Review of Psychology, 52,* 471–499.

Salas, E., Nichols, D., & Driskell, J. E. (2007). Testing three team training strategies in intact teams: A meta-analysis. *Small Group Research, 38,* 471–488.

Salas, E., Sims, D. E., & Burke, C. S. (2005). Is there a "Big Five" in teamwork? *Small Group Research, 36,* 555–599.

Saucier, G., & Goldberg, L. R. (1998). What is beyond the Big Five? *Journal of Personality, 66,* 495–524.

Saucier, G. & Ostendorf, F. (1999). Hierarchical subcomponents of the Big Five personality factors: A cross-cultural replication. *Journal of Personality and Social Psychology, 76,* 613–627.

Saunders, T., Driskell, J. E., Johnston, J., & Salas, E. (1996). The effect of stress inoculation training on anxiety and performance. *Journal of Occupational Health Psychology, 1,* 170–186.

Sitzmann, T., Brown, K. G., Casper, W. J., Ely, K., & Zimmerman, R. D. (2008). A review and meta-analysis of the nomological network of trainee reactions. *Journal of Applied Psychology, 93,* 280–295.

Stagl, K.C., Salas, E., & Day, D.V. (2008). Assessment of team learning outcomes: Improving team learning and performance. In V. I. Sessa & M. London (Eds.), *Work Group Learning* (pp. 369–392). Mahwah, NJ: Erlbaum.

Stewart, G. L. (1999). Trait bandwidth and stages of job performance: Assessing differential effects of conscientiousness and its subtraits. *Journal of Applied Psychology, 84,* 959–968.

Sutherland, L., & Watkins, J. (1997). *The role of personality in training performance in two military samples.* Paper presented at the International Military Testing Association Conference, Sydney, Australia.

Thoresen, C. J., Kaplan, S. A., Barsky, A. P., Warren, C. R., & de Chermont, K. (2003). The affective underpinnings of job perceptions and attitudes: A meta-analytic review and integration. *Psychological Bulletin, 129,* 914–945.

Tupes, E. C., & Christal, R. E. (1961). *Recurrent personality factors based on trait ratings* (Rep. No. ASD-TR-61-97). San Antonio, TX: Personnel Laboratory USAF, Lakeland Air Force Base.

Watson, D., Clark, L. A., & Tellegen, A. (1988). Development and validation of brief measures of positive and negative affect: The PANAS scales. *Journal of Personality and Social Psychology, 54,* 1063–1070.

Yik, M., Russell, J., & Barrett, L. (1999). Structure of self-reported current affect: Integration and beyond. *Journal of Personality and Social Psychology, 77,* 600–619.

Chapter 21

PROVIDING TIMELY ASSISTANCE: TEMPORAL MEASUREMENT GUIDELINES FOR THE STUDY OF VIRTUAL TEAMS

Susan Mohammed and Yang Zhang

Because the time based features of distributed teams have generally been treated as an afterthought, we argue that temporal dynamics should be brought into the forefront of virtual team research. Summarized in the form of 10 temporal measurement guidelines, we address the key issues of what, how frequently, and when to measure in a virtual team context. Through adopting a multilevel approach and highlighting several exciting opportunities for future research, we hope to stimulate a more systematic and comprehensive approach to capturing temporal dimensions in distributed team studies.

A common feature in many organizations is increasing levels of team virtuality, in which members utilize technology to work interdependently across locational, temporal, and relational boundaries (Martins, Gilson, & Maynard, 2004). In response, there has been a proliferation of research on virtual teams in the last decade, and a recent monograph on team effectiveness identified geographically dispersed teams as one of two emerging trends most likely to affect critical team processes (Kozlowski & Ilgen, 2006). Despite this burgeoning literature, however, "there is little theory and comparatively few deliberate studies of the effects of temporal dimensions on computer-mediated communication. Yet, temporal effects are crucial" (Walther, 2002, p. 251). Therefore, adopting a multilevel approach, this chapter offers 10 temporal measurement guidelines that address the key issues of what, how frequently, and when to measure in a distributed team context.

Virtual teams are characterized by distinct temporal qualities, including differences in working hours and time zones, the choice of synchronous (same time) or asynchronous (different time) technologies, and increased time pressure resulting from the rigid time limits imposed on temporary teams. Despite the multifaceted time based qualities of distributed teams, however, temporal dynamics have not been comprehensively incorporated into the conceptualization or measurement of virtuality. For example, although temporal boundaries are discussed by a small subset of researchers (for example, Espinosa, Cummings, Wilson, & Pearce, 2003; Martins et al., 2004), electronic dependence and geographic dispersion have received the most emphasis in defining virtuality (Gibson & Gibbs, 2006). Nevertheless, the distinction between teams that are distributed across space and time versus those that are distributed across space, but collocated in time, can be substantive (Bell & Kozlowski, 2002), although both would be subsumed under the general heading of "virtual."

Likewise, teams differ significantly with respect to lifecycle, history of interaction, and the timing and frequency of face-to-face (FTF) interactions, even though the current conceptualization of virtuality is not nuanced enough to take these temporal dimensions into account. Failure to recognize the critical role of time in virtual team functioning could cause root problems to go undiagnosed or be misattributed to other factors. For example, unattended synchronous meetings, unexpected delays, or missed deadlines may be interpreted as a lack of team dedication when the real source may be confusion resulting from lean communication media, diverse time based orientations, and/or time zone differences.

Therefore, in a virtual context, temporal factors should be taken into account in determining what and how to measure. Below, time based features are identified and phrased as questions that researchers should ask as they embark on virtual team studies. Adopting a multilevel approach, this chapter starts with the individual level of analysis and addresses temporal characteristics (time based individual differences and cultural background). At the team level, the discussion includes the team temporal mindset (prior history and life span), time based implications of technology use (synchronous and asynchronous), and temporal process mechanisms (coordination and leadership). At the macrolevel, the focus is on the role of the temporal context (time zone differences, deadlines, and time pressure). After addressing *what to measure,* attention shifts to *how frequently* and *when to measure*. Ten guidelines for measurement are discussed in the text and summarized in Table 21.1.

WHAT TO MEASURE: TEMPORAL DIMENSIONS OF VIRTUAL TEAMS

To What Extent Do Team Members Differ on Temporal Characteristics?

What is the configuration of time based individual differences in the team? Because they are so deeply ingrained, time based characteristics have been recognized as one of the fundamental parameters of individual differences (Bluedorn & Denhardt, 1988). For example, time urgency relates to the need to have control over deadlines, as well as the feeling of being driven and chronically hurried (for example, Conte, Mathieu, & Landy, 1998). Time perspective refers to the tendency to be past, present, or future oriented (for example, Zimbardo & Boyd, 1999).

Member diversity on time based individual differences may have potent influences in virtual teams. For example, because time-urgent members are

Table 21.1. Summary of Temporal Measurement Guidelines for the Study of Virtual Teams

Measurement Focus	**Level of Analysis**	**Temporal Measurement Guideline**
What to Measure		
Temporal Characteristics	Individual	1. Recognize that diversity of time based individual differences (for example, time urgency and time perspective) can exert considerable influence on virtual team processes and outcomes.
		2. Recognize that time is defined differently, depending on the cultural background of team members and that these time based differences may significantly affect group processes and performance.
Team Temporal Mindset	Team	3. Assess and report the length of time the group has been together prior to the measurement of study variables, including whether this contact has been FTF or distributed in nature.
		4. Ensure alignment between the life span of virtual teams and the nature of variables to be investigated (for example, use temporary virtual teams for constructs that emerge quickly and ongoing teams for processes that evolve over time).
Technology Use	Team	5. Measure the frequency of asynchronous and synchronous media use in virtual teams, as well as the match between the technology and the task type (for example, asynchronous communication for less complex tasks and synchronous communication for more complex tasks). Also consider assessing the timing and rhythm of FTF meetings in ongoing teams.
Temporal Process Mechanisms	Team	6. Explore temporal process mechanisms (for example, temporal coordination and leadership) as moderators in virtual team studies because they may ameliorate many of the problems caused by differing time zones, variability on time based individual differences and cultural backgrounds, as well as heavy reliance on asynchronous communication.

Temporal Context	Macro	7. Account for the type and degree of time separation within teams, as well as the mechanisms employed to handle time zone differences.
		8. Expand performance measurement (beyond quality and quantity) to include the timeliness of work completion and whether external deadlines have been met. Consider perceptions of time pressure as additional temporal measures.
How Frequently to Measure		
Virtual Teams over Time	Various	9. Make every effort to measure constructs over time, taking theoretical (for example, how often the variable is predicted to change) and logistical (for example, managing participant burden) concerns into account in determining the frequency of measurement.
When to Measure		
Determining Appropriate Intervals for Measurement	Various	10. Carefully plan the timing of measurement in virtual teams and align assessment with the time when critical processes are occurring.

chronically hurried and time-patient members underestimate the passage of time, the mix of the two within a team may generate dysfunctional conflict (Mohammed & Angell, 2004). In addition, individuals with future time perspectives may perceive collaborators with present time perspectives as undisciplined, whereas individuals with present time perspectives may perceive collaborators with future time perspectives as uptight and demanding (Waller, Conte, Gibson, & Carpenter, 2001). Because temporal individual differences are subtle and often remain in the background of thought processes, it is likely that they will be misattributed to more explicitly addressed personality traits and stereotypes, even when they spark serious conflict in the team (Mohammed & Harrison, 2007). Failure to identify the underlying source of team difficulties can cause teams to apply incorrect solutions to problems.

Heavy use of mediated communication and the distribution of collaborators across locations can exacerbate misperceptions about team members concerning time based individual differences. According to social identity/deindividuation theory, the reduced number and quality of cues available to communicators in lean media causes an overreliance on a few social cues (Lea & Spears, 1991). Therefore, scanty information provides the basis for social categorizations that exert considerable influence on how team members are perceived and treated.

Although temporal characteristics likely operate beneath conscious awareness even in collocated teams, the potential problems caused by this form of diversity in virtual contexts are multiplied. Nevertheless, the way a team resolves the asynchronies resulting from temporal diversity may be a potentially important determinant of performance (Mohammed & Harrison, 2007). Although the effect of individual differences on virtual team performance is often ignored (for example, Powell, Piccoli, & Ives, 2004), examining diversity of personality traits has been identified as a promising area for future research (for example, Martins et al., 2004). The study of temporal characteristics in distributed contexts is likely to be especially fruitful. Guideline 1 is stated in Table 21.1.

What are the temporal implications of diverse cultural backgrounds in virtual teams? Temporal individual differences derive, in part, from culture, and one of the defining characteristics of virtuality is nationally diverse members (Gibson & Gibbs, 2006). Because some of the most significant nonlanguage difficulties in cross-cultural interactions arise from temporal differences (Bluedorn, Kaufman, & Lane, 1992), members of globally dispersed teams are almost guaranteed to encounter divergence in how time and schedules are interpreted (Saunders, Slyke, & Vogel, 2004). For example, perceptions of how much margin there is around deadlines vary from culture to culture. In a series of three cross-national studies, Levine, West, and Reis (1980) found that public clocks and watches were less accurate in Brazil, and Brazilians expressed less regret over being late than Americans. Whereas Americans tend to apologize if they are 5 minutes late, Saudi Arabians do not feel the need to apologize unless they are 20 minutes late (Brislin & Kim, 2003).

Quick service is often equivalent to good service in the United States, but many other countries do not place as great an emphasis on speed (Brislin & Kim, 2003). Although Latin America and southern Europe subscribe to event time in which schedules are fluid and meetings take as long as needed, North America and northern Europe subscribe to clock time in which events follow prespecified schedules, and time is tightly allocated (Saunders et al., 2004). In addition, cultures with a short-term orientation (for example, the United States and Russia) focus on the present and immediate gratification, but cultures with a long-term orientation (for example, Japan and China) are more concerned with persistence (Hofstede, 2001). Furthermore, Asians and Pacific Islanders are comfortable with silence because it allows them to carefully plan the next step, but Americans and Western Europeans are often annoyed by lengthy gaps in interaction (Brislin & Kim, 2003)

As these examples illustrate, time is culturally bound (Saunders et al., 2004), and cultural orientation shapes the beliefs, preferences, and values of group members toward time. Nevertheless, because failure to communicate important contextual information has been identified as a key hindrance to establishing mutual knowledge in distributed teams (Cramton, 2001), it is unlikely that team members will spontaneously discuss differences in temporal orientations. Therefore, Saunders and colleagues (2004) advocate creating an awareness of differences and developing team norms on punctuality as solutions to handling variability on temporal perceptions among remote team members. In their review on multinational and multicultural (MNMC) virtual teams, Connaughton and Shuffler (2007) state that although culture is frequently examined in terms of nationality, race, and sex, there is a need to "move beyond unidimensional views of culture and beyond static, dichotomous views of distribution to reflect the complexities of MNMC distributed team characteristics and processes" (p. 408). Examining the time based features of culture would begin to address this research call. Guideline 2 is stated in Table 21.1.

What Is the Team Temporal Mindset Concerning Past and Present Interaction?

How much prior history does the team have? Familiarity facilitates interpersonal interaction and influences performance (for example, Harrison, Mohammed, McGrath, Florey, & Vanderstoep, 2003). Clearly, knowledge of the habits and abilities of team members implies a different level of development than teams without a prior history. Because participants may sign up for experiments with friends or be in classroom teams with previous work partners, researchers should take previous history into account. In the case of field studies with ongoing virtual teams, group (and not just organizational) tenure should be reported. In addition, as the level of team maturity affects measurement choices, it would also be useful to identify where the groups are in terms of social development (for example, forming, storming, norming, performing, and adjourning; Tuckman & Jensen, 1977), as well as task progress in relation to the deadline (Gersick, 1988).

While past member interaction is important for all team types, of particular interest in the virtual context is the extent to which FTF interaction is part of the distributed team's history (Connaughton & Shuffler, 2007). As there is substantial evidence that computer-mediated groups are less efficient, requiring more time and effort to achieve a common understanding (for example., Hightower & Sayeed, 1996), team members that have FTF contact, particularly at the beginning of the work cycle, are at a substantial advantage over those who have never met in person. For example, Furst, Reeves, Rosen, and Blackburn (2004) reported that working virtually slowed progress through the formation stage of project team development by reducing opportunities to communicate. Guideline 3 is listed in Table 21.1.

What is the expected life span of the team? Although the prototypical virtual team is generally characterized as an ad hoc group that adjourns after tasks with finite time limits are completed, longer-term virtual teams are predicted to become more common as globalization dictates their necessity (Zakaria, Amelinckx, & Wilemon, 2004). Longevity has been identified as an important characteristic in the distributed context because temporary and ongoing teams have diverse structures and processes (Saunders & Ahuja, 2006). Specifically, anticipating future interaction causes electronic partners to alter their communication strategies, resulting in improved interpersonal relationships (Walther, 2002).

Indeed, no significant differences in relational communication (Walther, 1994) resulted between computer-mediated versus FTF groups when there was the ongoing expectation of continued interaction. Therefore, "temporal effects can outweigh media effects on decision quality when groups have a chance to develop common experiences" (Walther, 2002, p. 251).

In contrast to the primary focus on task accomplishment in temporary teams, ongoing virtual teams are more socially oriented and have time to develop norms, establish deeper trust, and resolve conflict (Saunders & Ahuja, 2006). Therefore, a mismatch occurs when longer-term processes that take time to unfold are examined in shorter-term teams that have no expectation of future interaction. However, reviews of the virtual team literature concluded that most studies utilized short-term student samples that met an average of four to five weeks (Powell, Piccoli, & Ives, 2004), yet interpersonal processes of trust building and conflict resolution received the most emphasis (Martins et al., 2004). Certain processes emerge only after the team has worked together for an extended period and may be salient only in teams that expect continued contact. Clearly, the temporary versus ongoing nature of virtual teams permits the meaningful measurement of some variables, while constraining other possible choices. For example, long-term virtual teams are better suited to the investigation of variables such as group identity, social integration, group norms, and psychological safety. Guideline 4 is listed in Table 21.1.

What Are the Temporal Implications of Technology Use and Task Type?

Temporal differences are embedded in the technology that is used, with asynchronous (delayed or different time) teams allowing for more response time than synchronous ("real" or same time) teams (for example, Warkentin, Sayeed, & Hightower, 1997). Specifically, synchronous media (for example, FTF communication, telephone, and chat) facilitates turn taking, allows for subtle cues to be conveyed, and provides instantaneous feedback, but imposes several constraints on when and where members can participate (Montoya-Weiss, Massey, & Song, 2001). In contrast, information exchange takes longer with asynchronous technology (for example, e-mail, Internet newsgroups, and electronic bulletin boards) because members can reflect on received messages and carefully compose and edit responses (Warkentin et al., 1997). However, communication may become disjointed when feedback is delayed and interruptions or long pauses occur (Montoya-Weiss et al., 2001). Cramton (2001) identified differences in speed of access to information and difficulty interpreting the meaning of silence as key hindrances to establishing mutual knowledge in distributed teams.

Although both asynchronous and synchronous communication have strengths and weaknesses, their effectiveness is determined, in part, by the nature of the team task in which they are employed. For example, asynchronous technology is particularly well suited for straightforward tasks, such as idea generation, because it overcomes the limitation of only one person being able to speak at a time (for example, Valachich, Dennis, & Connolly, 1994). However, due to the increase in media richness, synchronous technology is generally recommended for tasks requiring detailed information sharing, reciprocal interdependence, and high coordination (for example, Bell & Kozlowski, 2002). To illustrate, Maznevski and Chudoba (2000) found that effective global virtual teams sequenced FTF coordination meetings at various intervals when there was a need to conduct complex decision making that required intensive interaction. In contrast, ineffective teams utilized expensive in-person contact to collect simple data. Therefore, the correspondence between the technology and the task type is a significant factor in determining virtual team effectiveness.

Whereas temporary virtual teams with straightforward objectives may achieve task accomplishment by relying solely on electronic communication, FTF exchanges can significantly enhance the performance of ongoing teams (Saunders & Ahuja, 2006). In-person contact provides opportunities to "clear the air" interpersonally, deal with long-standing conflicts, and handle complex issues, as well as rejuvenate motivation on extended projects (for example, Furst et al., 2004). The timing of FTF meetings is a significant consideration, with more frequent in-person contact advised when the task requires high interdependence and when members have not developed shared mental models of teamwork (Maznevski & Chudoba, 2000). Indeed, a longitudinal study of global virtual teams concluded that FTF contact set the basic temporal rhythm for team interaction. Specifically, FTF "coordination meetings served as a heartbeat, rhythmically pumping new life into the team's processes before members circulated to different parts of the world and task" (Maznevski & Chudoba , 2000, p. 486). Guideline 5 is stated in Table 21.1.

What Temporal Process Mechanisms Are Employed in Virtual Teams?

Whereas temporal patterns surface naturally in synchronous groups, asynchronous groups necessitate an explicit focus on synchronization (Massey, Montoya-Weiss, & Hung, 2003). Therefore, recent attention has been given to temporal coordination as a mechanism for improving asynchronous collaboration (for example, Im, Yates, & Orlikowski, 2005; Montoya-Weiss et al., 2001). In a study on global virtual project teams, temporal coordination, defined as a process intervention for directing the "pattern, timing, and content of interaction incidents in a team," enhanced convergence oriented behaviors and was associated with higher performance (for example, Massey et al., 2003, p. 131). In addition, there has been increased discussion on the role of leadership in helping distributed collaborators to manage performance in the virtual team literature (for example, Bell & Kozlowski, 2002). The notion of temporal leadership reflects the extent to which leaders prioritize and set milestones, as well as pace the team so that work is finished on time (Nadkarni & Mohammed, 2007). Indeed, temporal coordination and leadership may play a pivotal role in determining virtual team success or failure, especially in the face of time constraints, asynchronous communication, and diverse time based characteristics. Therefore, it is expected that interest in these constructs will continue to grow in the distributed team literature. Guideline 6 is stated in Table 21.1.

What Is the Temporal Context?

How do differences in time separation impact virtual team functioning? The challenges presented by varying time zones are substantial in internationally dispersed teams, including restricted possibility for synchronous interaction, miscommunication from reliance on asynchronous interaction, as well as delay and rework costs (for example, Carmel, 2006). Even a time zone difference of just one hour can reduce overlapping time by several hours because of divergence in when the workday starts and ends, as well as when lunch is taken (Espinosa & Carmel, 2003). Time zone differences can be exploited by operating "round-the-clock," in which dispersed collaborators pass off their work at the end of the day to team members around the world who continue to work while they sleep (Carmel, 2006). However, because near-flawless coordination and communication are required before these benefits can be realized, many teams are not able to sustain round-the-clock operations over the long term (Espinosa & Carmel, 2003). Indeed, most research reports that work in teams is disrupted rather than assisted by varying time zones (for example, Carmel, 2006; Munkvold & Zigurs, 2007; Sarker & Sahay, 2004).

Although receiving the most emphasis, it is important to note that time zone differences are only one type of time separation. Differences in work schedules, shifts, and lunch breaks, as well as nonoverlapping weekends and holidays, can also be significant hurdles to overcome in virtual teams (Espinosa & Carmel, 2003). For example, while a nine-to-five schedule in a Monday to Friday workweek is standard for Americans, Friday is not a workday in Arab countries, and Spaniards work until after 7 P.M. because they start later and take a longer lunch break (Espinosa & Carmel, 2003). In addition, the diversity in national holidays also presents coordination challenges, especially when remote partners fail to communicate this kind of contextual information (Cramton, 2001).

The extent to which global virtual teams will be disadvantaged depends on the type and degree of time separation. For example, having multiple collaborators widely dispersed across several time zones is more difficult than members being collocated within two sites that are working in different time zones (Espinosa & Carmel, 2003). In addition, the lower the number of overlapping hours, the greater the restriction of when synchronous technology can be utilized. For team members in India, coordinating synchronous communication with remote partners in New York is far more taxing than with partners in Europe or Asia because of the magnitude of the time separation (Carmel, 2006). Therefore, many global virtual teams may rely on suboptimal media such as e-mail for complex forms of collaboration when a simple phone call would provide much needed clarification. Significant project delays can ensue when the lack of media richness in asynchronous communication perpetuates miscommunication and misinterpretation (Espinosa & Carmel, 2003).

In addition to the configuration of time zone differences and the number of overlapping hours, another factor determining the extent to which time separation will affect team functioning is the effectiveness of strategies used to overcome these difficulties. Existing interventions include stating the clock time for all involved countries for each task, establishing liaison roles to facilitate interaction across locations, and instituting messaging norms to enhance communication (Espinosa & Carmel, 2003; Sarker & Sahay, 2004). Global information technology firms have also invested in technological tools to better structure data and have created organizational cultures that encourage employees to work longer hours in order to maximize overlapping hours with remote partners (Carmel, 2006). Guideline 7 is listed in Table 21.1.

What are the team's external deadlines, and how much time pressure are members experiencing? Deadlines are a significant component of a group's temporal context, and virtual team success is heavily dependent on the timeliness of work completion (for example, Sarker & Sahay, 2004). Nevertheless, the length of time taken to finish projects is commonly omitted as a criterion variable in virtual team research (for example, Carte, Chidambaram, & Becker, 2006; Montoya-Weiss, Massey, & Song, 2001), with more attention given to the quality and quantity of performance.

In addition, insufficient attention has been paid to subjective assessments of time pressure, despite its importance in the virtual context. One of the most robust results in this literature is that computer-mediated communication takes longer than FTF interaction (for example, Walther, 2002). This finding, coupled with the fact that temporary distributed teams are often required to execute tasks with short time limits, highlights the heightened time pressure in many distributed teams. Furthermore, time limits have been shown to have a significant impact on the interpersonal dynamics of dispersed collaborators (for example, Walther, 2002). Therefore, subjective assessments of workload intensity that capture whether team members perceive that they do not have enough time, have just enough time, or have more than enough time for team tasks should be collected. Guideline 8 is listed in Table 21.1.

HOW FREQUENTLY TO MEASURE: VIRTUAL TEAMS OVER TIME

Having discussed what to measure in terms of the temporal dimensions of virtual teams, we now turn our attention to the issue of how frequently to measure. Some of the concerns with cross-sectional studies stem from the potential for type I and type II temporal errors (McGrath, Arrow, Gruenfeld, Hollingshead, & O'Connor, 1993). Type I temporal errors occur when the conclusions from short-lived teams are not sustained over a longer term. For example, global virtual teams experience a pattern of "swift," but fragile, trust that erodes over time (Jarvenpaa & Leidner, 1999). Similarly, McGrath and colleagues (1993) found that performance losses commonly ascribed to the use of computer-mediated versus FTF communication disappeared by the third or fourth week that a team worked together. Type II temporal errors occur when the effects from longer-term teams do not occur in short-lived teams. To illustrate, Yoo and Kanawattanachai (2001) indicated that a virtual team's collective mind (social cognitive system in which individuals heedfully interrelate their actions) developed only in the later stages of a project's life after a transactive memory was in place. Given the compelling nature of these results, researchers should aggressively seek opportunities to conduct longitudinal studies.

Despite the empirical demonstration that observing teams over time is critical to uncovering team effects, however, a recent review concluded that "research on virtual teams has been predominantly conducted using single work sessions, thus ignoring the role of time on group processes and outcomes" (Martins et al., 2004, p. 819). Clearly, one reason for the prevalence of cross-sectional studies is the formidable logistical obstacles encountered in doing longitudinal research on distributed teams. The greater the frequency of data collection, the greater the likelihood of participant fatigue and attrition. Therefore, the theoretical need to capture changes over time in dynamic variables must be balanced with the logistical likelihood of gaining sufficient participation from a majority of members in each team for each wave of data collection. Guideline 9 is stated in Table 21.1.

WHEN TO MEASURE: DETERMINING APPROPRIATE TIME INTERVALS FOR MEASUREMENT

Based on the convenience of data collection, the timing of measurement may be somewhat arbitrary in many virtual team studies. Nevertheless, researchers are increasingly advocating for greater specificity regarding *when* measures are collected and whether data are gathered at appropriate times (for example, Marks, Mathieu, & Zaccaro, 2001). Because computer-mediated teams require more time and effort by members to achieve the same level of shared understanding in FTF teams, measurement must allow sufficient time for distributed members to adapt to one another and the communication medium. For example, it is possible to assess team processes too early, before the team has had adequate time for meaningful interaction, as well as too late, well after key communications have already occurred. A time-sensitive team task analysis, as well as qualitative research, can assist in identifying appropriate times for measurement.

When timing data collection, researchers should also be cognizant of critical events in the distributed team context (for example, project deadlines and major team conflicts). As the rhythm of FTF meetings has been shown to play a pivotal role in global virtual teams (Maznevski & Chudoba, 2000), it is imperative that investigators understand the implications for measurement. Whether data should be collected before, during, and/or after FTF interactions will depend on the purpose of the study and the nature of the variables under investigation. Guideline 10 is stated in Table 21.1.

CONCLUSION

Summarized in Table 21.1, the 10 temporal guidelines derived above alert investigators to the temporal variables that could potentially influence virtual team measurement. As an initial step, researchers should catalog the various ways in which time could play a role in distributed team functioning for their particular study so that assessment tools can be designed to tap key temporal dimensions. Process-related guidelines 4 (alignment between team life span and study variables), 9 (frequency of measurement), and 10 (when to measure) should be explicitly considered in the design of all virtual team studies. At a minimum, variables such as team tenure (guideline 3), the type and degree of time separation (7), the frequency of asynchronous and synchronous media use (5), and the timeliness of work completion (8) should be descriptively reported as a matter of course in studies of ongoing virtual teams. Content-related guidelines 1 (time based individual differences), 2 (member cultural background), and 6 (temporal process mechanisms) are more contingent on a study's nature and purpose, but are recommended as promising avenues for future investigation. In addition, several guidelines, including matching technology and task type (5), as well as assessing temporal coordination, leadership (6), and time pressure (8), are practically oriented to facilitate virtual team success. Despite the added complexities and costs incurred by more comprehensively incorporating time into measurement, it is predicted that some of the most fruitful research streams in the coming years will occur at the intersection of virtual teams and temporal dynamics.

REFERENCES

Bell, B. S., & Kozlowski, S. W. J. (2002). A typology of virtual teams: Implications for effective leadership. *Group & Organization Management, 27*(1), 14–49.

Bluedorn, A. C., & Denhardt, R. B. (1988). Time and organizations. *Journal of Management, 14*(2), 299–320.

Bluedorn, A. C., Kaufman, C. F., & Lane, P. M. (1992). How many things do you like to do at once? An introduction to monochronic and polychronic time. *Academy of Management Executive, 6,* 17–26.

Brislin, R. W., & Kim, E. S. (2003). Cultural diversity in people's understanding and uses of time. *Applied Psychology, 52*(3), 363–382.

Carmel, E. (2006). Building your information systems from the other side of the world: How Infosys manages time-zone differences. *MIS Quarterly Executive, 5*(1), 43–53.

Carte, T. A., Chidambaram, L., & Becker, A. (2006). Emergent leadership in self-managed virtual teams: A longitudinal study of concentrated and shared leadership behaviors. *Group Decision and Negotiation, 15,* 323–342.

Connaughton, S. L., & Shuffler, M. (2007). Multinational and multicultural distributed teams: A review and future agenda. *Small Group Research, 38*(3), 387–412.

Conte, J. M., Mathieu, J. E., & Landy, F. J. (1998). The nomological and predictive validity of time urgency. *Journal of Organizational Behavior, 19,* 1–13.

Cramton, C. D. (2001). The mutual knowledge problem and its consequences for dispersed collaboration. *Organization Science, 12*(3), 346–371.

Espinosa, J. A., & Carmel, E. (2003). The impact of time separation on coordination in global software teams: A conceptual foundation. *Software Improvement and Practice, 8,* 249–266.

Espinosa, J. A., Cummings, J. N., Wilson, J. M., & Pearce, B. M. (2003). Team boundary issues across multiple global firms. *Journal of Management Information Systems, 19* (4), 157–190.

Furst, S. A., Reeves, M., Rosen, B., & Blackburn, R. S. (2004). Managing the life cycle of virtual teams. *Academy of Management Executive, 18*(2), 6–20.

Gersick, C. J. G. (1988). Time and transition in work teams: Toward a new model of group development. *Academy of Management Journal, 31,* 9–41.

Gibson, C. B., & Gibbs, J. L. (2006). Unpacking the concept of virtuality: The effects of geographic dispersion, electronic dependence, dynamic structure, and national diversity on team innovation. *Administrative Science Quarterly, 51*(3), 451–495.

Harrison, D. A., Mohammed, S., McGrath, J. E., Florey, A. T., & Vanderstoep, S. W. (2003). Time matters in team performance: Effects of member familiarity, entrainment, and task discontinuity on speed and quality. *Personnel Psychology, 56*(3), 633–669.

Hightower, R. T., & Sayeed, L. (1996). Effects of communication mode and prediscussion information distribution characteristics on information exchange in groups. *Information Systems Research, 7*(4), 451–465.

Hofstede, G. (2001). *Culture's consequences: Comparing values, behaviors, institutions, and organizations across nations.* Thousand Oaks, CA: Sage.

Im, H., Yates, J., & Orlikowski, W. (2005). Temporal coordination through communication: Using genres in a virtual start-up organization. *Information Technology & People, 18*(2), 89–119.

Jarvenpaa, S. L., & Leidner, D. E. (1999). Communication and trust in global virtual teams. *Organization Science, 10,* 791–865.

Kozlowski, S. W. J., & Ilgen, D. R. (2006). Enhancing the effectiveness of work groups and teams. *Psychological Science in the Public Interest, 7*(3), 77–124.

Lea, M. R., & Spears, R. (1991). Computer-mediated communication, deindividuation and group decision making. *International Journal of Man-Machine Studies, 34,* 283–301.

Levine, R. V., West, L. J., & Reis, H. T. (1980). Perceptions of time and punctuality in the United States and Brazil. *Journal of Personality and Social Psychology, 38,* 541–550.

Marks, M. A., Mathieu, J. E., & Zaccaro, S. J. (2001). A temporally based framework and taxonomy of team processes. *Academy of Management Review, 26*(3), 356–376.

Martins, L. L., Gilson, L. L., & Maynard, M. T. (2004). Virtual teams: What do we know and where do we go from here? *Journal of Management, 30*(6), 805–835.

Massey, A. P., Montoya-Weiss, M. M., & Hung, Y. (2003). Because time matters: Temporal coordination in global virtual project teams. *Journal of Management Information Systems, 19*(4), 129–155.

Maznevski, M. L., & Chudoba, K. M. (2000). Bridging space over time: Global virtual team dynamics and effectiveness. *Organization Science, 11*(5), 473–492.

McGrath, J. E., Arrow, H., Gruenfeld, D. H., Hollingshead, A. B., & O'Connor, K. M. (1993). Groups, tasks, and technology: The effects of experience and change. *Small Group Research, 24,* 406–420.

Mohammed, S., & Angell, L. (2004). Surface- and deep-level diversity in workgroups: Examining the moderating effects of team orientation and team process on relationship conflict. *Journal of Organizational Behavior, 25,* 1015–1039.

Mohammed, S., & Harrison, D. (2007, August). *Diversity in temporal portfolios: How time-based individual differences can affect team performance.* Paper presented at the Academy of Management Conference, Philadelphia, PA.

Montoya-Weiss, M. M., Massey, A. P., & Song, M. (2001). Getting it together: Temporal coordination and conflict management in global virtual teams. *The Academy of Management Journal, 44*(6), 1251–1262.

Munkvold, B. E., & Zigurs, I. (2007). Process and technology challenges in swift-starting virtual teams. *Information & Management, 44*(3), 287–299.

Nadkarni, S., & Mohammed, S. (2007, December). *Diversity on temporal individual differences and team performance: The moderating role of temporal leadership.* Paper presented at the 21st annual meeting of the Australian and New Zealand Academy of Management, Sydney, Australia.

Powell, A., Piccoli, G., & Ives, B. (2004). Virtual teams: A review of current literature and directions for future research. *The DATA BASE for Advances in Information Systems, 35*(1), 6–36.

Sarker, S., & Sahay, S. (2004). Implications of space and time for distributed work: An interpretive study of US-Norwegian systems development teams. *European Journal of Information Systems, 13*(1), 3–20.

Saunders, C., Slyke, C. V., & Vogel, D. (2004). My time or yours? Managing time visions in global virtual teams. *The Academy of Management Executive, 18*(1), 19–31.

Saunders, C. S., & Ahuja, M. K. (2006). Are all distributed teams the same? Differentiating between temporary and ongoing distributed teams. *Small Group Research, 37*(6), 662–700.

Tuckman, B. W., & Jensen, M. A. C. (1977). Stages of small-group development revisited. *Group and Organization Studies, 2,* 419–427.

Valachich, J. S., Dennis, A. R., & Connolly, T. (1994). Idea generation in computer-based groups: A new ending to an old story. *Organizational Behavior and Human Decision Processes, 57,* 448–467.

Waller, M. J., Conte, J. M., Gibson, C. B., & Carpenter, M. A. (2001). The effect of individual perceptions of deadlines on team performance. *Academy of Management Review, 26*(4), 586–600.

Walther, J. B. (1994). Anticipated ongoing interaction versus channel effects on relational communication in computer-mediated interaction. *Human Communication Research, 20,* 473–501.

Walther, J. B. (2002). Time effects in computer-mediated groups: Past, present and future. In P. Hinds & S. Kiesler (Eds.), *Distributed work* (pp. 236–257). Cambridge, MA: MIT Press.

Warkentin, M. E., Sayeed, L., & Hightower, R. (1997). Virtual teams versus face-to-face teams: An exploratory study of a Web-based conference system. *Decision Sciences, 28*(4), 975–996.

Yoo, Y., & Kanawattanachai, P. (2001). Developments of transactive memory systems and collective mind in virtual teams. *International Journal of Organizational Analysis, 9* (2), 187–208.

Zakaria, N., Amelinckx, A., & Wilemon, D. (2004). Working together or apart? Building a knowledge-sharing culture for global virtual teams. *Creativity and Innovation Management, 13,* 15–29.

Zimbardo, P. G., & Boyd J. N. (1999). Putting time in perspective: A valid, reliable individual-differences metric. *Journal of Personality and Social Psychology, 77*(6), 1271–1288.

ACRONYMS

AAR	after action review
ACC	anterior cingulate cortex
ADDS	Assessment Design and Delivery System
AO	advance organizer
APGO	avoid performance goal orientation
APT	Amusement Park Theoretical
ARI	Army Research Institute
ARL	Army Research Laboratory
ASR	automatic speech recognition
ATM	automated teller machine
CAS	close air support
CAST	coordinated awareness of situation by teams
CC	command and control
CDMTS	Common Distributed Mission Training Station
CFF	call for fire
CONOPS	concept of operations
CTA	Cognitive Task Analysis
CTF	coalition task force
CTGV	Cognition and Technology Group at Vanderbilt
C2	command and control
CTT	Cognitive Transformation Theory
CVE	collaborative virtual environment
DARPA	Defense Advanced Research Projects Agency
DIS	distributed interactive simulation
EAD	enemy air defense
EEG	electroencephalogram/electroencephalography
EPIC	executive process/interactive control
FAB	First Australian Bank
FAC	forward air controller
FiST	Fire Support Team
fMRI	functional magnetic resonance imaging
FO	forward observer
FTF	face-to-face

GFS	Global Forecast System
GOMS	goals, operators, methods, and selection rules
GPS	global positioning system
HCIP	human-centered information processing
HCIP-R	human-centered information processing-revised
HF	human factors
HLA	high level architecture
HPRA	human performance requirements analysis
HQ	headquarters
HTA	hierarchical task analysis
ICT	information and communication technology
ID	identify/identification
IRR	inter-rater reliability
IS	Information Systems
IV	intravenous
JAD	Joint Application Development
K&S	knowledge and skills
KR	knowledge of results
KSAs	knowledge, skills, and abilities
KSAOs	knowledge, skills, abilities, and other characteristics
LGO	learning goal orientation
LIAN	lateral inferior anterior negativity
LOs	learning outcomes
LOS	learning objective statement
LSA	Latent Semantic Analysis
MFN	medial or mediofrontal negativity
MIT	Massachusetts Institute of Technology
MNMC	multinational and multicultural
MOEs	measures of effectiveness
MOPs	measures of performance
MOT^{2}IVE	Multi-Platform Operational Team Training Immersive Virtual Environment
NASA	National Aeronautics and Space Administration
NTC	National Training Center, Fort Irwin
OC	observer/controller
ONR	Office of Naval Research
ORA	operational requirements analysis
OSD	operational sequence diagram
PAST	performance assessment and diagnostic tool
PC	personal computer
PCC	posterior cingulate cortex
PPGO	prove performance goal orientation
ProMES	productivity measurement and enhancement system
RAD	Rapid Applications Development
RE	requirements engineering
ROI	return on investment
SA	situation awareness

SAGAT	Situation Awareness Global Assessment Technique
SAM	surface-to-air missile
SARS	Situational Awareness Rating Scale
SART	Situational Awareness Rating Technique
SBT	scenario/simulation based training
SEAD	suppression of enemy air defense
SET	stress exposure training
SME	subject matter expert
SMMs	shared mental models
SOPs	standard operating procedures
SPAM	Situation-Present Assessment Method
STE	synthetic task environment
TA	task analysis
TADMUS	Tactical Decision Making Under Stress
TDT	team dimensional training
TER	training effectiveness ratio
TI	training intervention
TIM	team interaction model
TIMx	Training Intervention Matrix
TLX	NASA Task Load Index
TMM	team mental model
TMS	transactive memory system
TNA	training needs analysis
ToT	transfer of training
TP	team performance
TSA	team situation awareness
UAV	unmanned aerial vehicle
UAV-STE	unmanned aerial vehicle synthetic task environment
UML	unified modeling language
USMC	U.S. Marine Corps
VR	virtual reality
VRISE	virtual reality induced symptoms and effects
XML	Extensible Markup Language

INDEX

ABOUT THE EDITORS AND CONTRIBUTORS

THE EDITORS

DYLAN SCHMORROW, Ph.D., is an international leader in advancing virtual environment science and technology for training and education applications. He has received both the Human Factors and Ergonomics Society Leland S. Kollmorgen Spirit of Innovation Award for his contributions to the field of Augmented Cognition, and the Society of United States Naval Flight Surgeons Sonny Carter Memorial Award in recognition of his career improving the health, safety, and welfare of military operational forces. Schmorrow is a Commander in the U.S. Navy and has served at the Office of the Secretary of Defense, the Office of Naval Research, the Defense Advanced Research Projects Agency, the Naval Research Laboratory, the Naval Air Systems Command, and the Naval Postgraduate School. He is the only naval officer to have received the Navy's Top Scientist and Engineers Award.

JOSEPH COHN, Ph.D., is a Lieutenant Commander in the U.S. Navy, a full member of the Human Factors and Ergonomics Society, the American Psychological Association, and the Aerospace Medical Association. Selected as the Potomac Institute for Policy Studies' 2006 Lewis and Clark Fellow, Cohn has more than 60 publications in scientific journals, edited books, and conference proceedings and has given numerous invited lectures and presentations.

DENISE NICHOLSON, Ph.D., is Director of Applied Cognition and Training in the Immersive Virtual Environments Laboratory at the University of Central Florida's Institute for Simulation and Training. She holds joint appointments in UCF's Modeling and Simulation Graduate Program, Industrial Engineering and Management Department, and the College of Optics and Photonics. In recognition of her contributions to the field of Virtual Environments, Nicholson received the Innovation Award in Science and Technology from the Naval Air Warfare Center and has served as an appointed member of the international NATO Panel on "Advances of Virtual Environments for Human Systems Interaction." She joined UCF in 2005, with more than 18 years of government experience ranging from bench level research at the Air Force Research Lab to leadership as Deputy Director for Science and Technology at NAVAIR Training Systems Division.

THE CONTRIBUTORS

G. VINCENT AMICO, Ph.D., is one of the pioneers of simulation—with over 50 years of involvement in the industry. He is one of the principal agents behind the growth of the simulation industry, both in Central Florida and nationwide. He began his simulation career in 1948 as a project engineer in the flight trainers branch of the Special Devices Center, a facility now known as NAVAIR Orlando. During this time, he made significant contributions to simulation science. He was one of the first to use commercial digital computers for simulation, and in 1966, he chaired the first I/ITSEC Conference, the now well-established annual simulation, training, and education meeting. By the time he retired in 1981, he had held both the Director of Engineering and the Direct of Research positions within NAVAIR Orlando. Amico has been the recipient of many professional honors, including the I/ITSEC Lifetime Achievement Award, the Society for Computer Simulation Presidential Award, and an honorary Ph.D. in Modeling and Simulation from the University of Central Florida. The NCS created "The Vince Amico Scholarship" for deserving high school seniors interested in pursuing study in simulation, and in 2001, in recognition of his unselfish commitment to simulation technology and training, Orlando mayor Glenda Hood designated December 12, 2001, as "Vince Amico Day."

RANDOLPH ASTWOOD is a Research Psychologist at NAWCTSD Orlando and is currently a doctoral candidate in the Industrial/Organizational Psychology program at the University of Central Florida. In addition, he holds an M.S. degree in Industrial/Organizational Psychology from the University of Central Florida. His primary research interests include training and teams.

EVA L. BAKER is Director at CRESST and a UCLA Distinguished Professor. Her research has focused on the design and validity of assessment models that integrate research from learning and psychometrics, exploring effectiveness of technology for assessment and instruction. Her interests traverse subject matter fields, learner ages, and education and training purposes.

HOLLY C. BAXTER, Ph.D., Co-Founder and Chief Scientist of Strategic Knowledge Solutions, has spent the past decade specializing in cognitively based instructional design, assessment metrics, and training in both military and commercial environments. She has published numerous articles in the field and has been an invited speaker at multiple conferences and events.

WILLIAM BECKER, Ph.D., is a member of the research faculty in the MOVES Institute at the Naval Postgraduate School. His specialty is the development of hardware and software to support advanced training for military personnel. He is currently working with the U.S. Marine Corps.

WENDY L. BEDWELL is a doctoral student in the University of Central Florida Industrial/Organizational Psychology program. Her research interests include motivation, technology, training, and distributed teams. She earned a B.A. in Psychology from James Madison University and a master's degree in Distance Education from the University of Maryland, University College.

WILLIAM L. BEWLEY, Ph.D., is Assistant Director at CRESST. His research focuses on applications of advanced technology to instruction and assessment of performance on complex tasks. He is an experimental psychologist with a background in education and training, computer science, software development, program management, and product management.

ELIZABETH BIDDLE, Ph.D., is the Instructional Systems Site Lead at Boeing Training Systems & Services in Orlando, Florida. She has led human performance and training research and development activities in the area of human performance, adaptive learning, and training simulation. She is currently leading new business initiatives for advanced training capabilities.

DEBORAH BOEHM-DAVIS, Ph.D., is currently Professor and Chair of Psychology at George Mason University and has worked previously at General Electric, NASA Ames, and Bell Laboratories. She has been president and secretary-treasurer of the Human Factors and Ergonomics Society and president of the Applied Experimental and Engineering Psychology Division of the American Psychological Association.

CLINT BOWERS is a Professor of Psychology and Digital Media at the University of Central Florida. His research interests include the use of technology for individual and team learning.

C. SHAWN BURKE is a Research Scientist at the Institute for Simulation and Training, University of Central Florida. She is currently investigating team adaptability, multicultural team performance, multiteam systems, leadership, measurement, and training of such teams. Dr. Burke received her doctorate in Industrial/Organizational Psychology from George Mason University in 2000.

GWENDOLYN CAMPBELL is a Senior Research Psychologist at NAWCTSD. She holds an M.S. and a Ph.D. in Experimental Psychology from the University of South Florida and a B.A. in Mathematics from Youngstown State University. Her research interests include human performance modeling and a cognitively based science of instruction.

JAN CANNON-BOWERS is a Senior Research Associate at the UCF's Institute for Simulation and Training and Director for Simulation Initiatives at the College of Medicine. Her research interests include the application of technology to the learning process. In particular, she has been active in developing synthetic learning environments for a variety of task environments.

MEREDITH BELL CARROLL is a senior research associate at Design Interactive, Inc. She is currently a doctoral candidate in Human Factors and Applied Experimental Psychology at the University of Central Florida. Her research interests include human/team performance and training in complex systems with focuses on performance measurement and virtual training technology.

GREGORY K. W. K. CHUNG, Ph.D., is a senior research associate at CRESST. He has experience embedding advanced computational tools in computer based assessments to measure problem solving and content knowledge in K16 and military domains. Dr. Chung earned a Ph.D. in Educational Psychology, an M.S. degree in Educational Computing, and a B.S. degree in Electrical Engineering.

JOSEPH COHN received his Ph.D. in Neuroscience from Brandeis University's Ashton Graybiel Spatial Orientation Laboratory and continued his postdoctoral studies with Dr. J. A. Scott Kelso. His research interests focus on maintaining human performance/human effectiveness in real world environments by optimizing the symbiosis of humans and machines.

NANCY COOKE received her Ph.D. in Cognitive Psychology from New Mexico State University in 1987. She is a Professor of Applied Psychology at Arizona State University Polytechnic and Science Director of the Cognitive Engineering Research Institute in Mesa, Arizona. She is currently Editor-in-Chief of *Human Factors*. Her research focuses on team cognition.

JESSICA CORNEJO received her Ph.D. in Industrial & Organizational Psychology from the University of Central Florida in 2007. She is currently an Organizational Development Project Leader at CVS Caremark. Her professional interests lie in the areas of selection, organizational development, and diversity.

JACOB CYBULSKI is Associate Professor in the School of Information Systems at Deakin University. His research includes IS theory, methodology and strategy, with a focus on business/IT alignment. His projects range from engineering and telecommunications to business applications and recently also e-commerce and Web systems, educational video, and e-simulation.

JOSEPH DALTON is currently a graduate student in the Industrial Engineering and Management Systems program at the University of Central Florida. He received his B.S. in Liberal Studies focusing in the areas of Mathematics, Computer Science, and Psychology. His research interests include human performance and applied usability.

SUSAN EITELMAN DEAN is a Senior Scientist at Applied Research Associates. Her work focuses on the development of intelligent agents, including instructor support and role-player simulations for military training. Susan holds a M.S. in Industrial Engineering from the University of Central Florida.

GIRLIE C. DELACRUZ is a Research Associate at CRESST. Her current work involves researching the use of technology, simulations, and games to improve assessment and learning in both military and educational contexts. Ms. Delacruz is currently a doctoral student in Psychological Studies in Education at UCLA.

DEBORAH DIAZGRANADOS is a doctoral candidate in the Industrial/Organizational Psychology program at the University of Central Florida. Ms. DiazGranados received a B.S. in Psychology and Management from the University of Houston, and her M.S. is in Industrial/Organizational Psychology from the University of Central Florida.

DAVID DORSEY is employed by the National Security Agency. Dr. Dorsey holds a Ph.D. in Industrial-Organizational Psychology and a graduate minor in Computer Science from the University of South Florida. His professional interests include performance measurement, testing and assessment, training and training technologies, and computational modeling.

JAMES DRISKELL is president of Florida Maxima Corporation and adjunct professor of psychology at Rollins College. At Florida Maxima, he has conducted research on training, selection, and performance under stress for the U.S. Army, the U.S. Navy, the U.S. Air Force, NASA, the FAA, the National Science Foundation, the Department of Homeland Security, and others.

JASMINE DURAN is a first year Applied Psychology graduate student at Arizona State University Polytechnic and is employed by the Cognitive Engineering Research Institute as a research assistant. She received a B.S. in Psychology from Arizona State University in 2005. Miss Duran's research interests include team performance related to coordination, communication, and decision making.

PETER FOLTZ is founder and Vice President for Research at Pearson Knowledge Technologies and Senior Research Associate at the University of Colorado, Institute of Cognitive Science. His work focuses on cognitive science approaches to measuring individual and team knowledge. Peter has served as principal investigator for research for the U.S. Army, the U.S. Air Force, the U.S. Navy, DARPA, and NSF.

JENNIFER FOWLKES, Ph.D., is a Managing Cognitive Engineer at CHI Systems, Inc. She enjoys working with other scientists to imbue scenario based training systems with sound training infrastructures that link training to real world needs, learning science, and emerging training technologies.

JARED FREEMAN, Ph.D., is Sr. Vice President for Research at Aptima. His research and development efforts focus on performance and communications measurement, training, and the design of organizational structures and processes. Dr. Freeman is the author of more than 80 articles and book chapters concerning these and related topics.

JAMIE GORMAN received his Ph.D. in Psychology from New Mexico State University in 2006. Dr. Gorman is currently a postdoctoral researcher at the Cognitive Engineering Research Institute and Arizona State University Polytechnic in Mesa, Arizona. His research includes dynamical systems theory of team coordination, team collaboration, and communications research.

MELISSA M. HARRELL is a doctoral student in Industrial/Organizational Psychology at the University of Central Florida. Her primary research interest is performance measurement and improvement. She earned a B.S. in Psychology from the University of Florida and her M.S. in Industrial/Organizational Psychology from the University of Central Florida.

GARY KLEIN, Ph.D., helped found the field of naturalistic decision making. He has written *Sources of Power: How People Make Decisions* (1998, MIT Press), *The Power of Intuition* (2004, A Currency Book/Doubleday), and *Working Minds: A Practitioner's Guide to Cognitive Task Analysis* (Crandall, Klein, and Hoffman, 2006, MIT Press).

NOELLE LAVOIE is a founder of Parallel Consulting, LLC, where she is the lead Cognitive Psychologist, and a former Senior Member of Technical Staff at Pearson Knowledge Technologies. Her work includes studying online collaborative learning, visualization tools to support multinational collaboration, tacit knowledge based assessment, and development of military leadership.

HEATHER LUM is a doctoral student pursuing a degree in Applied Experimental & Human Factors Psychology from the University of Central Florida. She is also a graduate research fellow at the UCF Institute for Simulation & Training. Her research currently includes psychophysiological assessment of team processes and cognition.

PHAN LUU, Ph.D., is Chief Technology Officer and Scientist at EGI. His research interests include learning and memory, affect, personality, and neural mechanisms of self-regulation.

LINDA MALONE, Ph.D., is a Professor in Industrial Engineering. She is the coauthor of a statistics text and has authored or coauthored over 75 refereed papers. She has been an associate editor of several journals. She is a Fellow of the American Statistical Association.

TOM MAYFIELD has been a Human Factors engineer for 35 years with experience in nuclear, military, and marine ergonomics. At Rolls-Royce plc, he was responsible for introducing virtual reality as an HF tool in the design for operability of nuclear submarine control systems. He is a FErgS and full member of the HFES, a past President of the Potomac Chapter of the HFES, and was an Adjunct Associate Professor at GMU.

DENNIS MCBRIDE, Ph.D., MPA, is Interim Director of the Center for Neurotechnology Studies and President of the Potomac Institute for Policy Studies. His Ph.D. in experimental psychology from the University of Georgia and his post-doctoral master's degree in systems from the University of Southern California focused on mathematical learning theory and cybernetics.

LAURA MILHAM received her doctorate from the Applied Experimental and Human Factors Psychology program at the University of Central Florida. At Design Interactive, she is the Training Systems Director and Principal Investigator of numerous projects in support of the development and assessment of the effectiveness of training systems and training management systems.

SUSAN MOHAMMED is an associate professor of Industrial and Organizational Psychology at The Pennsylvania State University. She received her Ph.D. from The Ohio State University. Her research focuses on teams and decision making, with a special emphasis on team mental models, team composition/diversity, and the role of time in team research.

RAZIA NAYEEM, Ph.D., is a Research Psychologist with SA Technologies. Her research interests include development of training requirements using knowledge elicitation techniques and the study of stress effects. Razia's recent work includes examining the effects of stress on the performance of combat medic tasks.

KELLY NEVILLE, Ph.D., is an Associate Professor of Human Factors and Systems at Embry-Riddle Aeronautical University and a cognitive engineer with CHI Systems, Inc. She is interested in the ways humans interact with each other and with technology in work and training contexts, and in influences that shape these interactions.

LEMAI NGUYEN, Ph.D., is a Senior Lecturer at School of Information Systems, Deakin University. Her primary contributions are to the study of creativity and problem solving activities in Requirements Engineering. Other areas of her research contributions include health care informatics, sociotechnical aspects of virtual communities, online learning, and information systems in general.

ROB OBERBRECKLING has worked with Pearson as a Senior Member of Technical Staff and currently performs research and engineering in natural language processing, cognitive science, and machine learning systems.

ORLANDO OLIVARES, Ph.D., is a Senior Industrial-Organizational Psychologist at Aptima. He provides a full range of services related to enhancing organizational effectiveness, namely, organizational assessment, problem identification and resolution, personnel selection, and the development of performance and business metrics and incentive systems. Dr. Olivares holds a Ph.D. in Industrial/Organizational Psychology from Texas A&M.

JENNIFER PHILLIPS is the President and Senior Scientist of Cognitive Training Solutions, LLC. She has over 12 years of experience conducting research and developing applications in the area of human cognition and naturalistic decision making. Her research interests include skill acquisition, cognitive performance improvement, and the nature of expertise.

CATHERINE POULSEN, Ph.D., is a Scientist at EGI, where she conducts research using cognitive experimental and dense-array EEG methods. Her primary line of research examines the neural dynamics underlying the adaptive control of learning and performance, and its modulation by goals, incentives, and feedback.

ROBERT D. PRITCHARD is Professor of Psychology and Management at the University of Central Florida. He received the Distinguished Scientific Contribution Award from the Society for Industrial and Organizational Psychology and is a Fellow in the Society for Industrial and Organizational Psychology, the American Psychological Association, and the American Psychological Society.

MICHAEL A. ROSEN is a doctoral candidate at the University of Central Florida and a graduate researcher at the Institute for Simulation and Training. He has co-authored over 60 peer reviewed journal articles, book chapters, and conference papers related to teams, decision making and problem solving, performance measurement, and simulation.

MARK ROSENSTEIN is a Senior Member of Technical Staff at Pearson applying machine learning and natural language processing techniques to problems involving understanding and assessing language and the activities connected with the use of language.

KAROL ROSS, Ph.D., is a research psychologist at the Institute for Simulation & Training at the University of Central Florida. She is also the Chief Scientist for the Cognitive Performance Group, a small business in Orlando, Florida. She has over 20 years of experience in military training research.

STEVEN RUSSELL is a Research Scientist at Personnel Decisions Research Institutes in Arlington, Virginia. He holds a Ph.D. in Industrial-Organizational Psychology from Bowling Green State University. His professional interests include the design and evaluation of training programs, criterion measurement, and test development and validation, including item response theory techniques.

EDUARDO SALAS is Trustee Chair, Pegasus Professor, and Professor of Psychology at the University of Central Florida and Program Director for the Human Systems Integration Research Department at the Institute for Simulation and Training. His expertise includes teamwork, designing/implementing team training strategies, training effectiveness, and developing performance measurement tools.

SHANNON SCIELZO is an Assistant Professor at the University of Texas–Arlington. She received her Ph.D. in Industrial/Organizational Psychology from the University of Central Florida in 2008 where she worked as a graduate fellow for the Multidisciplinary University Research Initiative and a graduate research assistant in the Team Training and Workforce Development Laboratory.

KIMBERLY SMITH-JENTSCH, Ph.D., has been a member of the psychology department at the University of Central Florida since 2003. Her research on teams, training, performance assessment, and mentoring is published in journals such as the *Journal of Applied Psychology, Personnel Psychology, Journal of Organizational Behavior,* and *Journal of Vocational Behavior.*

WEBB STACY, Ph.D., is Vice President for Technology at Aptima. Dr. Stacy is responsible for Aptima's current and future strategic technology portfolio and is involved in creating intelligent systems for modeling and assessing human performance. Dr. Stacy received a Ph.D. in Cognitive Science from SUNY/Buffalo and a B.A. in Psychology from the University of Michigan.

KAY STANNEY is President of Design Interactive, Inc. She received her Ph.D. in Industrial Engineering from Purdue University, after which time she spent 15 years as a professor at the University of Central Florida. She has over 15 years of experience in the design, development, and evaluation of human-interactive systems.

J. GREGORY TRAFTON, Ph.D., is a cognitive scientist at the Naval Research Laboratory. He is interested in building and applying theories in complex, real world situations, including meteorology, scientific visualization, interruptions and resumptions, and robotics. He has collected data in both naturalistic settings and laboratory environments to build computational and mathematical theories.

SUSAN TRICKETT is a cognitive scientist in Denver, Colorado. She is interested in spatial cognition, the use of simple and complex visualizations, learning, and the acquisition of expertise, particularly in complex, real world domains.

DON TUCKER, Ph.D., is CEO and Chief Scientist at EGI, Professor of Psychology and Associate Director of the NeuroInformatics Institute at University of Oregon. His research examines self-regulatory mechanisms of the human brain. These include motivational and emotional control of cognition, as well as neurophysiological control of arousal, sleep, and seizures.

WENDI VAN BUSKIRK is a Research Psychologist at NAWCTSD. She is currently a doctoral candidate in the Applied Experimental and Human Factors Psychology Ph.D. program at the University of Central Florida. Her research interests include human cognition and performance, instructional strategies, human performance modeling, and human-computer interaction.

JENNIFER VOGEL-WALCUTT, Ph.D., is a researcher at UCF's Institute for Simulation and Training. She leads the Learning Initiatives team within the ACTIVE Lab focusing on learning efficiency in complex environments. She is currently applying this research to military domains by creating strategies and methodologies for more efficient training.

SALLIE J. WEAVER is a doctoral student in the Industrial and Organizational Psychology program at the University of Central Florida. She earned her M.S. in Industrial/Organizational Psychology from UCF. Her research interests include individual and team training, performance measurement, motivation, and simulation, with an emphasis in health care.

YANG ZHANG obtained a Bachelor of Arts degree from Grinnell College and her Master of Science degree at Penn State. She is currently pursuing her Ph.D. in Industrial/Organizational Psychology at Penn State. Her research interests include teams and cross-cultural issues in organizations. Her dissertation focuses on temporal issues on teams.